NATURAL SCIENCES IN AMERICA

NATURAL SCIENCES IN AMERICA

THE

GEOGRAPHICAL AND GEOLOGICAL

DISTRIBUTION OF ANIMALS

BY

ANGELO HEILPRIN

ARNO PRESS
A New York Times Company
New York, N. Y. • 1974

Reprint Edition 1974 by Arno Press Inc.

Reprinted from a copy in The American
Museum of Natural History Library

NATURAL SCIENCES IN AMERICA
ISBN for complete set: 0-405-05700-8
See last pages of this volume for titles.

Manufactured in the United States of America

Publisher's Note: The frontispiece of this
edition has been reproduced in black and white.
The key has been coded accordingly.

————◆————

Library of Congress Cataloging in Publication Data

Heilprin, Angelo, 1853-1907.
    The geographical and geological distribution of
animals.

    (Natural sciences in America)
    Reprint of the 1887 ed. published by D. Appleton, New
York as v. 57 of the International scientific series.
    1. Zoogeography. 2. Paleontology. I. Title.
II. Series. III. Series: International scientific
series (New York) v. 57.
QL101.H46 1974   591.9   73-17824
ISBN 0-405-05742-3

THE

GEOGRAPHICAL AND GEOLOGICAL

DISTRIBUTION OF ANIMALS

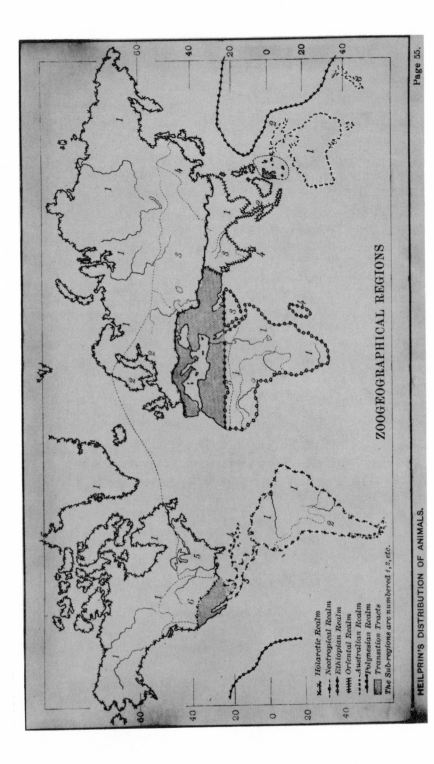

ZOOGEOGRAPHICAL REGIONS

HEILPRIN'S DISTRIBUTION OF ANIMALS.

Page 55.

✕✕✕ Holarctic Realm
✕✕✕ Neotropical Realm
✕✕✕ Ethiopian Realm
✕✕✕ Oriental Realm
...... Australian Realm
✕✕✕ Polynesian Realm
▨▨▨ Transation Tracts
The Sub-regions are numbered 1, 2, etc.

THE INTERNATIONAL SCIENTIFIC SERIES

THE

GEOGRAPHICAL AND GEOLOGICAL

DISTRIBUTION OF ANIMALS

BY

ANGELO HEILPRIN

PROFESSOR OF INVERTEBRATE PALEONTOLOGY AT, AND CURATOR-IN-CHARGE OF,
THE ACADEMY OF NATURAL SCIENCES OF PHILADELPHIA ; PROFESSOR OF
GEOLOGY AT THE WAGNER FREE INSTITUTE OF SCIENCE, PHILADELPHIA;
MEMBER OF THE AMERICAN PHILOSOPHICAL SOCIETY, &C.

NEW YORK
D. APPLETON AND COMPANY
1887

TO

PROFESSOR JOSEPH LEIDY, M. D., LL. D.,

PRESIDENT OF THE ACADEMY OF NATURAL SCIENCES OF PHILADELPHIA, &C.,

WHOSE PROFOUND RESEARCHES HAVE SO LARGELY TENDED TO DEVELOP

THE SCIENCE OF BIOLOGY,

THIS VOLUME IS RESPECTFULLY DEDICATED.

A. H.

PREFACE.

In the preparation of the following pages the author has had two objects in view : that of presenting to his readers such of the more significant facts connected with the past and present distribution of animal life as might lead to a proper conception of the relations of existing faunas; and, secondly, that of furnishing to the student a work of general reference, wherein the more salient features of the geography and geology of animal forms could be sought after and readily found. The need of such a work has been frequently felt and expressed. As far as he is aware, no work of that kind has as yet appeared, and therefore, to a certain extent, this publication stands alone in the field it is intended to cover. Necessarily, much that it embraces can be found elsewhere, and treated even at considerably greater length; but the matter is not contained under a single cover, and where a special subject is expounded *in extenso* the treatment is usually too exhaustive to permit of immediate use by the general reader. This applies particularly to zoogeography. With reference to geological distribution there is little connectedly written—indeed, beyond what is found in text-books largely devoted to cognate subjects, practically nothing. Moreover, what little of connected literature on the subject we do possess is almost entirely out of date, and in no way represents the present status of the science.

The subject of geographical and geological distribution is so vast that no full treatment of it could be expected in the limited number of pages set apart for it in the present work. The author has, therefore, been obliged to omit, or at least largely ignore, the consideration of some of the less important animal groups, and, while recognising the deficiencies resulting from such omission, trusts that it will not detract much from the general usefulness of the publication. The plan of treatment followed in the early part of the book (geographical distribution) is largely that so admirably unfolded by Mr. Wallace, to whom, for the constant use of his works, the author is under great obligations. He also wishes to express his special indebtedness to the pioneer workers in this field, Schmarda and Murray, whose writings have laid the foundation of much of our existing knowledge in the premises. No special mention need be made of the numerous other authors who have contributed more or less extensively to the subject under consideration, and whose works have aided in the preparation of the present volume; to those, collectively, the author likewise desires to acknowledge his indebtedness.

A few words need be said in relation to the zoogeographical regions that are recognised in this work, which differ essentially from those generally adopted by naturalists. The reasons for uniting the "Nearctic" and "Palæarctic" regions of zoogeographers into a single realm, designated, in accordance with a suggestion by Professor Alfred Newton, of Cambridge, the "Holarctic," are fully set forth in my paper "On the Value of the Nearctic as one of the Primary Zoological Regions," published in the "Proceedings of the Academy of Natural Sciences of Philadelphia," for December, 1882. Objections by Mr. Wallace and Professor Gill to the views there formulated appear in "Nature" of March 22, and June 7, 1883, respectively, and my rejoinders to the criticisms of these gentlemen in "Nature" of April 26, and the "Proceedings" of the Philadelphia Academy for November, 1883. To these papers I must refer the reader for a purely technical statement of the case. The classification of the "Transition" tracts is largely that which has been proposed by Forsyth Major, in "Kosmos" for 1884.

ACADEMY OF NATURAL SCIENCES, PHILADELPHIA, *October, 1886.*

CONTENTS

PART I.

GEOGRAPHICAL DISTRIBUTION.

PART II.

GEOLOGICAL DISTRIBUTION.

I.

II.

PART III.

GEOGRAPHICAL AND GEOLOGICAL DISTRIBUTION.

I.

II.

PART I.

GEOGRAPHICAL DISTRIBUTION.

I.

General principles of zoogeography.—Faunas of isolation.—Relations of past and present faunas.

EVERYWHERE upon the surface of the earth we meet with manifestations of animal life. The desert wastes, no less than the tropical jungles, the bleak ice-fields of the frozen north, and the most elevated mountain-summits—all have their faunas. The abyss of the sea, no less than its surface, contributes its quota to the animal world ; and in the atmosphere all around us, from the lowest stratum not unlikely to the highest, the germs of the organic universe lie everywhere scattered about. In what precise form or guise this life first manifested itself, or how inert matter became endowed with that potentiality which we recognise in vital energy, it seems hopeless to attempt to determine. True science takes cognisance of both fact and theory, but illusory speculation, whose groundwork is a simple outgrowth of the imagination, must find a resting-place without its domain.

No one who has paid the smallest amount of attention to the facts of nature as they present themselves can have failed to notice certain peculiarities in the way of the distribution of life, which do not always admit of an immediate or of a satisfactory interpretation. Why, for example, one piece of country should differ so essentially in its faunal aspects from another whose physical characteristics are practically identical with its own ; why the second should differ from a third, and this, again, from a fourth—may not appear comprehensible. Nor any the more comprehensible may appear the circumstance that, in most cases, island faunas are so eminently marked out from those of continental areas.

Another peculiarity in faunal distribution is presented in the fact that, while certain animal assemblages enjoy an almost limitless or universal extension, others, again, without apparent reason, are circumscribed within limits of the opposite extreme. The traveller to the most distant shores not infrequently recognises objects that are familiar to him as those of his native home, although possibly, in the interval of his journey, he has completely lost sight of their existence, so different might have been the creatures that successively met his gaze. "When an Englishman travels by the nearest sea-route from Great Britain to Northern Japan, he passes by countries very unlike his own, both in aspect and natural productions. The sunny isles of the Mediterranean, the sands and date-palms of Egypt, the arid rocks of Aden, the cocoa-groves of Ceylon, the tiger-haunted jungles of Malacca and Singapore, the fertile plains and volcanic peaks of Luzon, the forest-clad mountains of Formosa, and the bare hills of China, pass successively in review; till after a circuitous voyage of thirteen thousand miles he finds himself at Hakodadi, in Japan. He is now separated from his starting-point by the whole width of Europe and Northern Asia, by an almost endless succession of plains and mountains, arid deserts or icy plateaux, yet when he visits the interior of the country he sees so many familiar natural objects that he can hardly help fancying he is close to his home. He finds the woods and fields tenanted by tits, hedge-sparrows, wrens, wagtails, larks, redbreasts, thrushes, buntings, and house-sparrows, some absolutely identical with our own feathered friends, others so closely resembling them that it requires a practised ornithologist to tell the difference. If he is fond of insects he notices many butterflies and a host of beetles which, though on close examination they are found to be distinct from ours, are yet of the same general aspect, and seem just what might be expected in any part of Europe. There are also, of course, many birds and insects which are quite new and peculiar, but these are by no means so numerous or conspicuous as to remove the general impression of a wonderful resemblance between the productions of such remote islands as Britain and Yesso." *

* Wallace, " Island Life," p. 8.

On the other hand, a journey of only very moderate duration will frequently disclose the greatest diversity existing between contiguous faunas. The traveller who starts east from the African coast, and who has familiarised himself with the strange productions of the African continent, its elephants, giraffes, rhinoceroses, hippopotami, lions, and antelopes, finds none of these in the island of Madagascar ; the true monkeys have also disappeared, and in their place he meets with forms of half-monkeys (lemurs), a group of animals with which he will have already become acquainted before leaving the mainland. Strange creatures, wholly unlike anything previously known to him, now arrest his attention, and he finds himself in the midst of what might be termed a peculiar fauna. Likewise, if he leave the shores of Central America or Florida for the Great Antilles, the same marked isolation of the new fauna manifests itself. The larger forms of quadrupeds, such as the jaguar, couguar, tapir, and peccary, are wholly wanting, and even among the smaller and more numerously represented mammalian types many of the more prominent forms will be sought for in vain. On the other hand, he will make the acquaintance of entirely new groups of animals, some of which, like the Centetidæ, have their nearest foreign representatives in regions removed by nearly one-half the circumference of the globe. And this diversity in the faunal type is found to permeate to a greater or less extent all the individual groups, birds, reptiles, &c., of the animal kingdom.

It might be rashly supposed that the distance separating the regions under comparison would sufficiently account for the peculiarities of their respective faunas, or the disparities separating them; but distance alone, without a special relation binding together the principals between which it is supposed to act, can effect nothing. We have, indeed, seen upon what a vast extent of territory the British faunal facies is stamped, and were any further proof needed of the inefficacy of distance, pure and simple, as a prime factor in geographical distribution, we have but to transport ourselves to the Malay Archipelago, and observe how wonderfully diverse are the respective faunas on either side of the very narrow (but deep) channel separating the islands of Bali and Lombok from each other.

Mysterious as these various phenomena of distribution may appear, they yet have all their logical explanation. A quarter of a century ago, when the doctrine of independent creation still held sway over the minds of most naturalists, and when the organic universe was reflected in the eye of the investigator as an incongruous agglomeration of disjointed parts, there was, indeed, no necessity for specially accounting for the facts, since they were conceived to be such by reason of a previous ordination. Now, however, when the full value of the evolutionary process is recognised, and animate nature has come to be looked upon as a concrete whole, bearing special relations to its numberless parts, each individual fact seeks its own explanation, which explanation must of necessity stand in direct harmony with some previously observed fact. When, therefore, we seek to unravel the tangle of zoogeography, and to harmonise its apparent incongruities, we must at the outset admit that distribution, such as it is, is the outcome of definite interacting laws — laws which stand in relation to each other as absolutely as they do in any other field of action—and not a hap-hazard disposition, as some would lead us to suppose, setting all enquiry at defiance.

The naturalist who in the Western Hemisphere journeys southward from the ice-covered fields of British America fails to notice any very sudden or marked alternation in the character of the faunas that successively meet his view. New features are being constantly added, and old ones eliminated, but the interchange is effected so gradually that it becomes difficult to determine the limitations that properly define one fauna from another. The fur-bearing animals of the far north send their representatives into regions which border the habitats of the more exclusively tropical species, or are succeeded by forms which differ but little from them. The skunk, many of whose associates are animals of a distinctively Arctic character, finds its way into Mexico, and the ermine, which penetrates to the farthest northern point reached by mammals generally, still lingers on in some of the Southern United States. The Arctic fox is succeeded by the equally abundant types of the grey and the red fox; and similarly, the polar bear is followed on the one side of the continent by the grizzly, and on the other by the black bear. Having descended into the middle temperate regions, the traveller still finds about him mostly the forms with which he has already become acquainted. But many of the more familiar types have either wholly disappeared, or are fast disappearing. Such may be the musk-sheep, moose, stag, and reindeer, which will have left as their successors the bisons and the various species of smaller deer which range throughout the remainder of the continent. The grey wolf of the northern forests breaks up into a number of varietal forms more or less distinct from the typical one, and is carried by the coyote into the heart of Mexico.

Farther to the south the traveller observes entirely new features gradually appearing. In Arkansas he possibly meets with the peccary, the first indigenous member of the pig family with which he will have become acquainted; in Texas, with the armadillo, the first of that group of animals, the Edentata, which, in the past and present history of the South American continent, constitutes such an important element in its fauna; and, in the States adjoining the Mexican Republic, with an abundant representation of the iguanid lizards, which, by their numbers, so eminently typify the following region of the tropics. There are as yet neither monkeys, tapirs, nor guinea-pigs, but the first appear in Southern Mexico, the second in Central America, and the last in Venezuela or Guiana. The traveller is now in the region of the Equator, and surrounded by an association of animal forms most of which were unknown to him when he entered upon his journey, and which in many respects depart so widely from those with which he was familiar at his starting-point as to constitute a distinct fauna. There is no longer either wolf, fox, or catamount, beaver or musk-rat, and of the specifically important group of the hares or rabbits but a single species remains. The solitary species of bear is so different from its northern cousins as to be regarded by some naturalists as the type of a distinct genus.

The contrast between the successive faunal changes observed on the north and south journey and the faunal identity which so astonishes the traveller whose journey is directed eastward from England to Japan is very great. And yet if the traveller from Britain, instead of proceeding due eastward, were to shape his course a few degrees to the south, much the same kinds of changes as he noticed on his American trip would again present themselves. Along the shores of the Mediterranean he would no longer, or only at rare intervals, meet with his associates of the Arctic north; on the southern slopes of the Caucasus the tiger, and in Arabia the

camel, gazelle, and ostrich, would present to him certain features
of a fauna which was in the main unknown to him; in India the elephant, lion, and rhinoceros, and other curious denizens of the jun-
gle, the python and crocodile, and the numerous birds of resplendent
plumage, would probably crowd from his memory the forms of the
creatures ordinarily most familiar to him, and lead a passage to the
ultimate goal of his journey, Australia, where he would meet with
the most singular and most distinctive fauna on the surface of the
earth.

Much nearer to his northern home—on opposite sides of the
Mediterranean—and with much less travelling, the naturalist will
discern scarcely less well-marked faunal differences or peculiarities.
To account for the anomalies which the facts of distribution present
is the still unsolved problem that is put before the zoogeographer.

Granting, with the doctrine of evolution, that all the complex
assemblages of existing animal forms are modified derivatives from
previously existing forms, and that these are ultimately to be traced
back to some common ancestor, it must of necessity follow that any
given fauna will depend for the degree of its peculiarity, whether
great or small, upon the amount of modification, relative to any
other fauna, which it will have undergone. And this modification
can be effected in two ways: by inherent modification of the individual types composing the fauna, and by intermixture with, or
immigration from, contiguous or neighbouring faunas. In both
cases, manifestly, isolation or its opposite, union of habitation, will
constitute the governing factor in determining the amount of variation. A region that is broadly separated from all others will, naturally, tend to develop a fauna distinct from any other, since the
progressive modifications in its constituent faunal elements must ultimately lead to divergence; and the greater the period of isolation
the greater, of necessity, will be the amount of this divergence, or
the more pronounced the faunal individualisation. Hence it is that
in the greater number of the more distantly removed island groups,
or in those which are separated by more or less impassable barriers
from the nearest land-mass, we meet with such highly specialised
faunas. The Galapagos Islands, for example, as will be more fully
illustrated farther on, have a fauna very distinct from that of any
part of South America, although removed from it by a distance of less
than seven hundred miles. The birds are quite distinct, and so are

the reptiles, insects, and land mollusks. The island of St. Helena, in the South Atlantic, and the Sandwich Islands, in the North Pacific, present us with similar instances of faunal specialisation, and to a less extent, also, the group of the Azores. In the case of these last, which lie in the course of the storm-winds, a considerable intermixture has been effected with the faunas of Western Africa and Europe, for we find that by far the greater number of the resident land-birds are inhabitants of those two continents as well. The fact that there are so very few peculiar forms is proof either of a recent separation of the islands from the mainland—not sufficient time having been allowed for the development of new species—or of a recent or repeated peopling with old forms from the continents. Even irrespective of considerations connected with the physical geography or geology of the region, it would naturally be inferred, from the prevalence of in-blowing storm-winds, and the known fact that certain birds are transported hither, that the second supposition is the correct one; and that this is the true explanation is proved by evidence of a very positive character furnished by some of the other groups of animals. Thus, the land-Mollusca, which in their distribution are not so readily affected by aerial currents, are eminently distinguished from those of either Europe or Africa, or of any other continental land-mass, proving in their case a long-protracted period of isolation. Further, there is not a single species of fresh-water mollusk known in the entire group! The Bermuda Islands, which are about equally distant from the mainland, occupy a nearly analogous position with respect of their fauna; that is, partial interchanges have been effected with the fauna of the American continent.

In all these cases, necessarily, the amount of faunal specialisation will be the index of the period of isolation. Where faunal immigration from a foreign region takes place it not only checks the development of a newly-forming fauna, by infusing into it an element that does not properly belong there, but also prevents in a measure that variation among individuals which might otherwise obtain. The case of the bobolink of the Galapagos Islands is a well-known example of this kind. It alone, of about thirty species of land-birds inhabiting those islands, is considered to be indisputably identical with any form occurring on the mainland; hence it is concluded that this is about the only species of South American bird that ever visits the islands, for, if the case were otherwise, it would be incredible that no more common forms should have been detected there. But the fact that the bobolink has remained absolutely identical with the common form of South America, whence, doubtless, most of the species of Galapagos birds have been derived, while all the other birds of the island group have undergone more or less modification since the islands were first tenanted, proves that variation in its case has been prevented by the perpetuation of normal characters through interbreeding with the continental migrants. In other words, the breed has been kept true. Were the migrations of the visitors checked or interrupted, there can be little question that the island breed of bobolinks would undergo the same kind of modification which distinguishes the other birds, and which has developed in them new specific or varietal types. In the continent of Australia, again, we meet with the most remarkable example of a highly specialised fauna being developed as the result of long-continued isolation. Of all the varied mammalian forms which elsewhere crowd the surface of the earth we have here but the merest trace, for, with the exception of the rodents and bats, none of the ordinary orders—Carnivora, Ungulata, Insectivora, &c.—are represented.* And even of the rodents there is but a single family, that of the mice (Muridæ). On the other hand, the implacental mammals—kangaroos, wombats, duck-bill—whose only non-Australian representatives are the American family of opossums (Didelphidæ), acquire here a wonderful development, and exhibit a diversity of type-structure not met with in any other order of mammals. Now, the animals of this class, or such as might be considered most nearly allied to the marsupials, are the first of the Mammalia to appear in geological time, and they alone have thus far been detected in any of the deposits (Triassic, Jurassic) of the middle geological period, or Mesozoic era. They constitute the most primitively organised members of their class, and probably stand not far removed from what may ultimately be proved to be the bottom of the mammalian series.

In order to explain the anomalies of the Australian mammalian fauna we must have recourse to the hypothesis of isolation, for in

---

* The Australian wild-dog, or dingo, may prove to be indigenous, in which case it would represent the Carnivora.

no other way could we satisfactorily account for the remarkable development of the marsupial types, and the almost total absence of the commoner forms that are elsewhere so abundant. The oceanic barriers have evidently prevented that diffusion of species which would otherwise have sufficed to render the Australian fauna cosmopolitan in character. That this isolation, further, of the continent has been of very great duration is proved by the long period of time, dating from the Cretaceous epoch, during which the most diverse forms of mammalians have existed, and the high specialisation that its own fauna has acquired. It may appear not a little surprising, in view of what has preceded, that two groups of animals, so widely removed from the rest of the Australian mammalian fauna as are the mice and bats, should yet constitute a part of this fauna. In the case of the bats it is not difficult to account for their occurrence in the region in question, since their powers of flight have enabled them to overcome such obstacles as to other animals might have proved true barriers to migration. The mice, on the other hand, whose disposition to gnaw into, and conceal themselves among, timber of all kinds, is well known, may have found their way hither from the Asiatic continent or its adjoining islands through the intermedium of floating masses of vegetation. Much more inexplicable is the occurrence of the single non-Australian family of marsupials, the opossums, on the American continent, which is removed by a continuous water-way of several thousands of miles, when not a single member of the entire sub-class of implacental mammals is found on any other part of the earth's surface outside the Australian region. The hypothesis that land connection by way of the Antarctic region at one time existed between Australia and South America, and, possibly, also Africa, may or may not be true, but the evidence that has thus far been adduced tending to show that by such connection a transferrence of one section of the Marsupialia has been effected from one continent to the other is certainly very slim. Yet it is by no means impossible that such may have been the case. The Edentata—armadillos, ant-eaters, pangolins—whose home is preeminently the two great continents of the Southern Hemisphere, and which barely trespass north of the Tropic of Cancer, and the struthious birds, like the rhea, ostrich, and cassowary, offer equal perplexities in the way of an explanation of their anomalous distribution with the

2

marsupials, and they have likewise been considered to afford proof
of a land connection such as has been indicated. A serious difficulty, however, that lies in the way of this explanation is the
important fact that none of the characteristic African or South
American mammals are found in Australia, for it might justly be
contended that if a migration or transferrence was effected in one
direction, it could have been effected in the opposite direction as
well. But that such reciprocal distribution did not obtain is very
nearly certain. It may, indeed, be assumed that at the time of a
possible Australian migration the extremities of the southern continents were not yet inhabited; but this is very unlikely. Or, it
may be further assumed, with Rütimeyer, that the animals under
consideration had a polar origin, and that they were distributed
northward along continental lines that possibly now lie buried
beneath the sea ; but positive evidence in this direction is still
wholly wanting. An element in the problem which very materially narrows the issue is the circumstance that marsupial remains
have been found in the temperate regions of the Northern Hemisphere, and in both Europe and North America in deposits as ancient as the Triassic period. In this upper tract, therefore, we
find a possible and more probable clue towards the explanation of
the existing distribution of the animals in question ; and if it be
objected that some such living forms ought still to be found in
the connecting region, the fact, nevertheless, remains that they did
there once exist, but have since become largely extinct.

It will be evident that the key to the solution of the more
marked peculiarities of modern distribution must be sought in the
records of the past, for in the comparison between existing and
preexisting faunas alone can we expect to determine the conditions upon which present faunas were established, and to ascertain
the dates of their respective appearances or antiquity. In most
regions of the earth's surface a most intimate relationship links
together the existing fauna and the fauna of the geological period
or periods immediately preceding. The Pliocene and Post-Pliocene
marine shell-fish faunas of the Western United States are practically
identical with the equivalent fauna of the (modern) adjoining seas;
the Post-Pliocene mammals of Britain are such as still roam about
the land, although they include numerous forms which no longer
exist there; in India a large proportion of the mammalian types

that inhabit the region are already represented in deposits of the early Pliocene period; and in Australia the abundant remains of Marsupialia amply testify to the identity of character which unites the faunas of the past and present periods. A certain amount of antiquity is thus established for the several regional faunas. The farther back in time we proceed, however, the less pronounced appear the common characteristics of past and present periods; and, finally, they disappear almost altogether. Thus, the Eocene shell-fish fauna of the Atlantic coast of the United States and of France and Great Britain is very unlike that of the seas adjoining those regions at the present day, although, in a measure, it finds its analogue in the corresponding fauna of the eastern tropical seas. The Miocene mammals of the American continent are almost wholly unlike those which now inhabit the region, and what little similarity still remains completely vanishes with the animals of the more ancient Eocene period. And the same holds good with the European Tertiary fauna. Yet there are a number of existing types which in their own region can be traced through a series of progressive modifications to ancestral forms more or less unlike them, which belong to a comparatively remote geological epoch. The horse of the Old World, for example, has been traced through a number of intermediate forms to the Old Tertiary Palæotherium, one of the most abundantly represented mammalian genera of the deposits of Western Europe. The deer of the same region finds early ancestors in the horned and hornless species which occur fossil in the Miocene deposits of France and Germany; and not unlikely the wolf and fox see their progenitors among the early members of the canine race, whose remains have been traced to the Oligocene, and not impossibly also to the Eocene period. In so far as these animals are concerned, therefore, we have direct evidence of a fauna of considerable antiquity developing in place. In other cases, however, evidence of a very opposite character is often presented; that is to say, faunas, or their components, are very frequently shown to be in a given region of only brief duration. Thus, although bears are very plentiful at the present time in the North American continent, they are not known to have existed there before the last geological period, the Post-Pliocene. And the same is true of the members of the ox-family (Bovidæ)—most of which are, indeed, not represented at all as fossils—of which North America possesses five, in the main, widely-distributed species : two antelopes, two sheep (including the musk-ox), and the bison. The question as to how these animals obtained a foothold in the region which they now inhabit, whether they originated there as derivatives from previously-existing forms, or were introduced as migrants from some land-mass lying without their domain, can only be determined by a reference to the still earlier fauna of not only this, but of other regions as well. In the case of the bears, for example, no immediate ancestors of the tribe have thus far been discovered in the Western Hemisphere antedating the Post-Pliocene epoch; on the other hand, in the Eastern Hemisphere—Europe—the remains of such animals, and of the true bears themselves, are abundant in deposits of the earlier Pliocene age. Hence, the assumption appears almost unavoidable that the North American fauna received its ursine contingent from the Old World. The same may or may not be also true of the American Bovidæ; but the determination of this question is made difficult, or impossible, through the fact that at least two of the genera—Ovibos and Bison—occur fossil in the Post-Pliocene deposits, and there only, of both the Old and the New World, and consequently appear in the two hemispheres as being of approximately equivalent age. Yet the fact that neither goats, sheep, oxen, nor antelopes have thus far been discovered fossil on the North American continent, while their remains are sufficiently abundant in the deposits of Eurasia (Europe-Asia) of Post-Pliocene or even much older age, would seem to indicate that the true home of the Bovidæ is the Old World, whence, by gradually spreading, and through the facilities afforded them in the way of a northern land connection, they eventually came to occupy a considerable portion of the New World as well. The giant sioth-like forms, such as the Megalonyx, Megatherium, and Mylodon, which in North America are associated with the remains of animals of indisputably Post-Pliocene age, occur in South America in an older formation, the Pliocene, and thus seemingly represent an invasion of the north from the latter continent. This conclusion appears further borne out by the circumstance that the Southern Hemisphere is the home of the animals of this class, and that, with scarcely a single exception (Moropus, ? Morotherium) no edentate form has thus far been discovered in any North American deposit antedating the period which represents the development of the

South American forms. Similarly, the extinct proboscideans, mammoth and mastodon, are of later date in America than in Eurasia, and are in all probability to be traced back to the latter region for the place of their birth.

The countries of the Old World present to us perhaps no less direct evidence as to the origination of, or the lines of migration taken by, specific groups of organisms. The European mammalian fauna is at the present time not very unlike in its general features that of North America, but in the geological period immediately preceding the present one it numbered a host of forms wholly differing from anything known to have existed in the corresponding period of American history, and, indeed, quite different from anything now inhabiting Europe. Such, for example, were the mammoth, African elephant, hippopotamus, African lion, leopard, the spotted and the striped hyena, several species of rhinoceros, &c , forms the greater number of which are at the present day associated with the region lying south of the Mediterranean. The question that here presents itself is one, perhaps, that cannot be fully answered, but yet one whose partial solution is made very nearly certain. Did this fauna become suddenly exterminated, through some agency or other, in the region inhabited by it, or did it migrate elsewhere? There can be but little doubt that both conditions took place. The mammoth and the several species of (fossil) rhinoceros are now all extinct, and there is every reason to believe that their tribes perished gradually, without their having accomplished much migration immediately preceding final extermination. The case is, however, different with the other forms, for the fact of their inhabiting the African continent leads one to suspect that they may have found their way thither by way of some land connection no longer remaining. That such a connection uniting the two continents did exist within a comparatively recent geological period, permitting of an interchange of the respective faunas, is certain, as is proved by the numerous ties which bind together the faunas of the opposite shores of the Mediterranean. The Barbary ape of the Rock of Gibraltar inhabits Morocco, while the ichneumon of Spain, the porcupine of Italy, and the fallow-deer of the south of Europe generally, are all forms inhabiting the north of Africa as well. These animals evidently crossed over the intervening sea by some route or other, and, as has already been stated, in comparatively recent times, otherwise

while the type-forms represented on the opposing shores might have
been alike, the species would have almost undoubtedly differed.
Equally positive proof in this direction is furnished by the similari-
ties presented in the reptile and amphibian faunas. The shallow-
ness of the channel separating Spain from Morocco renders it practically certain that one such connecting land-mass occupied the
position of the present Straits of Gibraltar. On the other hand, the
finding of remains of several species of elephant in Sicily and Malta
is almost proof positive of a second connection having been formed
between Italy and Tunis. An elevation of the present bed of the
sea a few hundred fathoms would bring about this result. The
Mediterranean would then consist of two land-locked basins. But,
doubtless, many of the other islands besides Sicily and Malta were
united with the mainland, for otherwise it would be impossible to
explain the distribution of several modern animals, the moufflon,
for example, which is found in Sardinia, Corsica, Crete, and the
mountains of Greece.

Granting this connection between Africa and Europe, it appears
more than likely that the principal disturbing element which reacted
upon the Post-Pliocene European fauna, the great northern ice-sheet
and the accompanying cold of the glacial period, rather than causing the complete or sudden extermination of the receding fauna,
compelled it to migrate over into regions of a more congenial climate. That such was the fate of many of the forms there can be
no reasonable doubt. The African continent thus became stocked
with its existing fauna largely from the more temperate northern
regions. But there is every reason to believe that these same southward retreating forms were in great part primarily introduced into
Europe from Africa, and over the same routes by which the later
southerly migration was effected. Concerning the origin of the
African fauna itself we possess little precise information. The
paleontology and geology of the region are so imperfectly known
that we possess as yet no basis for satisfactory deductions. The
absence of sufficient data naturally renders uncertain all speculation
relating to the late European fauna as well. It may be considered
highly probable, however, that many of its characteristic elements
have been derived from the region about India, where a considerable
antiquity, extending back to the Miocene or early Pliocene period,
is proved for at least a number of the more prominent types. Seve-

ral of the antelopes have related, and apparently ancestral, forms in the Miocene deposits of Greece (Pikermi), which also contain a form not very far removed from the giraffe (Helladotherium), and a species of true giraffe itself (Camelopardalis Attica), so that possibly a contingent of the African fauna may have been derived from this region. Whether the southern or Ethiopian portion of the continent was at one time since the introduction of the placental Mammalia completely severed from the northern part or not there are as yet no means for determining. That Madagascar at one time formed part of the continent is indisputably proved by the character of its fauna ; but that its subsequent isolation is of very ancient date is conclusively shown by the complete absence of all the more distinctive Ethiopian placental mammals.

The few examples that have been cited in illustration of the appearance and disappearance of faunas are sufficient to show the character of the investigation that is open to the zoogeographer. While from the data that we now possess much can be done towards shaping our suppositions, it must be confessed that our knowledge is still much too limited to permit of very satisfactory conclusions being drawn therefrom. The principal danger that besets any investigation in the direction here outlined arises from the very natural assumption that the greater antiquity in any one region over another of a given type of animal indicates its prior appearance there, and migration thence to one or more secondary regions. This assumption might be well founded if we were only half conversant with the past paleontological histories of the regions under consideration; but where at best our knowledge is still very imperfect, as it is in the case of Africa, Asia, and South America, it would be, to say the least, highly injudicious. For what evidence have we that animal types not yet found, or dating back only to a comparatively recent period, might not some day be turned up in abundance, and in deposits of such age as to completely overthrow any deductions that may have been based upon their supposed non-occurrence ? A single illustration of this kind will suffice. Paleontologists are in the habit of considering the camels a New World family, which by migration finally occupied the region which it now inhabits. This conclusion is based upon the circumstance that numerous cameloid forms (Pliauchenia, Procamelus, Protolabis, Poebrotherium) carry this line of animals back in the North American continent to the early Miocene period, whereas such types are almost wholly wanting in the range of equivalent deposits of the Old World. Yet, if this is the true history of the family, it is certainly a surprising fact that the true camel itself (Camelus), which is entirely unknown on the American continent, should already be found fossil in the Miocene (or older Pliocene) deposits of India. Nor is it at all unlikely that ancestral forms leading up to this type may yet be found in deposits of still older age hereafter to be discovered.

II.

It is a fact of general observation that a given species of animal is so restricted in its range as to entitle the geographical area principally occupied by it to be considered as its home. This home may be limited in its extent to a very narrowly circumscribed area, possibly not embracing more than a few square miles, or even less, or it may spread out to dimensions coextensive (or nearly so) with the continental boundaries; or, finally, it may comprise considerable portions of two or more continental areas combined. As examples of animals having a very restricted geographical distribution may be cited the Pyrenean water-mole (Myogale Pyrenaica), a small insectivore found only in a very few localities of the northern valleys of the Pyrenees, and a species of buschbok (antelope, Cephalophus Natalensis), whose habitat is the region about Port Natal, South Africa. Arctomys caudata, one of the Asiatic marmots, is confined to the elevated valley of Gombur, in India, and to heights exceeding 12,000 feet. Of birds, whose powers for self-distribution are much more fully developed than among mammals, we have equally pointed examples of localisation. The brown-and-white cactus-wren (Campylorhynchus albibrunneus) is confined exclusively to the Isthmus of Panama, where its range is also somewhat limited; the Bornean yellow-bulbul (Otocampsa montis) has only been met with on the peak of Kina-Balu, in Borneo; and the red bird-of-paradise (Paradisea rubra) only within the narrow limits of the island of Waigiou, lying to the northwest of New Guinea. The most remarkable instances of localisation are probably afforded by the humming-birds, several species of which would seem to be restricted respectively to the volcanic peaks of Chimborazo and Pichincha, in the equatorial

Andes, and to the extinct crater of Chiriqui, in the province of Panama, Colombia. The Loddigesia mirabilis, one of the most beautiful of the Trochilidæ, has been observed thus far only at Chachapoyas, in the Peruvian Andes, and even there it occurs so rarely as to have been obtained but once during the period of forty years following its first discovery.[1]

Too much stress should not, however, be laid upon what would appear to be the absolute localisation of a species, since such supposed localisation is frequently only the expression of our defective knowledge in the premises. In the case of the famous South American oil-bird, or guacharo (Steatornis Caripensis), for example, which was for a long time considered to inhabit solely a cave near Caripé, in the province of Cumaná, Venezuela, more recent research has revealed a comparatively broad area of distribution, which embraces Sarayacu and Caxamarca in Peru, Antioquia in Colombia, and the Island of Trinidad. The garden-mouse (Mus hortulanus), which for some twenty years was known only from the botanic gardens of Odessa, Russia, has been found in abundance in Kaschau and several other towns of Northern Hungary.[2] So, likewise, in the case of the anthropoid apes of the genus Troglodytes, which were formerly supposed to be restricted to the western regions of the African continent, but which the more recent explorations of Schweinfurth, Von Heuglin, and others have shown to inhabit East Central Africa as well.

Of species having a very broad distribution—excluding such as have been transplanted through the agency of man—may be cited the African elephant, whose domain extends over the greater part of the African continent south of the Sahara Desert ; the tiger, whose habitat embraces the entire east and west extent of Asia, from the Caucasus to the Island of Saghalien; and the ermine, which is found throughout the greater portion of the temperate and boreal regions of the Northern Hemisphere. The leopard ranges over entire Africa and throughout most of Southern Asia, having, with perhaps the exception of the common European wolf, whose identity with the various forms of American wolves is conceded by many naturalists, and some of the smaller carnivores, the most wide-spread distribution of any mammalian species. There is but little question as to the identity of the North American and European species of brown-bear, Arctic fox, glutton, ermine, weasel, elk, reindeer, and beaver,[2] all of which have, consequently, a very extended range. The American panther or couguar (Felis concolor) inhabits the territory included between Canada and Patagonia, an extent covering upwards of one hundred degrees of latitude, which probably represents the greatest north and south range of any mammal.

As might naturally have been expected from the greater facilities for dispersion, we find many more marked instances of broad specific distribution among birds than among mammals. Indeed, when we consider with what apparent facility certain birds accommodate themselves to the varying conditions of atmospheric pressure and climatic changes, and the readiness with which they traverse broad expanses of the oceanic waters—e. g., the North Atlantic between Ireland and Labrador—it might at first sight appear as though there ought to be, at least in many cases, no absolute limit to their distribution ; yet, from our present knowledge, it may safely be affirmed that there exist but very few species of birds which are in any way cosmopolitan. The fish-hawk (Pandion haliaëtus), with probably the most extensive range of any known bird, inhabits the greater portion of all the continents, with the possible exception of Australia, where its place appears to be supplied by a closely-allied (and by many ornithologists considered identical) species, the P. leucocephalus. Scarcely, if at all, less extensive is the range of the common peregrine falcon (Falco communis or peregrinus) and the barn-owl (Strix flammea), the former of which is distributed, according to Professor Newton, from " Port Kennedy, the most northern part of the American continent, to Tasmania, and from the shores of the Sea of Okhotsk to Mendoza, in the Argentine territory," and the latter, according to Sharpe, over the entire world, with the exception of New Zealand, and many island groups of Oceania, Malaysia, &c. The common American raven (Corvus corax) has, likewise, a very broad distribution, its range extending from Mexico into the far north, over the whole of Europe and Northern and Central Asia, as far east as the Island of Saghalien.

The fishes present scarcely less well-marked examples of broad distribution; but in such aquatic forms the physical conditions of the medium which they inhabit offer far less obstacles to a very general diffusion than are to be encountered in the case of terrestrial animals. The same holds true with other aquatic animals capable of self-locomotion, and, indeed, in the case of those pelagic forms whose dispersion or "migration" is less a matter of volition than the result of an interaction of extraneous physical causes there would seem to be no barriers set to a practically universal distribution. But here, too, Nature has set a limit to the possibilities of migration, and, therefore, even among those lower forms which might be considered best adapted for withstanding the varying physical vicissitudes of their surroundings we meet with but very few species whose distribution might be said to be in any way cosmopolitan. The free-swimming pteropods, or winged - Mollusca, and medusoids, although exhibiting individual examples of very broad distribution, are still more or less restricted specifically to well-defined oceanic areas, whose boundaries may in a measure be dependent upon the prevalent surrounding water-currents. Shells of the Spirula Peronii, a member of the two-gilled order of cephalopods, are met with almost all over the oceanic borders, as well in the temperate as in the tropical zones, but, owing to the extreme rarity of the animal itself, which has been observed, perhaps, but a half-dozen times, it is impossible to say what the exact, or even approximate, range of the species is, and, consequently, of how much of the area of the distribution of the shell it partakes. The common form of argonaut (Argonauta argo) is found in the tropical parts of the Atlantic, Pacific, and Indian oceans, and in the Mediterranean Sea, and it has been met with as far north in the Atlantic as the New Jersey coast, and as far south as the Cape of Good Hope. The animal might, therefore, be said to be almost cosmopolitan.

It may be laid down as a fundamental law in geographical distribution that the areas inhabited by a given species are continuous with each other; in other words, we do not find, except at rare intervals, and under peculiar circumstances, the same species of animal inhabiting distantly-separated localities, in the interval between which no individual of the species is to be met with. Thus, in the entire range of the leopard there occurs no district of any significance where the animal may not be confidently looked for, and which would negatively tend to render its distribution discontinuous. And the same may be said of the hundred or more degrees of latitude prowled over by the couguar, an animal whose home is at one place the lowland forests, at another the elevated mountain plateaus, and at a third the grassy savannas and rolling plains. Naturally, in the case of such animals as are dependent for their existence upon certain physical peculiarities of their environment, or upon particular conditions of food and climate, we shall meet with local areas scattered through the region of distribution of a given species where no individuals of that species are to be met with, an apparent discontinuity being thus presented. For instance, such denizens of the forest as the South American monkeys and the sloths will but very exceptionally be found anywhere else than in their forest homes, and, therefore, the partial destruction of this forest, or its invasion by a grassy savanna, will tend to render the "home" of those animals discontinuous. Possibilities of such, or a similar, discontinuity may likewise arise in the case of the animals of the plains, marshes, and deserts, since the physical aspects of the earth's surface are constantly subjected to vicissitudes of greater or less magnitude, and, as a matter of fact, we find numerous instances where, in an extensive range, particular animals are restricted in their habitats to certain favoured spots or localities. But in all or most of such instances a former, and comparatively recent, continuity of area, or possibility of migration from one locality to another, can be proved. The chamois, whose range embraces the entire east and west extent of Southern Europe, is found almost exclusively on the higher mountain summits—the Pyrenees, Alps, Carpathians, Caucasus, and the mountains of Greece—and would appear, therefore, to occupy several widely-removed habitats. But there can be no reasonable doubt that the peculiar distribution of this animal is the outcome of migration from a central home. The hippopotamus is found in the Nile, Niger, Senegal, and most of the larger rivers of South Africa, between which stretch vast areas where no individuals of the animal have ever been found—regions untenantable by reason of their aridity; but here, as in the case of the chamois, there can be no doubt that a migration or diffusion did take place at a time when the physical aspects of the country were favourable for such a dispersion, and were, consequently, different from what they are at present. One of the most remarkable instances of areal discontinuity among mammals is that exhibited by the variable hare, whose home, in the Old World, is Eurasia north of the fifty-fifth parallel of latitude. The animal reappears, after skipping the lowlands of Central Europe, in the Pyrenees, Alps, and the Bavarian Highlands, and again in the Caucasus, the last region isolated by fully one thousand miles of non-inhabited country. Equally striking examples were supposed to be afforded by the fresh-water seals of Lake Baikal and the brackish-water species of the Caspian, which were considered to be identical with the northern Phoca fœtida and P. vitulina respectively, but more careful study has shown this identification to be erroneous.[4] The critical studies made by Mr. Seebohm of the Central and East Asiatic faunas have disclosed a number of extraordinary instances of discontinuous habitation among birds. One of these is exemplified in the case of a South European variety of the common marsh-tit (Parus palustris), which reappears in an undistinguishable guise in China, although in an intervening tract of some four thousand miles (east of Asia Minor) the variety is entirely wanting, being replaced by one or more closely related forms. Ceryle guttata, a spotted king-fisher, appears to be confined to Japan and the Himalaya Mountains, being completely wanting in China; and the same is true of a species of crested eagle (Spizaëtus orientalis), with the exception that its range embraces the Island of Formosa. Similarly, we have two species of birds, the rufous-breasted fly-catcher (Siphia superciliaris), and the Darjeeling wood-pigeon (Palumbus pulchricollis), which are absolutely confined to the Himalayas and the Island of Formosa.

But while individual cases of species inhabiting discontinuous areas do present themselves, they are of comparatively rare occurrence, and the general law of regional continuity may be recognised. In a region occupied by a given species of animal there is usually an area which is *par excellence* more thickly inhabited than any other, and which may, consequently, be termed the "metropolis" of that species. From this metropolis there is in most cases a radial distribution of the individuals of the species, with a thinning out towards the periphery. Distinct species of the same genus rarely have coincident geographical distributions; in other words, they rarely occupy precisely the same areas, but more generally these areas, if at all continuous, overlap each other to a greater or less extent. This fact is beautifully exemplified in the case of the American hares, which are represented by some eleven species, and about as many well-marked varieties. Commencing at the far north, we have the polar or variable hare (Lepus variabilis or L.

timidus, var. Arcticus), whose range extends from the Arctic coast southward to Newfoundland, and in the interior to Fort Churchill, on Hudson's Bay. Along its southern confines it meets and slightly overlaps the boundaries of the northern varying hare (L. Americanus), which, in its several geographical varieties, is distributed from the Barren Grounds in the north southward to a zone which corresponds generally with the isotherm of 52° F. On the Atlantic coast region, the southern limit of this species appears to be Connecticut; along the line of the Appalachian highlands, Virginia (or possibly North Carolina); and in the Rocky Mountain region, New Mexico. Lepus Americanus is found throughout the northern parts of nearly all the northern tier of States interposed between the Missouri and the Atlantic coast, and over the greater portion of this vast area of distribution, which is continued westward to the Pacific, it forms the sole representative of the family. In the south its habitat overlaps the range of the wood-hare (L. sylvaticus), which, in its several varietal forms, is distributed along the Atlantic coast from Southern New England to Yucatan. Westward, the range of this species extends quite, or very nearly, to the Pacific, keeping, however, to a course south of the isotherm of 45° F. The prairie-hare (L. campestris) is found in the interior region, principally between the isotherms of 56° and 36°, its range being consequently overlapped on the north by that of Lepus Americanus, and on the south by L. sylvaticus. In the Southeastern United States there are two distinct species, L. palustris and L. aquaticus, which are almost exclusively confined to the marshy lowlands, and whose habitats, extending to Yucatan on the south, are partially comprised in those of the wood-hare and jackass-hare (L. callotis), the last a western species, whose range descends into the arid interior of the Republic of Mexico. Finally, we have a solitary species of South American hare (L. Brasiliensis), whose reputed range embraces a considerable portion of the continent from Patagonia to Panama, continuing thence into Central America.[5]

It frequently happens that the boundaries of a given species are sharply defined against those of another, stopping just where the others begin, and where, consequently, no overlapping takes place. Such cases of specific limitation occur where natural obstacles to a free migration are suddenly encountered, as where mountain or water barriers project themselves into a given region. Thus, it will not rarely be found that a genus of animals is represented by one or several species on one side of a long mountain-slope, and by entirely distinct species on the other. And, similarly, distinct species of a genus may be encountered on opposite sides of a river-bed, although instances of such a nature among the higher animals are probably not of very frequent occurrence. Mr. Wallace cites the case of certain species of Saki monkey (Pithecia), found on either side of the Amazon River, whose range either southward or northward appears to be limited by that stream. The same naturalist instances among birds species of jacamar (Galbula) and trumpeter (Psophia) which exhibit a similar limitation, particularly the latter, where five distinct species are relegated to as many distinct, but contiguous, geographical areas, separated from each other by the Amazon and some of its tributaries (Negro, Madeira, Tocantins). Of about twelve species of armadillo (separated by some naturalists into several distinct genera), most of which are inhabitants of Brazil, it would seem that not a single species is common to Brazil and the Argentine Republic, or the Argentine Republic and Paraguay, the Parana River, with its tributaries, evidently forming an insurmountable barrier to the passage of this animal. The Uruguay River appears in the same way to limit the eastward progression of the viscacha (Lagostomys trichodactylus), an animal allied to the chinchilla, although, as has been pointed out by Mr. Darwin, the trans-Uruguayan plains are fully as well adapted to the animal as those of its native home.

Just as the boundaries of land-animals are in many instances defined by the dominant river-courses, so, in a like manner, but in a much more marked degree, the domains of fresh-water forms are frequently circumscribed by the land surfaces bordering the waters inhabited by them. This fact is beautifully exemplified in the geographical distribution of two American families of fluviatile mollusks, the Strepomatidæ, or American melanians, and the Unionidæ, the fresh-water mussels, where the species of several genera, at least in the Southern United States, are restricted in their habitats to certain individual streams, to the exclusion of all others. Indeed, it would appear that even in such aquatic forms a large river may constitute an almost insuperable barrier to migration, as is shown in the case of the Strepomatidæ by the Mississippi (south of the line of the Ohio River), which but very few members of the family have been able to surmount. According to Tryon, only one species of the family, Goniobasis sordida, is positively known to be common to the region on both sides of that great stream.[6]

Probably no group of animals, as Mr. Wallace well observes, illustrates in a more striking manner the extreme features of specific distribution than the true jays, birds of the genus Garrulus. About fourteen species are recognised by ornithologists, whose combined domain embraces the entire east and west extent of the continent of Eurasia, from the Bay of Biscay to the Sea of Okhotsk, and also includes the continental British Isles on the west, and the Japanese group on the east. Most of these species occupy independent areas of their own, or areas which but barely overlap on their contiguous borders. Thus, the common jay (Garrulus glandarius) inhabits the greater portion of the semi-continent of Europe, ranging from the Barbary States in Africa northward to about the sixty-fourth parallel of latitude (in Scandinavia and Russia), and east to the Ural Mountains. Along its southern border it meets the Algerian jay (G. cervicalis), a distinctly-marked species, and one having but a very limited range. On the southeast, again, its confines meet those of the black-headed jay (G. Krynicki), which occupies a somewhat circular district extending some distance on all sides of the Black Sea. Contiguous with this last is the region inhabited by the Syrian jay (G. atricapillus), a species very closely allied to the preceding, whose domain extends through Syria, Palestine, and Southern Persia. North of this we have the limited area occupied by the Persian jay (G. hyrcanus), which has thus far been found only on the Elbruz Mountains. In an almost direct line east of this region, but separated from it by a considerable area where no jays are to be met with, we pass consecutively over the haunts of the black-throated jay (G. lanceolatus), from the Northwestern Himalayas, the Himalayan jay (G. bispecularis), from the Himalaya Mountains to the eastward of Cashmere, the Chinese jay (G. Sinensis), from South and Central China (and, occasionally, Japan), and the Formosan jay (G. Taivanus). The home of the Burmese jay (G. leucotis) adjoins that of the Himalayan jay on the southeast. North of the belt occupied by the species of southern jay we have a vast region —the desert area of Central Asia, with Thibet, Turkestan, Mongolia, and Gobi—throughout the greater part of which no jays have as yet been discovered. Bounding this area on the north, and extending from beyond the Ural Mountains (Kazan) to the northern island of the Japanese group, there exists an almost continuous and comparatively broad belt which is tenanted throughout its entire extent, except where it overlaps the habitat of the common European G. glandarius, by a solitary species, known as Brandt's jay (G. Brandti). Finally, in the southern island of Japan there are found two species, G. Japonicus and G. Lidthi, the former of which, singularly enough, is the species which is most nearly allied to the common European jay, although separated by the greatest distance from it.[7]

Generic Distribution.—The laws governing specific distribution are in considerable measure likewise applicable to the distribution of genera. Thus, we have genera that are restricted to very limited areas, and, as a necessary consequence resulting from specific distribution, those whose areas are coextensive with continental boundaries, or embrace portions of two or more continents; and, again, we have genera of a given family which occupy contiguous, overlapping, or discontinuous provinces. The localisation of a genus to an exceptionally narrowly circumscribed area, such as we have seen in the case of the species of humming-birds of the volcanic peaks of South America, can almost necessarily obtain only there where the number of species belonging to the genus is also exceptionally limited, or, more nearly, when the genus is coextensive with a single species. Potamogale, which comprises the single species P. velox, a singular otter-like insectivore of the west coast of Africa, appears to be confined to the region included between Angola and the Gaboon ; Chœropsis, with the single species C. Liberiensis, an animal closely allied to the true hippopotamus, inhabits, as far as is yet known, only the wilds of Liberia; and, likewise, the singular carnivore constituting the genus Ailurus (A. fulgens) has been met with only in the Southeastern Himalayas. Instances of restriction are much more numerously presented in the case of insular than of continental faunas, whether the examples be taken from the class of birds or mammals.

Genera of very broad, or almost world-wide distribution, are of frequent occurrence, both among the lower and higher animals. Among the latter, in the class of birds, we have numerous examples among the swimmers, waders, and birds of prey, whose range covers the greater extent of the primary divisions of the earth's surface, and which may, consequently, be said to have a cosmopolitan distribution. Generic groups with a nearly world-wide distribution among the Mammalia are of much rarer, although of not exactly infrequent, occurrence, and if the Australian dingo, a species of wild dog, be not considered indigenous to the country which it inhabits, there would appear to be, if we except the bats, not a single altogether cosmopolitan genus among that class of animals. Leaving out of consideration the continent of Australia, whose mammalian fauna is deficient in nearly all the orders of the class, we have a considerable number of genera whose range comprises the greater portion of the habitable globe.* Thus, the members of the genus Felis (cats) are spread throughout the entire expanse of the continents of both the Eastern and the Western Hemisphere, through regions the extremes of whose temperature may be measured by probably no less than 225 degrees of the Fahrenheit scale. The genus Canis (dogs) has an almost equally broad distribution ; and the same range is exemplified in the case of the weasel genus (Mustela). Ursus, the bear, is met with throughout the greater part of the Northern Hemisphere, and in the continent of South America the genus has one or more representatives whose habitat is situated considerably to the south of the Equator.† The genus Cervus (deer), in its broader sense, has representatives in both North and South America, Europe, and Asia, with a very limited number of species (fallow-deer, stag) in Africa north of the Sahara.

Discontinuous generic areas, like specific areas, are of comparatively rare occurrence. Among the most remarkable instances of such discontinuity we have that exhibited in the case of the

---

* The only placental animals indigenous to the Australian continent, if we exclude the rather doubtful dingo, which is by most naturalists considered to have been introduced by man, are the Cheiroptera (bats) and Rodentia, the latter represented by the family of mice (Muridæ). The implacental mammals—kangaroos, wombats, phalangers—have, on the other hand, an extraordinary development.

† The solitary species of bear inhabiting the continent of Africa appears to be confined to the Atlas Mountains ; it constitutes the genus Helarctos of some authors (H. Crowtheri).

genus Myogale, the water-mole, already referred to, which embraces two species, one of which, M. Pyrenaica, is an inhabitant of
the northern valleys of the Pyrenean chain of mountains, and the
other, M. Muscovita, the plains of Southeastern Russia skirting the
Don and Volga rivers. The pikas (Lagomys), small rodent animals
having a rather near relationship with the hares, which are extensively distributed along the upper mountain heights from the Ural
to Cashmere and the eastern extremity of Siberia, have a single
outlier in the Rocky Mountains of North America. The members of
the genus Capra—the goats and ibexes—occupy disjointed patches
of territory in Europe, Asia, and Africa, mainly confined to the
elevated mountain regions, such as the Pyrenees, the Sierras of
Spain, the Alps, Caucasus, Himalayas, &c., the intervals between
which are deficient in the wild or indigenous representatives of the
genus. A similar discontinuity is exhibited in the case of the
snow-partridges of the genus Tetraogallus, a bird likewise partial
to the elevated mountain-slopes. Numerous other instances of
birds occupying discontinuous areas may be cited, and they appear
particularly noticeable among families of a more or less tropical
habit. Such, for example, are the jaçanas (Parra), which inhabit
the tropical regions of both the Old and the New World, the similarly distributed flamingoes (Phœnicopterus), the wood-ibises of
the genus Tantalus, the gerontics (Geronticus), and the marabou
storks (Mycteria). Among perching birds a most remarkable instance of generic discontinuity has been cited by Wallace in the case
of the blue magpies (Cyanopica), which comprise two species, one of
which, C. Cookei, inhabits the Spanish Peninsula, and the other, C.
cyanus, Eastern Siberia, Japan, and North China, the habitats of
the two being removed from each other by an interval of fully 5,000
miles. Still more marked is the case of the bluebirds constituting
the genus Sialia, all of whose members, with one exception, inhabit
temperate and tropical America; a solitary form, Sialia (Grandala)
cœlicolor, singularly enough, crops up again among the Himalaya
Mountains, and eastward throughout the mountainous region separating China from Thibet.* The most remarkable instance of a
mammalian genus occupying two widely - removed areas is furnished by Tapirus, the tapir, several species of which are natives
of the South American continent, and one, very distinct from the
others, of Malacca and Borneo, the group of animals, therefore,

appearing at localities separated from each other by nearly half of the earth's circumference.

Distribution of Families.—The restriction of families to certain local areas is of comparatively rare occurrence, an almost necessary consequence of the number of species and genera of which they are in most cases composed. Among mammals the Cheiromydæ, with one genus and one species, the aye-aye (Cheiromys Madagascariensis), are confined exclusively to the Island of Madagascar; the Protelidæ, likewise consisting of but a single genus and species, the aard-wolf (Proteles Lalandii), an animal in several respects intermediate between the cats and dogs, and considered by some as representing a greatly modified form of hyena, are confined to the extra-tropical regions of South Africa. Occupying pretty nearly the same region, and confined to it, are the Chrysochloridæ, or golden-moles, with a single genus and about five species. The Ailuridæ, a group of animals having their nearest allies in the coatis and bears, and consisting of one or two species, appear to be restricted to the forest region of Eastern Thibet and the Eastern Himalaya. Among the class Aves we have likewise families that are restricted both as to the number of species comprised by them and the region which they inhabit. The Paictidæ, a group of birds, considered by some ornithologists to have affinities with the American ant-thrushes (Formicariidæ), and by others with the Old-World pittas, consist of a single genus and two or more species, both of which are confined to the Island of Madagascar. Here, also, exclusively belong the Leptosomidæ, birds allied to the cuckoos and rollers. The Apterygidæ, with one genus (Apteryx) and four species, are strictly confined to the two larger islands of New Zealand; the Drepanidæ, with some four or five genera and ten species, are restricted to the Sandwich Island group; and, finally, the paradise-birds, excluding the bower-birds, which are classed together with them in one family by some ornithologists, with about eighteen genera and thirty species,[8a] are almost entirely confined to New Guinea and the surrounding islands, only four representatives of the group finding their way into the neighbouring continent of Australia. Mr. Wallace has emphasised the very remarkable case of localisation presented among reptiles by the Uropeltidæ, or rough-bellied, burrowing snakes, all of whose members appear to be strictly confined to Ceylon and the adjacent parts of the Peninsula of India.[9]

Families with restricted ranges, like genera and species, are of infrequent occurrence, broad distribution being with them the rule rather than the exception. Nevertheless, owing to the peculiarily isolated position of the Australian fauna, there are among the land Mammalia only two families which can lay claim to being strictly cosmopolitan. These are the mice (Muridæ) among rodents, and the Vespertilionidæ among bats, the former being universally distributed throughout the globe, if we except some of the island groups of Australasia. The Vespertilionidæ have representatives almost everywhere, being apparently limited, as stated by Wallace, only by the necessities of procuring insect food. Among birds, examples of practically cosmopolitan families are presented by the thrushes, warblers, crows, swallows, king-fishers, goatsuckers, and pigeons. The hawks, owls, ducks, and gulls are cosmopolitan *par excellence*, being found in almost every habitable locality throughout the globe, whether on the mainland or on the most distantly removed oceanic islands. The extensive family Fringillidæ (finches, buntings), as now generally constituted by ornithologists, with upwards of seventy genera and five hundred species, appears to have no representative in Australia, all the finch-like birds of that continent belonging to the family of the weavers (Ploceidæ).

As with genera and species, so likewise in the case of families, we have numerous instances of groups occupying discontinuous areas. In the class of Mammalia, for example, the swine (Suidæ), which are so extensively distributed throughout both the tropical and temperate regions of the Old World, have no representatives in the New World north of about the thirty-fourth parallel of latitude —the Red River, in Arkansas—although they have two species (of peccary) in the region south of that line. The Orycteropodidæ have a solitary representative in the Cape District, the aard-vark, or Cape ant-eater (Orycteropus Capensis), and another in the interior of Northeast Africa and in Senegal, the form occurring in the latter region being possibly a third species.[10] The tapirs, constituting the family Tapiridæ, have, as already stated, their representatives on opposite sides of the globe, one species inhabiting the Malay Peninsula and some of the adjacent islands, and the four or five others the tropical forests of Central and South America. The chevrotains, or deer-like animals of the family Tragulidæ, abound in India and some of the islands of the Malay Archipelago, where they constitute the genus Tragulus; a solitary representative of the same family, but belonging to a distinct genus (Hyomoschus), is a native of West Africa. The anthropoid apes (Simiidæ) are represented in Western (and probably also East Equatorial) Africa by one or more species of gorilla and chimpanzee (Troglodytes), which are almost exclusively confined to the forest region. The form most nearly allied to these man-like apes, and belonging to the same family, is the orang (Simia satyrus), which, as an inhabitant of the islands of Sumatra and Borneo, is encountered after an interval of not less than seventy degrees of longitude. Inhabiting the same region, but with a northward extension to China, and westward to Assam (south of the Brahmaputra River), we find the members of the genus Hylobates, the gibbons. Probably the most striking example of a divided family is furnished by the Camelidæ, which in the Old World are represented by the genus Camelus, with two species—the dromedary and the Bactrian camel—whose habitat extends from the Sahara through the desert regions of Western and Central Asia to Lake Baikal; and in the New World by the genus Auchenia (the llama, alpaca, vicuña, and guanaco), with about four species, all of them restricted to the mountainous and desert regions of Western and Southern South America. We have here, therefore, a family which is not only divided by a vast ocean and the greater mass of two continents, but the members of which, in one hemisphere, inhabit the region north of the Equator, and, in the other, the region south of it.

Instances of divided families among birds occur as in mammals, although probably to a less marked extent, owing naturally to their increased facilities for dispersion; such division obtains more especially among the so-called "tropicopolitan" forms, or those whose homes are properly the region of the tropics, or that immediately adjoining it. The flamingoes (Phœnicopteridæ), consisting of a solitary genus and about eight species, are about equally distributed as to the number of species throughout the warmer regions of America, Africa, and Asia, some of the forms extending their range to a considerable distance within the bounds of the Temperate Zone, as in Southern Europe and South America. The trogons (Trogonidæ), comprising many of the most beautifully-arrayed of birds, and with upwards of forty species, are more strictly confined to the tropical regions of the earth's surface, but few forms being found beyond the limits of that zone. They are fairly abundant in the forest region of South America, ranging from Paraguay to Mexico, and less so in South and Southeast Asia, and some of the islands of the Malay Archipelago. In Africa the family is represented by but two species. The Psittacidæ among parrots furnish us with another good example of a divided family, whose members are to be found only in the two great southern continents, Africa and South America, and in some of the adjacent islands. Still more remarkable is the case of the ostriches, of which there are two species (of the genus Struthio) pertaining to the desert regions of Africa and Western Asia (Arabia and Syria), and likewise two (of the genus Rhea, sometimes placed in a distinct family) belonging to temperate South America, whose range extends from Patagonia to the confines of Brazil.

Among reptiles similar instances are presented by the tropicopolitan groups. Thus, we have the Crocodilidæ inhabiting the tropical waters of both the Eastern and Western Hemispheres. The Pythonidæ, or giant constricting serpents—boas, anacondas, pythons—are, with the exception of the Californian genus Charina, sometimes referred to this family, distinctively tropical, but they have representatives in the South American continent, in Africa, Asia, Australia, and in several of the continental and oceanic islands. The family of iguanas (Iguanidæ), comprising upwards of fifty genera and some three hundred species, is almost distinctively American, being distributed from about the fiftieth parallel of south latitude, in Patagonia, to the Canadian boundary-line on the north. No member of the family is known from either of the continents of Eurasia or Africa, yet the family crops up again in a solitary genus—Brachylophus—in the Feejee Islands, and two (doubtfully placed) genera have also been described from Madagascar and Australia.

Distribution of Orders.—The principal features of geographical distribution exhibited by species, genera, and families repeat themselves in a measure in the case of the higher groups of the animal kingdom known as orders. Very narrowly circumscribed areas of habitation, at least among the orders of higher animals, do not exist; broad distribution is the rule. Among mammals the most marked instances of semi-localisation, if so it may be termed, are furnished by the Monotremata, comprising the two families of duck-bills and echidnas, both restricted to Australia and the Island of Tasmania, and the Hyracoidea, an order consisting of two genera, Hyrax (the coney) and Dendrohyrax, and about a dozen species, all of which are restricted to the continent of Africa and the immediately adjoining parts of Asia (Syria). The only orders of terrestrial mammals which can lay claim to being cosmopolitan are the Cheiroptera (bats) and Rodentia, none of the other orders, except the Marsupialia—unless the dingo, as a member of the Carnivora, be considered indigenous to the continent it now inhabits —having any representatives in Australia. Among birds we have no instance of an order being restricted to the limits of a single continent. Among reptiles the Crocodilia are almost entirely confined to the tropical and sub-tropical regions, and occur in both the Eastern and Western Hemispheres. The Ophidia (serpents) have what might be called a world-wide extension, although no member of the order has been met with farther to the north than the Arctic Circle. The order Anura (frogs and toads) among amphibians is very nearly cosmopolitan ; the Urodela, on the other hand, comprising the tailed forms, such as the newts, salamanders, &c., are almost strictly confined to the Northern Hemisphere, a few only of its representatives passing through Central America as far south as Colombia. The entire class of the Amphibia (as indigenous forms) is absent from the vast majority of oceanic islands—New Zealand, New Caledonia, and the Andaman Islands, and possibly the Solomon and Seychelles groups, almost alone, according to Darwin, presenting exceptional instances.

The marsupials afford a remarkable example of a comparatively large order of animals occupying widely separated and discontinuous areas. With the exception of the opossums, of which there are two genera and about twenty species, confined to the two continents of America, and more particularly to the tropical regions of these continents, all the members of this peculiar and lowly-organized order of animals are strictly limited in their range to the Australian continent and its dependent islands, and some of the islands of the Malay Archipelago. In all the broad intervening region—Europe, Africa, and Asia—no representative of the order is to be met with. The Edentata—ant-eaters, armadillos, &c.—are largely confined to the tropical and sub-tropical regions

3

of South America, Asia, and Africa, and are almost completely absent from the vast northern tracts which spread out towards the polar confines, and tend to bring together the terrestrial areas of the Old and the New World. The order is entirely wanting in Europe, and nearly so in North America, the genus Tatusia, an armadillo, alone penetrating within the boundaries of the last into the State of Texas. In Asia no member of the order is found to the north of the Himalaya Mountains. A remarkable example of discontinuous habitation among birds is furnished by the Struthiones, or ostrich-like birds, whose members are distributed throughout considerable reaches of tropical and sub-tropical South America, Africa, Asia, and Australia, some of the Australian islands, and New Zealand, and are entirely wanting in Europe and North America. It is a singular circumstance in connection with the distribution of the birds of this very limited order that two genera, so closely allied as are Rhea and Struthio, should occupy areas so distantly removed from each other as Africa and South America. The Psittaci, or parrots, inhabitants of both the New and the Old World, may likewise be considered as being preeminently tropical and sub-tropical, for although a few examples are found whose range in the Southern Hemisphere ascends to the fifty-fourth degree, yet the true home of the order is located in the zone embraced between the thirty-fifth parallels north and south of the Equator. Being absent from Europe and the greater portion of the continent of North America, the distribution of the order is necessarily discontinuous.

III.

Conditions affecting distribution.—Climate.—Food-supply.—Barriers to migration.—Migrations of mammals and birds.—Dispersal of insects and mollusks.

Of the Conditions which affect or limit Distribution among Animals.—Climate.—It is a common belief that the principal factor limiting or regulating the distribution of animals is constituted by climate; in other words, certain groups of animals are associated with certain grades or conditions of climate, beyond the reach of whose interacting influence they could no longer maintain an existence. Thus, among quadrupeds, the elephant, camel, and tiger are popularly associated with the hottest climates of the earth's surface; the reindeer and moose with climates of equal, but opposite, severity. And, similarly, among birds, the ostriches and hummers are considered to be particularly indicative of hot or tropical climates, and the auks, guillemots, puffins, and penguins, as products of the cold northern or southern climes. That climate does regulate distribution, or impose a bar upon the migration of certain forms of life, there can be no manner of doubt; but that it does not exercise the paramount influence that is generally attributed to it there can likewise be no question. Taking, for example, some of the instances that have just been mentioned as indicating the supposed association between animal distribution and given conditions of climate, we find that the tiger, while its home, *par excellence*, may be considered to be the hot districts of India and the Indian Archipelago, is in no way restricted in its range to those regions, or to regions having at all a similar climate. Thus, the animal is found in the elevated regions of the Caucasus and the Altai chain, and in the Himalaya range its footprints are not infrequently found impressed in the fields of snow. It is a permanent inhabitant of the cold plains of Manchuria and the Amoor region, as well as of the plains lying north of the Hindu-Kush, in Bokhara, prowling about even in winter along the icy margins of the Aral Sea. As a matter of fact, the range of the tiger extends to about the fifty-third parallel of north latitude—or what corresponds to the position of Lake Winnipeg, in British America—in the neighbourhood of Irkutsk and Lake Baikal. Nor can this northern range be taken to represent the range of simply stray individuals, since in the region of Southeast Siberia traversed by Radde that traveller affirms that tigers were uncommonly abundant.[11]

Although at the present time the lion is confined almost exclusively to the tropical and sub-tropical regions of Africa and Asia, there can be but little doubt, as appears from the writings of Herodotus and Aristotle, that as late as the beginning of the historic period that animal still inhabited in Europe a region lying as far north as about the fortieth parallel of latitude — or what corresponds in position to the State of Pennsylvania—namely, the region of Thessaly, in Greece. And even at the present day the Tunisian lion is occasionally found in the neighbourhood of the thirty-seventh parallel of north latitude, and until recently the Cape lion was abundant in or about the district of the Cape, extending to the thirty-fifth parallel of south latitude. Although the climate of these latitudes in Africa is of an unusually mild character, yet there are sudden changes of temperature, as between day and night, which may be likened to the changes in the temperature between the summer and winter climates of more temperate regions. We are informed by travellers that in the Kalahari Desert and other dry open districts of South Africa the nights are frequently unpleasantly cool, or even cold, the free and rapid radiation of heat from the soil not rarely being accompanied by a freezing of the surface. The formation of ice in the Desert of Sahara is, likewise, not exactly of exceptional occurrence, but in that region of the African continent, except on its immediate borders, lions are only rarely met with.

That a restriction to warm climates is likewise not the case with the elephant is almost conclusively proved by the readiness with which, in the Roman period, these animals were made to pass the barrier offered by the lofty Alpine chain. Still more indisputable evidence on this point is, however, afforded by the habits of the Indian elephant, which appears to be equally at home among the cool mountain heights as amidst the hot and jungly lowlands. In Ceylon, according to Sir Emerson Tennent, "the mountain-tops, and not the sultry valleys, are his favorite resort. In Oovah, where the elevated plains are often crisp with the morning frost, and on Pedro-Tella-Galla, at the height of upwards of eight thousand feet, they are found in herds, whilst the hunter may search for them without success in the jungles of the low country. No altitude, in fact, seems too lofty or too chill for the elephant, provided it affords the luxury of water in abundance; and, contrary to the general opinion that the elephant delights in sunshine, he seems at all times impatient of its glare, and spends the day in the thickest depths of the forest, devoting the night to excursions, and to the luxury of the bath, in which he also indulges occasionally by day." [12] Mr. Johnston, during his recent explorations of the Kilimanjaro region, encountered elephants, together with buffaloes, and one or more species of antelope (kudu), at an elevation of thirteen thousand feet.[12a]

The camel is an animal popularly associated with the burning desert regions of Africa and Asia, yet the two-humped or Bactrian species is found throughout the greater portion of Mongolia and Chinese Tartary, in the mountain region as well as in the lowlands, lying between the fortieth and fiftieth parallels of latitude, and it extends its range even considerably beyond the fiftieth parallel into Siberia, as along the borders of Lake Baikal, where it appears to pass the winter season without discomfort. It is a fact worthy of note that the only other existing representatives of the camel family —the llama and llama-like animals of the New World—are strictly adapted to a rigourous winter climate, as is shown by their partiality to the highly-elevated tracts of the South American Andes. The same adaptability to different extremes of climate likewise presents itself in the case of many of the so called Arctic animals. The reindeer, while it habitually prefers for its home a region that enjoys a more or less rigourous climate, and where the soil is for the greater part of the year covered with snow, does not appear to be impatient of the summer heat of comparatively low latitudes, as is proved by the circumstance that in the various zoological gardens of Central Europe it not only develops in good condition, but also breeds freely. Indeed, its restriction to the high northern latitudes appears to be in no way dependent on considerations connected with either cold or snow, but merely upon the presence there in the greatest abundance of its particular food, the reindeer-moss and various lichens, without which it seems incapable of flourishing. There can be little doubt that were individuals of the reindeer transplanted to an elevated mountain region, such as the European Alps, for example, where their own proper nourishment would be again met with, they would thrive very nearly, if not fully, as well as in their true homes north of the fifty-fifth or sixtieth parallel of latitude. Indeed, even in their northern haunts the animals, at least as is shown by the American species or variety, would seem to be impatient of too great a cold, since in the winter they seek the inner recesses of the forests for protection.

Turning now to the class of birds, we find that similar illustrations of climatic adaptation present themselves. Thus, the usually considered "tropical" or "equatorial" humming-birds are in reality not such at all. While it is true that by far the greater number of species belonging to this family are found within the region embraced within the tropics, yet the range of the family extends all the way from Cape Horn (Eustephanus galeritus) to Sitka (Selasphorus rufus), or over a territory covered by no less than one hundred and fifteen degrees of latitude. And even among the strictly tropical forms many of them extend their range to the limits of perpetual snow, some remaining in the cold region permanently. The Oreotrochilus Chimborazo and O. Pichincha have their abode in the equatorial peaks indicated by their respective specific names at an elevation of no less than sixteen thousand feet—or higher than the summit of the Mont Blanc—in a world of almost perpetual snow, hail, and sleet.[13] In fact, the elevated Andean slopes are much more thickly visited by humming-birds than the deep lowlands, no matter how luxuriantly these last may be clothed with vegetation.

The ostriches constitute another group of animals whose habitat is popularly associated with the burning deserts of the Torrid zone. While it is unquestionable that these birds do delight in just such districts, it may yet be doubted whether the matter of climate has very much to do with the selection of a region, since ostriches are, or have been until recently, equally abundant in all parts of the African continent, in the high table-lands as well as in the lowlands, from Algeria to the Cape, and from the east to the west coast, where the suitable desert conditions present themselves, and where, consequently, as has already been stated, the differences between the temperature of night and day are excessively marked. In the desert region of Western Asia—Persia and the Valley of the Euphrates—the bird ranges or ranged as far north as about the thirty-fifth parallel of latitude, and, indeed, it is not exactly improbable, as has been maintained by Vámbéry,[14] that even at the present day it exists in limited numbers along the shores of the Sea of Aral, in about the forty-fifth parallel, or what would correspond to the position of the southern portion of the State of Maine. In the case of this family—Struthionidæ—we also notice the singular fact, analogous to that which has been observed in relation to the distribution of the Camelidæ, that the only representatives of the group other than Struthio (the ostrich proper), constituting the American genus Rhea, are birds belonging almost strictly to the temperate regions, their range extending from Patagonia to the southern confines of Brazil. The parrots (Psittaci) may be considered to be preeminently tropical birds, the vast majority of the species being included in a zone bounded by the thirtieth parallel on each side of the Equator, but yet it may be doubted whether this limitation does not depend more upon the nature of the food-supply than upon the character of the climate. In South America a species of Conurus extends its range as far as the Strait of Magellan, and·in the Macquarie Islands, in the South Pacific, representatives of the family are met with as high as the fifty-fourth parallel of latitude, corresponding to a position removed by only six degrees from the southern extremity of Greenland. Wallace probably justly refers to the "almost universal distribution of parrots wherever the climate is sufficiently mild or uniform to furnish them with a perennial supply of food."[15]

But while in numerous, and perhaps the majority of, instances the limitation of animal groups to certain geographical regions is dependent more upon the physical character of the immediate environment and the nature of the food-supply than upon particular conditions of climate, yet it cannot be denied that in very many cases climate appears to exercise a paramount influence upon distribution. This influence is frequently considered to be nowhere more forcibly illustrated than in the migration of birds, both as regards the northern species and those inhabiting the southern climes. That the climatic explanation of the phenomenon of bird migration is a fallacy most ornithologists are now agreed. It is a well-ascertained fact that the vast majority of birds are migrants to a greater or less degree, and that non-migration with this class of animals is much more of an exception than the rule. Yet, by reason of their peculiar covering, birds generally, as compared with other vertebrates, are but slightly affected by extremes of either heat or cold, and indeed, as far as we are capable of judging, by most climatic influences, provided only that their food-supply is not affected thereby. The condor in its aerial flight within a few minutes of time accommodates itself to the most varying climatic conditions, the change from the freezing cold of the mountain heights to the scorching heat of the tropical lowland plains seemingly having no effect upon the vigour of the bird. There can be but little doubt, as has been insisted upon by Professor Newton, that a deficiency in the food-supply—the necessity for searching for new food—is the most obvious cause or impulse promoting bird migration. Migrations of a somewhat similar character, indisputably governed, at least in part, by considerations connected with the food-supply, but also in greater part by conditions of climate, manifest themselves among several other classes of animals. Thus, in India, the monkeys habitually ascend the Himalaya Mountains in summer to elevations of ten or twelve thousand feet, and again descend in winter. Semnopithecus schistaceus has been observed at a height of eleven thousand feet, leaping in fir-trees laden with snow wreaths! Wolves in severely cold weather descend from the mountain-slopes to the lowlands, and bears not infrequently migrate in great numbers to escape the rigours of an extreme winter. The migratory instincts of the northern hares and squirrels, and more particularly of the Norway rat and lemming, which in severe winters move in amazing numbers in direct lines over lake, river, and mountain, overcoming all obstacles that might be placed in their path, are well known. The Kamtchatka rats, under the pressure of numbers, are stated by Pennant to travel westward for a distance of eight hundred miles or more. Similar instances of the force of migration are presented by the hoofed animals. The vast herds of moving buffalo were until recently familiar sights to the traveller on the American plains ; in South Africa countless numbers of antelope, impelled by the necessities of food-supply, pour down upon the more favoured districts lying without the region of parched soils; and similar excursions, although in this case governed by reversed thermometric conditions, are practised by the onager or wild ass of Tartary. Even the reindeer is to an extent a migrant, since in both Russia and Chinese Tartary it descends far southward in advance of a rigourous winter, and, indeed, frequently reaches a lower latitude than any part of England, although in Scandinavia the animal is rarely seen south of the sixty-fifth parallel.

It is not alone among the higher animals that the migratory instinct is developed. Turtles, during the ovipositing season, move in considerable numbers from one part of the sea to another, and they are stated to find their way annually to the Island of Ascension, which is distant upwards of eight hundred miles from the nearest continental land-mass.[16] Fishes migrate in immense numbers, but the periodical shifting of the abodes of these animals is directly connected with the processes of reproduction. Certain fishes, as the salmon, shad, and smelt, ascend the waters of fresh-water streams for the purpose of depositing their eggs; others, again, as the herring and mackerel, frequent in immense shoals, during the breeding season, the neighbourhood of the coast-line. The young eel follows the line of the river-courses in myriads, ascending all the tributary streams, and frequently overcoming apparently impassable water - falls by squirming over the moss-covered ledges on either side. Among insects, the devastating migrations of the locust are proverbial, and similar illustrations of the wandering instinct could be cited from other members of the same class of animals. A remarkable example of migration has recently been observed in the case of a species of grapsoid crab (Sesarma ?) off Cape San Antonio, the western extremity of the Island of Cuba.

Barriers to Migration, and Facilities for Dispersion.—It has already been remarked that the interposition of extensive and elevated mountain-chains and of large bodies of water, and also sudden changes in the physical character of a country, are insurmountable obstacles in the way of the migration or dispersion of certain classes of animals. The most serious of these obstacles, as affecting the dispersion of the Mammalia, is of course that of large bodies of water. We are well aware that the most experienced swimmer among this class of animals can accomplish by the natatorial process but an insignificant journey, and, therefore, it would necessitate the interposition of but a very moderate expanse of water to effectually bar its progress in any given direction. Several members of the cat family are expert swimmers, the jaguar being known to cross the broadest of the South American rivers, the La Plata, as observed by Lieutenant Page. The tiger and elephant are both good swimmers. Deer are likewise prone to take to water, but it may be questioned whether animals of this kind would be apt to trust themselves beyond the sight of land. The domestic pig, even at a very young age, has been known to swim five or six miles, and it is not exactly impossible that the wild-hog, in cases of absolute necessity, might successfully attempt a passage of three or four times this distance. Probably the most remarkable exhibition of the natatorial powers of a land animal is that shown in the case of a polar bear, which was observed by Captain Parry vigourously paddling away in Barrow's Strait at a nearest distance of twenty miles from the shore, with no ice in sight on which it could have secured needed repose. It may safely be conceded, from our present knowledge on the subject, that while many of the land Mammalia can effect with safety, and even readiness, such water passages as are most generally to be met with on continental areas, none, probably, would be prompted to undertake a journey across an arm of the sea whose width measured fifty or more miles, or even one much exceeding half that extent.* To these difficulties or impossibilities in the way of dispersion must be attributed the circumstance that the vast number of oceanic islands are deficient, except where man has effected an introduction, in representatives of this particular class of animals. The fact that certain allied, or even identical, forms of mammals are found in regions widely removed from each other, and which at the present time are separated by impassable bodies of water of greater or less extent, is practically conclusive evi-

---

* In the case of the polar bear above cited, the absence from view of any ice need not necessarily, or even probably, indicate that there was no ice present nearer to the swimming subject than the ice of the land-border. From the mast of a vessel, elevated one hundred and fifty feet above the surface of the water, an iceberg rising to the same height could not, owing to the curvature of the earth, be distinguished at a greater distance than thirty-four miles; flat masses of pack-ice, rising but a few feet above the water, at only about half that distance.

dence that in the former periods of the earth's history the surface of the globe must have undergone such vicissitudes as to have at various times disturbed the general relations existing between land and water. In other words, much of the surface that at one time was occupied by water must have been replaced by land, and, *per contra*, what was at one time land must at another have been water. And evidences of such variations in terrestrial equilibrium are abundantly afforded by geological landmarks. Had the greater portion of the surface of the globe at one time since the introduction of the Mammalia consisted principally of dry land, or had there been since that period a general alternation in the relative positions of the land and water areas, the geographical distribution of the Mammalia would have been very different from what we actually find it to be. Hence, it must be assumed that a land and water alternation, such as could have brought about the present result, must have taken place in certain parts of the earth's surface only, and without affecting others. There would seem to be very strong grounds for concluding that the most recent connection uniting the principal land-areas of the globe was formed in the Northern Hemisphere, as a belt closing off the Arctic Sea (if it then existed) from the Pacific and Atlantic oceans.

The only class of terrestrial mammals to which a broad arm of water offers no impediment in the way of migration or dispersion is that of the bats; and, singularly enough, just in the case of these animals, as has already been remarked, are we furnished with an example of universal distribution, there being but very few of the habitable oceanic islands which are not tenanted by one or more representatives of the order. But even among the habitually terrestrial Mammalia there are certain exceptional methods by which dispersion to very considerable distances from the mainland can be effected. In the northern regions the frozen sea constitutes a connecting bridge between distantly-removed land-masses which is constantly taken advantage of by various forms of Arctic animals. By the breaking up and drifting away of fragments of the northern ice-masses animals that might be temporarily wandering over them could readily be transported to very considerable distances from their true homes; and, indeed, it is through such means that polar bears are periodically stranded upon the coast of Iceland. In one year alone twelve of such wandering animals made their appearance

upon the island.[17]   The reindeer is stated to cross the Behring
Straits by way of the Aleutian Islands and the frozen sea, and in a
somewhat similar manner the musk-ox finds its way to Melville
Island; it is, however, singular that the last named, despite its long
ice-journeys, never manages to reach either the continent of Asia or
Greenland.   In regions like the tropics, which support a luxuriant
vegetable growth, and which are subject to periodical fluminal
overflows, and, consequently, to the uprooting or outwashing action
of the inundating waters, it not infrequently happens that islands
or "rafts" of considerable magnitude, consisting mainly of inter-
laced or matted vegetation—tree-trunks held together by various
creepers and climbers, and containing a sufficient quantity of vegetable mould and soil bound together in the roots—are floated down
stream into the open sea, where they are at once placed at the mercy
of the prevailing oceanic and atmospheric currents.   These rafts
have been frequently noticed at the mouths of some of the larger
streams, as the Mississippi, Amazon, and Ganges, and, in the case
of the last named, at a distance of a hundred miles from its mouth.
Floating masses of wood, with upright trees growing over them,
were mistaken by Admiral Smyth in the Philippine seas for true
islands, until their motion made their real nature apparent.   Such
floating masses not rarely harbour various forms of animal life in
their midst, and among these the Mammalia with arboreal habits are not inadequately represented.   The South American travellers Spix and Martius assert that on different occasions they observed monkeys, tiger-cats, squirrels, crocodiles, and a variety of
birds, carried down stream (the Amazon) in this manner, and similar observations have been made by other travellers in the case of
the Rio Paraná.   It is asserted that no less than four pumas were
landed in one night from such rafts in the town of Montevideo.[18]
Some of the animals thus conveyed may travel unconcernedly, and
without any special disadvantage arising from a change of abode;
others, as the larger quadrupeds, will have been caught up and
transported through accident.   To what distance such a floating
raft with its living cargo may ultimately be carried in safety, and
without detriment to its inhabitants, over the oceanic surface there
are as yet no data for determining.   But there would appear to be
no reason for assuming that they could not be transported to a
distance of several hundreds of miles, seeing that the upright vege-

tation found on many of them would serve with powerful effect in the face of a wind. And while the majority of the animal inhabitants might be exterminated before the end of the voyage the safe arrival on an island or distant shore of a very limited number of individuals, embracing both males and females, would serve in a short period, under favourable conditions, to stock the new land with the species. That an absolute limit is set, however, to migration as effected in this manner is proved conclusively by the utter absence in most of the oceanic islands of indigenous mammals, excepting bats.

The same obstacle that is interposed by the ocean to the dispersion of the Mammalia presents itself in the case of the vast majority of other terrestrial animals in which the power of flight is not at all, or at best but feebly, developed. Thus, the serpents, although many of them are fairly good swimmers, are, if we except the marine forms, as incapable of passing oceanic barriers as are the quadrupeds, and their transportation from continental areas to regions far remote can only be effected by such or similar accidental means as that just described. As might have been expected, therefore, they are absent from nearly all oceanic islands. The Amphibia (frogs and toads) are no more fortunate in passing broad arms of the sea than are the serpents, despite the circumstance that in their young or larval condition they are strictly aquatic in their habits. Salt water proves fatal both to them and their eggs. Since moisture is a necessary condition for the early existence of this class of animals, it is evident that an extensive desert region will be an effectual barrier to their distribution—in fact, about as much so as an ocean. Lizards, in their adult condition, are as incapable of traversing an oceanic region as are the snakes and amphibians; but it would appear that in some special way—whether as effected by the oceanic currents themselves or through the agency of birds—their eggs may be transported to very considerable distances out to sea, since this order of animals is sufficiently represented in remote islands where neither snakes nor amphibians have as yet been encountered. That the ocean offers no insuperable obstacle to the broad dispersion of a very large body of birds is known from almost daily observation. Birds are known to pass several hundreds of miles on the wing without halting, and, indeed, it is not exactly impossible, or even improbable, that such unassisted flight may extend over one or more thousands of miles.   The flights of the wild-goose and the swallow have been estimated to be performed at the almost incredible velocity of from sixty to ninety miles per hour, and the flights of many of the smaller birds at not very much less.   A sustained flight of ten or more hours in duration, especially when assisted by a favourable wind, involving an amount of muscular exertion probably within easy command of many birds, would carry them over an enormous stretch of territory, during a period of time which, by its brevity, would render the question of food-supply comparatively unimportant.   Land-birds have been encountered in the North Atlantic at almost all points of the oceanic expanse; but to what extent these stragglers have received assistance in their flight, by taking temporary shelter on board the numerous vessels plying between Europe and America, can hardly be determined.   There is no question as to such assistance in numerous instances, but whether it is afforded in all or most cases is a matter of pure conjecture.   By whatever means or methods the oceanic travel of birds may be effected, it is a matter placed beyond all question that numerous American birds make their appearance at intervals along the European coast.   Upwards of sixty species of such foreigners, embracing examples from nearly all the orders of birds, have at different times been noted on the eastern coast of the Atlantic, principally in the British Isles and the Island of Heligoland.[19] Singularly enough, no distinctively European birds make their appearance on the American coast, except a few whose journey over is made by way of Greenland and Iceland.*   Despite the long-sustained flight of which birds are capable, it may be considered exceedingly doubtful whether many or any of them undertake these protracted journeys as a matter of their own pure choice or volition.   It seems hardly possible that an animal would subject itself to such an amount of exertion and privation as would appear to be involved in journeys of this length, when no material advantage could in the end be derived therefrom.   It therefore appears more than probable, as has been urged by Baird, Wallace,

---

* No account is here taken of the purely pelagic forms, which are found on the opposite borders of the oceanic expanse, and which find suitable resting-places on the surface of the waters.   The greenshanks (Totanus glottis) has been obtained once in Florida, and apparently nowhere else in the United States.

and Newton, that the oceanic wandering of land-birds must be
attributed in most, or nearly all, cases to accidental circumstances—
namely, storms, or the prevalence of certain winds—which may
have wafted the birds beyond their control off to sea.   Winds
from the west, as has been shown by Professor Baird, are prevalent between latitudes 32° and 58° N., and, hence, would be liable
to catch such birds as may be passing southward during their au-
tumnal migration, especially there where their flight would be
at some distance off from the shore, or across broad arms or inlets of the sea.   The dispersal would naturally be facilitated by
the interaction of a heavy storm, and it is a most noteworthy
confirmatory fact that the appearance of American birds on the
European coast is either presaged or accompanied by heavy westerly
winds blowing in that quarter.   North of the fifty-eighth parallel
of latitude the polar winds trend westward, and with them we
have the accompanying transferrence of European birds, by way of
Iceland and Greenland, to the American continent.   That storms
or heavy winds do influence the flight of birds in the manner here
described, is indisputably proved by the facts that present themselves in connection with the occurrence of marine birds over continental areas at some distance from the shore-line.   The stormy
petrel, during and after the prevalence of a northeast storm, has
been seen in considerable numbers in the Eastern United States
beyond the Alleghany Mountains; the Thalassidroma Leachii has
been abundantly killed at or about the city of Washington; and
Professor Baird instances the case of a Pomarine jäger (Cataractes
Pomarinus), which was killed on the Susquehanna, at Harrisburg,
in 1842.[20]   The golden plovers, in their southerly flight, start direct from Nova Scotia or Newfoundland for the West Indies,
whence they continue their journey along the South American
coast to Patagonia.   In this journey but comparatively few individuals touch or rest along the Atlantic States, yet it is known
that during heavy northeastern winds, in the month of August,
great numbers of the birds may be confidently expected along the
New England coast.   And it not infrequently happens that under similar conditions immense numbers of these and allied birds
are driven to very considerable distances in the interior of the
continents.   In a like manner, during the prevalence of heavy
storms, European birds are cast upon the Azores, situated about

one thousand miles from the nearest continental coast. Among
these are the kestrel, hoopoe, oriole, and snow-bunting, and not
improbably also swallows, larks, and grebes.* If, then, birds
may be drifted by accidental storms to a distance of one thousand miles in a direction contrary to that of the prevalent winds,
it may be asked, Why may they not be thus drifted, at least
after their first landing-place, another one thousand or two thousand miles further ? In other words, if European birds are carried
to the Azores, why are they not at intervals also transported from
there to the American coast? This question can, with our present
knowledge, not yet be answered. Three or four species of European
birds have been noticed in the Bermuda Islands—the wheat-ear (Saxi-
cola œnanthe), the sky-lark (Alauda arvensis), the snipe (Gallinago
media), and the land-rail (Crex pratensis); but three of these are
also found in Greenland or on the North American mainland, while
the fourth, the sky-lark, appears to have been brought over in, or
to have escaped from, a ship.† In an ocean studded with islands,

* Most of the resident land-birds of the Azores are identical with forms
found in Europe and North Africa, and it, therefore, becomes impossible to
ascertain how many of the individuals actually peopling the islands may *not*
have been recently transported from the mainland. It is only under exceptional circumstances — barring the case of recognised stragglers — that such
wanderers can be determined.

† The total number of European birds known to have found their way
across the Atlantic to the American shores (including Greenland) is, according to Freke ("Zoologist," 1881), thirty-seven, of which Greenland counts
about thirty, and the Eastern United States only twelve. This determination
naturally excludes all birds that have been artificially introduced. Of the
twelve species occurring in the Eastern United States, six are swimmers and
five waders, and only one (and that somewhat doubtful, Buteo vulgaris, reported to have been obtained in Michigan, in October, 1873) is a true land-
bird. The wheat-ear, referred to as occurring in the Bermudas, is considered
a member of the North American fauna. The number of species of American
birds crossing the Atlantic in the contrary direction is, according to the same
authority ("Proc. Royal Dublin Soc.," 1881), sixty-nine, of which twenty-two
are swimmers, sixteen waders, and no less than thirty-one land-birds. The
last include, among other forms, representatives of the genera Turdus (four
species), Galeoscoptes, Regulus, Dendrœca, Hirundo, Loxia, Zonotrichia,
Ceryle, Coccyzus, Picus, and several species of birds of prey. The bald-
headed eagle has been recorded from Sweden. It is significant that, of the
forty-seven species of waders and land-birds, only two are known from Iceland (Falco candicans and Numenius Hudsonicus) and none from the Faroe

which afford numerous resting-places, it would not seem difficult to account for the occurrence of land-birds at the remotest distances from the mainland, even without having recourse to the accessory transporting agency of prevalent winds and storms. But even with this favourable condition added, it would appear that most land birds are not disposed to undertake of their own free will extended oceanic journeys, as is proved by the avi-fauna of many of the oceanic islands. Thus, while, as we have already seen, nearly all the representatives of the bird-fauna of the Azores, situated more than one thousand miles from the mainland, are identical with forms inhabiting either Europe or Northern Africa, indicating that the islands were peopled in comparatively recent times from those continents, in the Galapagos, situated only six hundred miles off the west coast of the continent of South America, we meet with an entirely different state of things as regards the bird-fauna. Of about thirty species of indigenous land-birds, apparently only one, the common rice-bird (Dolichonyx oryzivorus), which ranges from Canada to Paraguay, is absolutely identical with a form found outside the limits of the island group. In addition to this a species of owl (Asio Galapagoensis) is considered by some authors to be but a mere variety of the cosmopolitan Asio brachyotus, or short-eared owl, which is distributed from China to Ireland, and from Greenland to Patagonia.[21] We have here, therefore, positive evidence that migrant stragglers from the South American continent are at the best of but very rare occurrence, and, on the other hand, visitors from the islands to the mainland appear to be equally rare. But since, from the resemblance which the fauna as a whole presents to that of the mainland, it is practically proved that the same was at one time derived by migration from the continental areas—the islands being of volcanic origin— it is manifest that this migration must have taken place at a period sufficiently remote to have permitted the differences separating the two faunas to have been brought about. On the other hand, the absolute identity of the rice-bird with the similar form from the continent, proves, as has been pointed out by Wallace, that the island breed has been kept unaltered only through repeated or fre-

Isles. The easterly dispersion is attributed to causes identical with those which have been assigned in explanation of the phenomenon by Professor Baird.

quent visits from the specific congeners on the mainland.   Not
only are, with the one or two exceptions above noted, all the
Galapagos land-birds specifically distinct from those found anywhere else,* but they also belong largely to distinct genera.   Of
the fourteen genera represented, four are peculiar to the islands.
The rarity of continental visitors to the Galapagos, as compared
with the Azores, is to be attributed to the circumstance that these
islands are situated in a zone characterised by an absence of storm
winds.   In the island of Juan Fernandez, situated in latitude 34°
S., and only four hundred miles from the Chilian coast, there are
but five species of land-birds, and of this number two are peculiar.
In the Keeling or Cocos Archipelago, situated in the Indian Ocean
at about the same distance from the Sumatran coast as are the
Galapagos from the coast of South America, there is not a single
species of true (indigenous) land-bird, although snipes and rails
of the common Malayan species are sufficiently abundant; and the
same is true in the case of the island of St. Helena, situated eleven
hundred miles from the nearest point of the continent of Africa.†
Of the twenty species of Passeres, or perching-birds, inhabiting the
Sandwich Islands—about the most strictly oceanic of any group of
oceanic islands so-called, being situated fully two thousand miles
from the nearest continental coast-line, and the same distance from
the nearest island groups (Marquesas and Aleutian), if we except the
small and almost tenantless shell and coral reefs—all the forms are
peculiar; and, furthermore, in all cases but one or two they belong
to genera which are likewise confined to the islands.   And even of
the twenty-four or more species of aquatic and wading birds that
have been observed on or about the islands, five—a coot (Fulica
alai), a moor-hen (Gallinula Sandvichensis), a rail (Pennula Millei),
and two ducks (Anas Wyvilliana and Bernicla Sandvichensis)—are
peculiar.[22]   All in all there are some fifty species of birds known
from the island group, of which about one-half are peculiar.
It is evident that migrants (true land-birds) from distantly removed countries but rarely arrive here.   In the case of the Ber-

---

* The Dendrœca aureola, a species of wood-warbler closely allied to the
"golden" or summer warbler of the United States (D. æstiva), is only doubtfully separable from the D. petechia of the Island of Jamaica.

† A small wading-bird of the genus Ægialitis (Æ. Sanctæ Helenæ), allied
to a species of plover common in South Africa, is found in the island.

muda Islands, which are distant from seven hundred to eight hundred miles from the nearest coast, we meet with a different order of things. The bird-fauna of these islands consists in all of about one hundred and eighty species, including both the land and aquatic forms, of which number, however, about thirty have been noticed only on one occasion. Of the eighty-five species of land-birds less than ten are permanent residents, the rest making their way principally from the North American continent and the West India islands.[23] It is a singular circumstance that most of the foreign invaders are strictly migrating birds, whose course of migration lies along the Atlantic coast, and which in their periodical wanderings frequently pass at some considerable distance out to sea. Entering the region of violent winds and hurricanes, they are liable to be snatched from their track, and to be forcibly transported to some remote shore, where, of necessity, they will be compelled to secure for themselves a new home, and where, through frequent visitations of a like character, the original breeds established will remain pure and unaltered. Such is the condition of the bird-fauna of the Bermudas at the present time. None of the strictly non-migratory birds are represented in those islands. Two or more species of bat, also North American forms, are, with the exception of rats and mice, the only indigenous mammals.

Dispersal of Insects.—It is a well-known fact that insects have been found in nearly all parts of the world that have thus far been trod by man, from the extreme limits of the Arctic and Antarctic regions to the Equator, and from the level of the sea to—and considerably above—the line of perpetual snow. Butterflies were observed by the naturalists of the "Alert" and "Discovery" nearly as far north as the eighty-third parallel of latitude; and Humboldt met with insects on Chimborazo, at an elevation of upwards of 18,000 feet. They are found in fresh and salt waters, freely swimming on the surface—and at very considerable distances from the mainland—as well as below it; in hot springs, where the water has attained to a moderately high temperature, and in subterranean caves. But, while the members of this class of animals, taken collectively, appear to be specially adapted to all the various conditions of existence that might be imposed upon them by accidental circumstances, the same does not hold for the individual members composing the class. Thus, certain insects are entirely dependent upon some special vegetable product for their existence, whether it be, as it may happen, the leaf, the flower, or the juice of the plant in question. Again, while in some cases the adult insect may be entirely independent of such a circumscribed food-supply, the larva may still be governed in its diet by a particular kind, without which, consequently, the prolonged reproduction of the species would be impossible. Such instances of limitation are exhibited by numerous forms of caterpillars. Hence, it is not difficult to comprehend why, in regions which are affected by similar conditions of climate, and which collectively show a general correspondence in the character of the vegetation, certain species of insects should be found at one locality and not at another, even where no physical barrier separating the two should be interposed. In fact, the barrier interposed by conditions of vegetable growth is fully as effective in restraining a broad specific distribution as are the barriers resulting from the physical conditions of the earth's surface, most of which they are able to overcome, either voluntarily or involuntarily. The mature insect, from its lightness, is frequently carried away in aerial currents from its native or favourite haunts to regions widely remote, in a manner precisely similar to what obtains in the case of birds. Hawk-moths have been caught on board ship at a distance of two hundred and fifty miles from shore, and a large Indian beetle (Chrysochroa ocellata) was captured some years ago, in the Bay of Bengal, at a distance of two hundred and seventy miles from the nearest land. During Captain King's expedition to the Straits of Magellan dragon-flies flew on board his vessel when still fifty miles out at sea (south of the Rio de la Plata); and Admiral Smyth reports that, in the Mediterranean, myriads of flies were brought to his ship by a southerly wind from a region fully one hundred miles distant. A beetle is recorded by Darwin as having been caught aboard the " Beagle " when the vessel was upwards of forty miles distant from the nearest shore; from what actual distance the insect may have come could, necessarily, not be determined. A locust was observed by the same naturalist three hundred and seventy miles from land; and in 1844 swarms of these insects, " several miles in extent, and as thick as the flakes in a heavy snow-storm, visited Madeira. These must have come with perfect safety more than three hundred miles, and, as they continued flying over the island for a long time, they could evidently have travelled to a much greater distance."[24]   In addition to this means of aerial dispersion, the distribution of insects may be to a great extent effected in the condition of eggs, which retain a considerable amount of vitality, and which are not infrequently laid in decaying timber and in the living tissues of various plants.   When, therefore, floating rafts or mats are apt to be formed, and to be floated out to sea, it is almost certain that with them will be carried out a host of insects—whether in the perfect form, as grubs, or as eggs—of different species, a fair proportion of which will, doubtless, have retained their vitality even after a protracted sea-voyage of several thousand miles.   It is in this manner that many or most of the tropical forms which periodically make their appearance on the British coast have been transported thither, the current of the Gulf Stream, which trends in a general northeasterly direction, being instrumental in drifting tropical log-wood to the trans-Atlantic temperate shores.

Dispersal of Mollusks.—The world-wide distribution of the fresh-water and terrestrial Mollusca, and the occurrence of identical or very nearly allied generic forms at opposite quarters of the globe, prove conclusively that the animals of this class are favoured with special instrumentalities by which a broad distribution is effected. Land-snails of the genus Helix are found in all the continental areas, from the polar regions to the Equator, and from the limit of perpetual snow on mountain summits to the level of the sea; they are also found in all the oceanic islands, even the most remote, that have thus far been visited.   The exact nature of this distribution has not yet been positively determined, and, in fact, there are several difficulties in the way of accounting for it.   It is well known that these animals cannot survive for any length of time the effects of salt water, and this water is almost immediately fatal to the vitality of the eggs.   Hence, only under exceptional conditions is it possible to account for a transference over a broad expanse of oceanic surface.   But it has been ascertained that such forms as are capable of secreting an epiphragm, and therewith closing up the entrance to the shell, are able to resist the injurious effects of salt water for a very considerable period, in some instances as much as two weeks, or more, as has actually been determined experimentally by the immersion of land-shells in the briny medium.   In regard to these, therefore, there will be no difficulty in accounting for a broad distribution, since they, and especially the genus Helix of all others, would be liable to be concealed in and transported away by floating timbers. In this manner they could be drifted away for several hundreds of miles, and, under exceptionally favourable circumstances, to possibly one or two thousand, the more readily since some of these animals possess an enormous amount of vital tenacity, even under the most adverse conditions of existence. Thus, a Helix from North Africa (H. desertorum), contained in the British Museum collection, and glued on to a tablet, was found by the conservators to be alive after a period of more than four years. A similar instance of resuscitation, although after a less protracted period, has been noted in the case of one of the tabulated snails of the Academy of Natural Sciences of Philadelphia. Again, it has been conclusively shown by Darwin and others that the eggs of pond and other fresh-water bivalve-mollusks are occasionally found attached to the feet of wading-birds—ducks, and the like—visiting such waters, and are by them liable to be carried to very considerable distances from their true homes, and thereby to have their range almost illimitably widened. Such a method of transport, although exercised to a much more limited extent, has been observed to be effected even by species of water-beetle, whose legs may have become entrapped between the valves of the shell, as well as by newts and other amphibians. The broad distribution of allied or identical generic and specific forms of fluviatile mollusks over the most extended or widely remote geographical areas receives a partial explanation in the circumstance that the physical forces operating upon the earth's crust, causing movements in it of a differential character—i. e., elevation at one point and subsidence at another—tend to destroy the permanency of river courses, turning them now to one side, then to another, and ultimately, possibly, uniting the basins of streams whose waters were at one time quite remote from each other. With this union or coalescence of the waters there will necessarily also be a union of the contained molluscan faunas, and, by a repetition of the process, a general transference may in course of time be effected of the same or but barely modified forms over the most distant portions of the earth's surface. Existence, under the new conditions of habitation, will be rendered possible or materially facilitated by the comparatively slight alteration in its physical properties to which the watery medium will in many or most cases be subjected.

IV.

ZOOLOGICAL REGIONS.

As an outcome of the laws governing distribution, and the
varying adaptabilities of animal organisms to overcoming the many
conditions of existence which present themselves on the surface of
the earth, it has resulted that different assemblages or groups of
animals have been thrown into different quarters of the habitable
globe, which may, accordingly, be said to be divided into a number of regions, of greater or less extent, each of which is characterised by its own particular fauna. To the more comprehensive of
such zoological divisions the term "region" or "realm" has been
applied by scientists. But just as the earth's surface taken collectively may be divided into zoological regions, so may these be
again further subdivided into minor regions, these still further, and
so on, until we have, as generally recognised, "regions," "sub-
regions," "provinces," and "sub-provinces."

By most naturalists the terrestrial portion of the earth's surface is recognised as consisting of six primary zoological regions,
which correspond in considerable part with the continental masses
of geographers. These six regions are: 1. The Palæarctic, which
comprises Europe, temperate Asia (with Japan), and Africa north
of the Atlas Mountains; also, the numerous oceanic islands, with
Iceland, of the North Atlantic. 2. The Ethiopian, embracing all
of Africa south of the Atlas Mountains, the southern portion of
the Arabian Peninsula, Madagascar, and the Mascarene Islands,
and which, consequently, nearly coincides in its entirety with the
Africa of geographers. 3. The Indian or Oriental, which embraces

India south of the Himalaya, Farther India, Southern China, Sumatra, Java, Bali, Borneo, and the Philippines. 4. Australian, comprising the continent of Australia, with Papua or New Guinea,
Celebes, Lombok, and the numerous oceanic islands of the Pacific.
5. The Nearctic, which embraces Greenland, and the greater portion of the continent of North America (excluding Mexico); and,
6. The Neotropical, corresponding to the continent of South America, with Central America, the West Indies, and the greater portion of Mexico. A seventh region has been established by some
authors to receive New Zealand; but there would seem not to be
sufficient reasons for isolating this island, or group of islands, from
the Australian region.

While the regions here designated are to a great extent clearly
defined by their zoological characters, it would, nevertheless, appear more in consonance with actual facts to depart somewhat
from their generally recognised limitations. Thus, the Palæarctic
and Nearctic tracts, in the absence of both positive and negative
faunal characters of sufficient importance to separate them from
each other, are indisputably linked together, and should constitute
but a single region (the Holarctic). On the other hand, the scattered island groups of the Pacific, which have been united with the
Australian realm, may with sufficient reason be constituted into an
independent region of their own; at any rate, they appear to bear
no special relationship with the Australian region, any more than
with the Oriental. Again, it seems advisable to separate from what
has hitherto been known as the Palæarctic region the tract that is
comprised within the "Mediterranean sub-region "—i. e., the peninsular portion of Southern Europe, North Africa, and, in Asia, Asia
Minor, Persia, Afghanistan, Beloochistan, and the northern half of
Arabia—and to consider it by reason of its faunal association as a
"connecting" or intermediate region between the Holarctic, Ethiopian, and Oriental. A similar, although not yet clearly defined,
intermediate region, comprising in a general way Lower California,
the province of Sonora in Mexico, Arizona, New Mexico, and parts
of Texas, Nevada, and California, with probably also the extremity
of the peninsula of Florida, connects the western division of the
Holarctic realm with the Neotropical; and in the Eastern Hemisphere, the Austro-Malaysian islands lying to the east of Bali and
Borneo, as far as, and inclusive of, the Solomon Islands, form a

transitionary tract between the Oriental, the Australian, and Polynesian realms.

The major faunal divisions of the globe are, therefore:

1. The Holarctic realm.
2. Neotropical realm.
3. Ethiopian realm.
4. Oriental realm.
5. Australian realm.
6. Polynesian realm.
      a. Tyrrhenian, or Mediterranean transition region.
      b. Sonoran, or American transition region.
      c. Papuan, or Austro-Malaysian transition region.

THE HOLARCTIC REALM.

This division comprises the greater portion of the continent of North America, the whole of Europe north of the Alpine chain of mountains, and by far the larger half of the continent of Asia. It is preeminently the region of the Temperate and Frigid zones, and is, in fact, the only one into the consideration of whose organic products a well-marked Arctic element enters. As here defined, it comprises both the Palæarctic and Nearctic regions of zoogeographers, which do not differ very essentially from each other in the general characters of their faunas, or, at any rate, not nearly to the extent that the other regions differ from each other, or these individually from any third. The southern limits of this Holarctic tract, owing to the intermingling along the several border-lines of its fauna with the faunas of the various other regions, is difficult of precise determination; and there can be no doubt that what at many points is considered to belong properly to one region belongs just as properly to another. But such "debatable grounds" between two regions will occur in the case of any other two regions, and likewise in the case of the minor divisions—sub-regions, provinces, etc. In the Western Hemisphere the debatable lands between the Holarctic and the Neotropical realms cover a considerable portion of the Southwestern United States—namely, Arizona, New Mexico, and parts of Texas, Nevada, and California, a tract of territory generally included in the Nearctic region of most zoogeographers. But there can be no question that the preponderating faunal element in this tract is that of the region farther to the

4

south, the Neotropical; and the same can probably be said of the
extremity of the peninsula of Florida.   With these limitations the
Holarctic in the Western Hemisphere embraces the whole of the
United States, and all the region stretching thence northward
towards and into the Arctic Sea.   In the Eastern Hemisphere the
southern boundary may in a general way be said to be the mountain complex which, as the Pyrenees, Alps, Balkans, and Caucasus,
traverses the south of Europe from the Bay of Biscay to the Caspian, the northern line of Persia and Afghanistan, the Himalaya Mountains, and the Nanling range in China, which forms the
southern water-shed to the Yangtse-Kiang.   These various boundaries are principally of a physical nature, and of such a character as to be insurmountable to most animals.

No other region can compare with the Holarctic in the manifold variety of its physical characteristics.   Every form of terrestrial configuration, or condition of soil or climate, that may be represented in any other region, is also represented here, and on an
imposing scale.   From the ice-bound fields of the far north to
the burning desert wastes of Turkestan on the south, and from the
deep forest-grown lowlands to mountain summits soaring thousands of feet above the level of perpetual snow, we pass through
all those various gradations of climate which respectively characterise the Frigid, Temperate, and Torrid zones.   Densely covered
forest tracts, supporting, as in the north, a sombre growth of pine
and other coniferous trees, or, as in the south, a vegetation of
almost tropical luxuriance, alternate with broadly open grass or
pasture lands (*tundras* of Siberia, American prairies and plains),
which in some cases support over enormous areas only a very scanty
vegetation, and in others display a profuse variety of vegetable
productions.   It is in this region that, in addition to a most boun-
tiful development of desert tracts, we meet with the most elevated
table-land (the Central-Asian), and, at the same time, with the
greatest expanse of lowland on the surface of the globe, the great
plain of Siberia and Northeastern Europe.

For convenience of treatment, and to facilitate comparison with
other zoogeographical publications, the Old and New World divisions of the Holarctic region will be considered separately.

The Old World or Eurasiatic Division (Palæarctic region [in
part] of most authors).—The southern boundaries of this region

have already been indicated. In the northwest and west it embraces Spitzbergen and Iceland, and the numerous larger and smaller islands which lie between these and the mainland.

Although this division has an east and west extent not far short of half the circumference of the globe, yet so great is its zoological unity "that the majority of the genera of animals in countries so far removed as Great Britain and Northern Japan are identical. Throughout its northern half the animal productions of the Palæarctic region are very uniform, except that the vast elevated desert regions of Central Asia possess some characteristic forms; but in its southern portion we find a warm district at each extremity with somewhat contrasted features." [25]

Zoology of the Eurasiatic Region.—Although the Eurasiatic fauna comprises representatives of thirty distinct families of Mammalia, not a single one of these is absolutely confined, or is peculiar, to that region. Perhaps on the whole its most distinctive group of quadrupeds is that of the sheep and goats, forming the sub-family Caprinæ of the Bovidæ (oxen). There are represented in this group some twenty-two or twenty-three species (belonging to the genera Capra and Ovibos), which, with four or five exceptions, are either absolutely confined within the limits of the region, or just pass beyond it. The genus Capra, comprising the goats and ibexes on one side, and the sheep on the other, have an outlying Old World representative—a goat—in the "Warryato" (Capra hylocrius) of the Neilgherries (Oriental realm), and another—a sheep, the moufflon (C. [Ovis] musimon)—in the larger islands (Corsica, Sardinia, Crete) of the Mediterranean, and the mountains of Greece and Persia. A species of ibex (C. beden) inhabits the elevated districts of Egypt, Syria, and Sinai, and another (C. Valie), possibly only a variety of the preceding, the highlands of Abyssinia, just within the boundaries of the Ethiopian realm. The two American representatives of the family, the Rocky Mountain big-horn (C. [Ovis] montana) and the musk-ox (Ovibos moschatus), are both absolutely confined to the Holarctic tract. One, at least, of the two generally recognised species of camel, the Bactrian or two-humped species (Camelus Bactrianus), is at the present time entirely, or almost entirely, restricted to the Eurasiatic region, and not unlikely the dromedary (C. dromedarius) was also at one time indigenous to it, although from the long-continued subjection under which it has been held by man, whose wanderings the animal has been forced to follow, it has become almost impossible to determine the precise region constituting its true home.

The extensive group of the antelopes, so highly indicative of the Ethiopian region, are but very sparingly represented, the most characteristic forms being the chamois (Rupicapra tragus), confined to the elevated mountain summits of Southern Europe, from the Pyrenees to the Caucasus, and the saiga (Antelope saiga), an inhabitant of the plains of Southeastern Russia and the adjoining country of Asia. These are the only forms of antelope found in Europe; two or three species inhabit the Thibetan plateau, and several goat-like forms, of the genus Nemorhedas, range from the Eastern Himalayas into China and Japan. The deer (Cervidæ) are sufficiently abundant, and comprise among the more distinctive genera of the region the roe-deer (Capreolus) and the eastern musk (Moschus), the latter considered by many authors to constitute the type of a distinct family (Moschidæ). The stag (Cervus elaphus) ranges over nearly the whole of Europe, and eastward in Asia to Lake Baikal and the Lena River. The only members of the Quadrumana, or monkeys, known to exist within the limits of the region under consideration, belong to the genera Semnopithecus and Macacus, one species of the former (S. Roxellana) occurring in the elevated mountain region of Eastern Thibet, in about latitude 32°, and several of the latter likewise in Eastern Thibet, and also in China and Japan. The Barbary monkey (Macacus inuus), a North African species, which inhabits the Rock of Gibraltar, is the only European representative of the order; but its habitat is located within what has been designated the Tyrrhenian transition region. The Carnivora constitute an important feature in the Eurasiatic fauna, both by the number and variety of the individual forms represented and by their broad geographical range. But the actual number of carnivore genera specially distinctive of this fauna is very limited. The badger (Meles), is found throughout Central and Northern Europe and Asia, in Japan and China, in the latter country extending its range as far south as Hong-Kong, or within the boundaries of the Oriental region. In brief, the most distinctive Eurasiatic mammalian genera may be said to be the following :

Talpa, the Mole.—Distributed throughout the entire region, and passing in Northern India beyond its limits into the Oriental region.

Meles, the Badger.—Temperate Eurasia, Palestine, Japan, and China.

Camelus, the Camel.—At present distributed from the Sahara northeastward throughout Western and Central Asia to the shores of Lake Baikal, and the region of the Amoor.

Capreolus, the Roe-deer.—An inhabitant of temperate and Southern Europe, and Western Asia, with a distinct species in North China.

Moschus, the Musk-deer.—Central Asia, from the Amoor and the district of Peking to the Himalayas and the elevated peaks of Siam.

Poephaga, the Yak.—The elevated plains of Western Thibet.

Rupicapra, the Chamois.—Elevated mountain slopes of the Pyrenees, Alps, Carpathians, Balkans, Caucasus.

Saiga.—The Steppes of Southeastern Russia, and Western Asia.

Capra, the Sheep and Goats.—The former are found in a natural state only in the mountain wilds of Corsica, Sardinia, and Crete, and in Greece, Asia Minor, Persia, and Central and Northeast Asia. The single American form, the big-horn, as above mentioned, is a native of the Rocky Mountains. The goats are found throughout nearly the whole of the South European Alpine region, from Spain to the Caucasus, whence they extend their range through Armenia and Persia to the Himalayas and China.

Myoxus, the Dormouse.—Found throughout the greater part of the region.

Lagomys, the Pika, or Tailless Hare.—A group of small rodents, whose distribution extends from the elevated slopes (11,000 to 14,000 feet) of the Central-Asian mountain system, and Southeastern Russia, north and northeastward to the Polar Sea, and the farthest extremity of Siberia. The genus has a solitary representative in North America.

Myogale, the Water-mole, or Desman.—A singular insectivorous animal, resembling the water-rat, of which there are but two species, one of them inhabiting the valleys along the northern face of the Pyrenees, and the other the river banks of Southern Russia.

Of other well-known types which may be said to be characteristic of, but which are not absolutely confined to, the Eurasiatic region, are the reindeer (Rangifer), the elk (Alces), aurochs or European bison (Bison)—now in a wild state confined to Lithuania and the Caucasus—the polar bear (Thalassarctos), and the beaver (Castor), all of which, comprising in each case but a single species, appear to be, with the possible exception of the bison, specifically identical with North American forms.

North American, or Nearctic Division.—The dominant features of the North American mammalian fauna are preeminently those which also stamp the character of the Eurasiatic fauna. Thus, among the commoner animals we have the deer, moose or elk, reindeer, bison (possibly identical with, or at least very closely allied to, the European aurochs), cats, lynxes, weasels, bears, wolves, foxes, the beaver, hares, squirrels, and marmots. Many of the forms embraced in these types, moreover, are, as has already been stated, specifically identical with their Eurasiatic congeners. But, while there are such striking resemblances between the two faunas—resemblances that penetrate to almost all parts of the regions that are under consideration—it cannot be denied that there are also a number of almost equally well-marked differences; but these are neither sufficiently numerous, nor sufficiently important, to invalidate the claims carried by the positive characters for uniting the two trans-Atlantic divisions into one region, the Holarctic. The preponderating element in the North American mammalian fauna (as, indeed, also in the Eurasiatic) is furnished by the group of the rodents, which here comprise nearly, or fully, one-half of all the recognised mammalian forms. Of about twenty-six genera represented, nearly one-half are restricted, or are peculiar, to this region; but the actual number of specific forms embraced in these peculiar genera scarcely numbers one-fourth of the total number of species. The most distinctively North American families are the Haploodontidæ, a very limited group (two species) of beaver-like animals inhabiting the west coast, and the Saccomyidæ, or pouched-rats and gophers (Saccomys, Geomys, Thomomys, &c.), animals characteristic of the fauna of the Western plains and elevated mountain regions. Among the rats and mice (Muridæ) we meet with, in addition to certain peculiar North American forms, the genus Arvicola, the field-mouse, or vole, which has an extensive representation throughout the temperate portions of the Eastern Hemisphere as well; along with this animal we find the lemming (Myodes), another Eurasiatic form. It is a singular fact, to be noted in this connection, that the typical genus Mus, which includes the common or domestic rats and mice, and which is represented on all the grand divisions of the Eastern Hemisphere, is completely wanting, not only in North America, but in the entire New World, where its place is taken by the closely-allied vesper-mice, constituting the genus Hesperomys. The musk-rat (Fiber), belonging to the same family, is not found outside the limits of the North American continent, although its range extends into the Neotropical realm (Mexico). The squirrels (Sciuridæ) are principally Old World forms; they comprise the true squirrels (Sciurus), flat-tailed flying squirrels (Sciuropterus), ground-shrews (Tamias), marmots (Arctomys), and pouched-marmots, or spermophiles (Spermophilus). In addition to these forms we are presented with the curious animal known as prairie-dog (Cynomys), whose range is confined to the central continental region. Among other rodents may be mentioned the jumping-rat (Jaculus, or Zapus), allied to the eastern jerboas, and the Canadian porcupine (Erethizon), belonging to a group of animals (Cercolabidæ) distinguished from the true or Old World porcupines both structurally and in their arboreal habits. The ungulates, or hoofed animals, have but a very feeble development in the Nearctic division of the Holarctic realm. The goats and sheep are, with two exceptions, the big-horn (Ovis montana), an inhabitant of the Rocky Mountains, and the musk-ox (Ovibos moschatus), from the Arctic district, completely wanting, a faunal characteristic which eminently serves to distinguish the western division of the Holarctic tract from the eastern, to which almost the whole of this group of animals is confined. The antelopes are limited to two species, representing two distinct types, both of them confined to the more temperate regions of the continent. The one is the "prong-horn" of the Western plains (Antilocapra), and the other the Rocky Mountain goat (Aplocerus laniger), which, as the name indicates, is partial to the mountain fastnesses. Two varieties of the bison, or American buffalo, are recognised—the buffalo of the plains, and the buffalo of the forests and mountains; but the variation observable between these is one pertaining to habit and not to structure, and therefore not of specific importance. The Carnivora present several distinctively American types, and notably so the raccoons (Procyonidæ), a small group of interesting quadrupeds, whose home is primarily the region of the tropics, and which appear to hold a somewhat inter-

mediate position between the weasels and bears.   To the same
family belong the South American coatis (Nasua) and the prehensile-tailed kinkajou (Cercoleptes).   The Mustelidæ, or weasels, comprise the weasels proper, marten, ermine, mink, glutton, American
badger (Taxidea), skunk, American otter (Latax), and the singular
sea-otter (Enhydris), from the California coast.   The most formidable carnivores are the grizzly bear (Ursus horribilis), not improbably identical with the European brown-bear (U. arctos), and
the couguar, or American panther (Felis concolor), whose range
extends from the sixtieth parallel of north latitude to the southern
extremity of Patagonia.   One species of implacental mammal—the
Virginian opossum (Didelphis Virginianus)—penetrates as far north
as the Canadian frontier.

Taking the Nearctic and Palæarctic divisions of the Holarctic
region collectively—i. e., the region as a whole—we find it to be
characterised by the exclusive, or almost exclusive, possession of
the following families: Talpidæ (moles), Trichechidæ (walruses),
Castoridæ (beavers), and Lagomydæ (pikas); and if the reindeer,
moose, and sheep and goats, be considered as distinct families, as
is maintained by many naturalists, then also by the Rangiferidæ,
Alcidæ, and Capridæ.   In addition to these seven families, we
have also the hares (Leporidæ) and bears (Ursidæ), which, though
not exclusively restricted to these regions, are by their numbers
and vast distribution, eminently characteristic of them.   Of about
one hundred and twenty genera represented, upwards of seventy
(or sixty per cent.) are found in no other region.   Among the most
characteristic forms are—

In the Old World :

Talpa, the mole.	Rupicapra, the chamois.
Meles, the badger.	Saiga, the saiga ʳ ɪtelope.
Camelus, the camel.	Capra, the goat.
Capreolus, the roe-deer.	Myoxus, the dormouse.
Moschus, the musk-deer.	Myogale, the water-mole.
Poephaga, the yak.	

In the New World:

Saccomys,	} pouched-rats or	Ovibos, the musk-ox.
Geomys,	} gophers.	Antilocapra, the prong-horn.

Thomomys, ⎫ pouched-rats or
Dipodomys, ⎬ gophers.
Perognathus, ⎭
Jaculus, the jumping-mouse.
Fiber, the musk-rat.
Cynomys, the prairie-dog.
Erethizon, the Canadian porcupine.

Aplocerus, the Rocky Mountain goat.
Procyon, the raccoon.
Mephitis, the skunk.
Latax, the American otter.

Common to both divisions:

Lagomys, the pika.
Arctomys, the marmot.
Spermophilus, the pouched-marmot.
Castor, the beaver.
Myodes, the lemming.
Arvicola, the field-mouse.
Ovis, the sheep.

Bison, the bison.
Rangifer, the reindeer.
Alces, the elk.
Thalassarctos, the polar bear.
Gulo, the glutton.
Lyncus, the lynx, and most of the seals and the walruses.

The bird-faunas of the Old and New World divisions of the Holarctic tract differ very materially from each other, a condition in great measure explained by the circumstance that in both a large representation is obtained through migration from extra-limital regions. Thus, the Eurasiatic or Palæarctic avifauna is largely made up of types which are equally Ethiopian or Oriental; and in like manner a very large proportion of the similar North American fauna is made up of forms which might with equal justice be considered Neotropical or Nearctic. But even in the case of the resident birds, or such as may be considered to be more properly belonging to the region, marked differences, sufficient to characterise the two divisions, present themselves. The preponderating Eurasiatic forms belong, among the perchers, to the families of thrushes (Turdidæ)—with the cosmopolitan genus Turdus; warblers (Sylviadæ), with the true warblers (Sylvia), red-start, robin, and nightingale (Luscinia); nuthatches (Sittidæ), tits (Paridæ), Muscicapidæ (Old World fly-catchers), shrikes (Laniidæ), crows (Corvidæ), with the pies, crows proper, and jays; swallows (Hirundinidæ), finches (Fringillidæ)—gold-finch, haw-finch, cross-bill, bull-finch, linnet, sparrow, grosbeak, lark-bunting, true finch

(Fringilla), and bunting (Emberiza), the last two almost exclusively confined to this region (and the adjoining debatable tracts)—starlings (Sturnidæ), larks (Alaudidæ), wag-tails (Motacillidæ), wood-peckers (Picidæ), king-fishers (Alcedinidæ), swifts (Cypselidæ), and pigeons (Columbidæ). All of these families, not a single one of which is restricted to the Holarctic region, are, with the exception of the starlings (Sturnidæ) and the fly-catchers (Muscicapidæ), likewise distributed throughout the Nearctic division, of whose avifauna they constitute a very important factor. In the New World the true starlings are replaced by the family of hang-nests (Icteridæ), to which the Baltimore bird (Icterus), bobolink (Dolichonyx), cow-bird (Molothrus), and red-wing (Agelaius) belong. The Old World fly-catchers have their representatives in the tyrant shrikes (Tyrannidæ), familiarly also known as fly-catchers. The true warblers (Sylviadæ) are but very feebly developed in the Nearctic division, where, of about ten species, three are kinglets (Regulus), and three blue-birds (Sialia); but their place is taken by a multitude of forms belonging to the preeminently South American family of wood-warblers (Mniotiltidæ). Of the Holarctic gallinaceous birds the most distinctive forms in the eastern division are the true partridge (Perdix), snow-partridge (Tetraogallus), capercaillie (Tetrao), true pheasant (Phasianus), golden-pheasant (Thaumalia), tragopan (Ceriornis), and impeyan (Lophophorus), forms either exclusively restricted to the region, or just passing beyond the boundaries; of the western division, the California quail (Oreortyx), cupido (Cupidonia), tree-grouse (Canace), sage-grouse (Centrocercus), and turkey (Meleagris). The ruffled-grouse (Bonasa) and ptarmigan (Lagopus) are common to the northern regions of both hemispheres. The birds of prey comprise, throughout both divisions of the region, a variety of eagles, falcons, hawks, buzzards, kites, and owls, and of forms nearly all of which are also found in other portions of the earth's surface. America has no representative of the Old World group of (true) vultures, forming the sub-family Vulturinæ, their place being filled by the carrion vultures, or so-called turkey-buzzards (Cathartinæ). Of the wading-birds the Eurasiatic region alone possesses the true bustard (Otis), the typical representative of a family whose members are spread throughout Africa, Asia, and Australia.

The Holarctic region is deficient in reptilian forms as compared with the warmer regions of the earth's surface, which appear to be more suitable to the habits of this class of animals. In the whole of Europe north of the Alps, or in what has been recognised as the "European province," naturalists recognise only about fifteen species of snakes, and a nearly equal number of lizards ; in the Nearctic division, while the number of lizards is not very much greater—about twenty species—that of serpents is very materially increased—to about eighty to ninety species—most of them belonging to the family of colubers, which includes the black constrictors. The headquarters of the rattlesnakes are situated in the debatable land bordering the Neotropical realm.

The entire reptile-fauna of Europe is, according to Schreiber ("Herpetologia Europæa," 1875), comprised in sixty-two species, of which twenty-five are serpents, thirty-two saurians, and five chelonians. Northern Europe, or the region lying to the north of the fifty-fifth parallel of latitude, is represented by but six species: Viperus (Pelias) berus, Tropidonotus natrix, Coronella Austriaca, Anguis fragilis, Lacerta vivipara, and L. agilis; the chelonians are completely wanting in this tract. Central Europe, including the Alpine system of mountains, has twenty-nine species, while the entire number is represented in the Mediterranean fauna. A number of additional species has been added to the list enumerated by Schreiber; but these do not materially affect the ratio for the different zones. The most northerly of all serpents appears to be the common European viper, Viperus (Pelias) berus, whose range in Scandinavia extends to about the sixty-seventh parallel of latitude. The species is distributed throughout nearly the whole of Europe, and eastward through Central Asia to the Japanese islands; it is also found in England and Scotland, and in some of the Scotch islands (Arran, Hebrides). Tropidonotus natrix (Natrix vulgaris), a species of equally broad distribution, which is stated to ascend mountains to a height of six thousand feet, is found in Norway as far north as the sixty-fifth parallel.

The most northerly, and at the same time most broadly distributed, species of European lizard is the Lacerta vivipara, whose range in Norway is extended by Collett to the seventieth parallel of latitude. It is found throughout most of Europe (wanting in the Iberian Peninsula, Southern Italy, and Greece), and is an inhabitant of the Alpine region, up to an elevation of nine thousand feet.

While the Holarctic region is relatively meagre in its reptilian fauna, it is preeminently the home of the tailed amphibians, newts, salamanders, &c., of which we have the blind proteus (Proteus anguinus), in the cavern-waters of Carinthia, Carniola, and Istria; the giant salamander, known as Sieboldia (Cryptobranchus), in Japan, and its allied American form, the menopoma, the eel-like sirens, mud-puppies (Necturus), and almost limbless amphiumes of the Eastern and Southern United States; the true salamander and triton in Europe and Asia, and their American representatives, the amblystomes, to which the singular form known as the axolotl belongs. The European tail-less amphibians (frogs and toads) number some dozen or more species of the genera Bombinator, Pelobates, Alytes, Hyla, Discoglossus, Rana, and Bufo, the most broadly distributed of which appears to be Rana temporaria and R. esculenta, the former extending its range eastward to Japan and America, and northward in Norway to beyond the seventieth parallel of latitude.

The fish-fauna of the Holarctic tract is characterised by the special development, among fresh-water forms, of the carps (Cyprinidæ), salmon (Salmonidæ), pikes (Esocidæ), perches (Percidæ), sculpins or bull-heads (Cottidæ), sticklebacks (Gasterosteidæ), sturgeons (Accipenseridæ), and lampreys (Petromyzon), which are distributed over both the eastern and western divisions of the region. The Cyprinoids are especially abundant, constituting, in the number of species, according to Günther (Ency. Brit., XII., p. 675), nearly two-thirds (two hundred and fifteen species) of the entire fish-fauna of temperate Europe (including the Mediterranean transition region) and Asia, and more than one-third (one hundred and thirty-five species) of the equivalent fauna of North America. The cat-fishes (Siluridæ), so eminently characteristic of the more southerly equatorial zone, are largely deficient in the number of species. Silurus occurs in some of the Eurasiatic waters as an immigrant from India; most of the North American forms belong to the genus Amiurus. Among the more distinctive ichthyic features separating the faunas of the eastern and western divisions of the Holarctic realm are the possession, by the former, of the barbels (Barbus) and cobitoids, and, by the latter, of the suckers (Catostomidæ), sun-fishes (Centrarchidæ, most abundant in the Mississippi Valley), and two genera of ganoid fishes, Amia and Lepidosteus, both of which occur as fossils in the Tertiary deposits of North America, and the latter also in Europe. On the other hand, the two regions exhibit a marked inter-relationship by the possession of a number of identical specific forms, as Accipenser sturio (sturgeon), Perca fluviatilis (perch), Salmo salar (salmon), Esox lucius (pike), Lota vulgaris (ling), &c.

The correspondence existing between the vertebrate faunas of the Old and the New World divisions of the Holarctic tract extends also to the Invertebrata, and is especially marked in the case of the beetles (Staphylinidæ, Carabidæ), butterflies, and the land and fresh-water mollusks (Limnæa, Planorbis, Physa, Paludina, Valvata). The land-snails (Helicidæ) and naiades (Unionidæ) are very largely developed, the latter more particularly in the American streams, where distinctive types appear to be relegated to the different water-courses. The eastern melanians are wholly wanting in America, where they are replaced (principally to the east of the Mississippi River) by the members of the allied family of the Strepomatidæ (Io, Goniobasis, &c.).

The Holarctic realm may be conveniently divided into the following sub-regions:

1. *The Boreal Sub-Region*, which extends northward into the Polar Sea, and whose southern confines are fixed approximately by the northern limits of the cultivation of the cereals, and the southern limits of the migration of the reindeer. In the Western Hemisphere it comprises most of the region lying to the north of the United States and Canada boundary-line, and in Eurasia the tract lying north of a line starting from about the sixty-sixth parallel of latitude, on the Norwegian coast, and passing southeastward to the East Asiatic coast, in about latitude fifty degrees north. The fauna of this region is a very homogeneous one, and, generally speaking, also a limited one. Among the more distinctive mammalian forms, which comprise almost exclusively only ruminants, carnivores, and rodents, are the Arctic fox (Canis lagopus), polar-bear, glutton, ermine, mink, sable, walrus, variable hare, lemming, and reindeer. The musk-ox, which occurs fossil in the Quaternary deposits of Europe, is at the present time found only in America. Among the more characteristic birds are the snow-partridges (Lagopus), snowy-owl (Surnia nivea), Iceland falcon (Falco candicans), eider-duck (Somateria mollissima), and various alks (Alca), divers (Colymbus), and guillemots (Uria, Lomvia). Captain Markham observed the footprints of the

polar-hare in the snow-bound ice in latitude 83° 10′, and the antlers
of a reindeer were picked up by the officers under Sir George Nares,
in latitude 82° 45′ (Grinnell Land). A skeleton of the latter animal, recently picked by wolves, was also obtained in latitude 80°
27′. Traces of the rock-ptarmigan (Lagopus rupestris) have been
met with as far north as latitude 83° 6′, and the snow-bunting (Plec-
trophanes nivalis) in latitude 82° 33′. The reptile-fauna is very
limited, no serpent, apparently, passing beyond the sixty-seventh
parallel of latitude, and no lizard above the seventieth. The fishes,
which include the common perch and pike, are mainly Salmonoids.
Insects are fairly numerous, and even in the far north the number
of species is considerable. Sir George Nares obtained no less than
forty-five species, representing nearly all the orders (including Lepidoptera) in Grinnell Land; Greenland has thus far yielded eighty
species, and Iceland three hundred. Among the most northerly
genera of non-marine mollusks are Helix, Pupa, Succinea, Limnæa,
and Planorbis.

2. *The European Sub-Region*, which includes practically the whole
of Europe lying between the Arctic tract and the Alpine system of
mountains, extending southeastward and eastward to the Caucasus
and the Caspian steppes. The fauna of this region is typically
that of temperate Eurasia, taken as a whole, and therefore requires
no elaborate analysis. Among its more or less characteristic Mammalia are the moose or elk (in the north), stag, roe, aurochs (Bison
Europæus, in the Caucasus and the forests of Lithuania), the Alpine
chamois and ibex (with the accompanying marmot, Arctomys marmotta), brown bear (Ursus arctos), badger, glutton, dormouse,
hamster (Cricetus frumentarius), mole (Talpa Europæa), and hedge-
hog (Erinaceus Europæus), many of which also form integral parts
of the Arctic and Central Asian faunas. The birds comprise several hundred species, of which Germany alone possesses nearly three
hundred; distinctive types are, however, not numerous—indeed,
they may be said to be almost wholly wanting—and even the number of restricted forms is very limited. Among these may be mentioned a number of song-birds of the finch tribe, as the chaffinch
(Fringilla cœlebs), siskin (F. spinus), goldfinch (F. carduelis), bull-
finch (Pyrrhula rubricilla), yellow-hammer (Emberiza citrinella),
and linnet (Linota linaria), and the nightingale (Luscinia). The
most distinctive bird of prey is the bearded vulture of the Alps, or

lämmergeier (Gypaëtus barbatus). The reptilian-fauna, as has already been remarked, when treating of the Eurasiatic region generally, is very limited. Among the poisonous serpents are the Pelias berus and the sand-viper (Vipera ammodytes), the range of the latter extending from the Mediterranean into Sweden.

3. *The Central Asian Sub-Region* comprises that portion of the Holarctic tract which lies south of the Arctic sub-region, and is included between the European sub-region on the west and China proper and Manchuria on the east. A steppe character prevails in the western portion of this region, to which desert features are also added, and hence we find a distinct individuality imparted to its fauna. The rodents, whose most distinctive forms are the spermophiles and jumping-mice (Dipus, Scirtetes), are more extensively developed than in the European tract; on the other hand, the larger Carnivora are almost completely wanting, although the tiger seems to occasionally reach the region about the Caspian and Aral seas. The deer are replaced by one or more forms of antelope (Saiga), which, with the exception of the camel and some wild equine representatives, are almost the sole hoofed animals of the region. A distinctive faunal feature is constituted by the seals of the Caspian (Phoca Caspica), which appear to be very closely related to the common Phoca vitulina of the North Atlantic. Among the birds may be mentioned the steppe partridges (Pterocles, Syrrhaptes), and, as an occasional visitor, not improbably also the ostrich. Reptiles are fairly abundant, and include among the lizards the agamid genus Stellio, and among serpents the sand-snake (Psammophis) and the poisonous Trigonocephalus.

In the eastern half of the Central Asian sub-region, plateau and desert features largely predominate, and the fauna acquires somewhat distinctive characters. The ungulate and carnivore types of Mammalia are more abundantly represented, the former comprising, in addition to the camel (Camelus Bactrianus), whose range extends to the shores of Lake Baikal, and two or more species of antelope, the dziggetai or kiang (Equus hemionus) of the plains of Thibet, the recently-discovered species of wild horse described by M. Poliakoff as Equus Przevalskii, the Thibetan yak, several wild goats, and the argali (Ovis argali), the last recalling the big-horn of the Rocky Mountains. The tiger and ounce (Felis uncia) both range across the region into Siberia, sharing the habitat of the

snow-leopard (Felis irbis). The grouse, partridges, steppe partridges, and bustards constitute important elements in the bird-
fauna, which exhibits the general features of the north temperate
region.

4. *The Manchurian Sub-Region* embraces Manchuria, Northern
China (with a westerly extension along the northern face of the
Himalayas), and the Japanese islands. The fauna, especially in
the more southern districts, exhibits characters drawn from both
the Tropical and Temperate zones. A distinguishing feature of
this sub-region is the presence of monkeys, which are represented
by the genera Macacus and Semnopithecus, the former penetrating
even into Japan. The Cervidæ, whose range extends into Japan
(where the antelope is also met with), are represented by both
horned (Elaphodus) and hornless species (Hydropotes, Lophotra-
gus), and by the diminutive musk-deer (Moschus moschiferus),
which inhabits the mountain-valleys from the Amur to the Himalayas. The edentates have their most northerly representative in
the scaly Manis (Japan). Among the more distinctive carnivore
forms are the genera Lutronectes, Ailuropus, and Nyctereutes.
The facies of the Japanese bird-fauna is distinctly European, but
in China, which is properly the home of the pheasants, there is a
considerable intermixture of tropical (Oriental, &c.) forms, such as
the babbling-thrushes, caterpillar-eaters, honey-suckers, and weaver-
finches.

5. *The Alleghanian Sub-Region* may be approximately defined as
comprising that portion of the Holarctic realm in the Western
Hemisphere which lies south of the Arctic tract and east of the
one hundredth meridian of west longitude (Greenwich). Its faunal
features are those of the North American or Nearctic division generally, although among the Mammalia a deficiency is brought about
by the absence of some of the pouched-rats or gophers, the big-
horn (Ovis montana), the Rocky Mountain goat (Aplocerus laniger), and the musk-ox, which, with the exception of the last, belong
to, or are most abundant in, the central or Rocky Mountain sub-
region. The majority of the Arctic Carnivora penetrate to within,
or far beyond, the northern boundary-line, and the same is true,
to a certain extent, of the moose and caribou (reindeer). The
more strictly southern forms are the opossum, raccoon, and peccary, the last, however, not penetrating farther north than the

Red River, in Arkansas. The bison, which was at one time very abundant on the western prairies, has now almost wholly disappeared, and impending destruction likewise threatens the pronghorn (Antilocapra Americana), which in the furcation and shedding of its horns may be considered to be the most divergent type of all antelopes. Characteristic rodent forms of the prairies are the prairie-dog (Cynomys Ludovicianus) and the gopher (Geomys bursarius). More properly belonging to the eastern wooded sections are the beaver (in the north), identical with the European species, the skunk, and Canada porcupine.

6. *The Rocky Mountain Sub-Region* comprises the mountainous tracts of the West-Central United States, and a similar region in Canada extending to about the fifty-fifth parallel of latitude. Its southern boundary is determined by the as yet not very exactly defined Sonoran transition region. Its general and distinctive faunal features have already been indicated.

7. *The Californian Sub-Region* is comprised within the borderland which stretches along the Pacific west of the main mountain axis, and extends northward to about the fiftieth parallel of latitude. Many of its faunal features are borrowed from the Neotropical realm, but in its entirety the fauna is distinctively North American. Among its characteristic mammalian forms are the sewellel (Haploodon) and the grizzly bear (Ursus horribilis). Other distinctive types are the California condor (Sarcorhamphus Californianus), the ground cuckoos (Geococcyx), and the singular bird known as Chamæa. The number of humming-birds is greater than in the Eastern United States.

THE NEOTROPICAL REALM.

This region, as usually recognised, comprises the continent of South America, the West India Islands, Central America, and the lowlands—*tierras calientes*—on either side of the Mexican plateau. While, therefore, it is in the main clearly circumscribed by its water-boundary, the northern portion, or that which borders on the Holarctic, is much less sharply defined, as must invariably be the case where two faunal regions overlap. The principal features of the greater portion of this vast tract are singularly uniform. An enormous expanse of forest, unequalled for its continuity and luxuriance of growth, occupies fully one-half of the surface area, covering not only the deep lowlands, but also the mountain-slopes to a very considerable elevation. Commencing on the Atlantic border, it stretches, through a north and south extent of thirty degrees of latitude, almost unbroken for nearly three thousand miles to the base of the Andes, harbouring in its dense recesses a host of the most varied animal forms. Beyond, and partially enclosed within, the limits of this vast forest region are the various forms of pasture-land, or grassy plains, the llanos or savannas of Venezuela, the campos of the highlands of Brazil, and the pampas of the Argentine Republic and Patagonia. In the great Andean chain, which traverses in one continuous sweep the entire north and south expanse of the region, we have all the more characteristic features of a mountain-system developed on a most gigantic scale—high plateaus, deep valleys, wooded and barren slopes—conditions affecting in a most marked degree the diversity of its animal and vegetable creations. Desert areas, or such as are rendered almost unfit for habitation by reason of extremes of climate, like the north of both the North American and Eurasiatic continents, are, if we except the most elevated mountain-summits, limited to a few scattered patches of small area in the Argentine Republic, and to a narrow tract of littoral lying in Peru and Chili, on the Pacific side of the mountain-axis.

Zoological Characters of the Neotropical Realm.—In comparing the fauna of the Neotropical with that of the other zoogeographical regions we are struck with two things: 1, its extraordinary richness; and 2, the very great preponderance of forms that are peculiar to the region and not met with anywhere else. According to Wallace, the region comprises no less than forty-five families and nine hundred genera of vertebrate animals which are strictly peculiar to it, while it has representatives of one hundred and sixty-eight out of the total of about three hundred and thirty families recognised by naturalists ; in other words, more than one-half of all the families scattered over the globe are here represented. Of about thirty-one mammalian families eight are almost completely confined to the region, as follows: The Cebidæ, or true South American monkeys; the Hapalidæ, or marmosets; the Phyllostomidæ, or simple leaf-nosed bats, which include the vampires (with one extra-limital species in California); the Chinchillidæ, comprising the chinchilla and vizcacha, a small group of animals confined principally to the Alpine slopes of the Andean chain, in Peru, Bolivia, and Chili, but also (Lagostomus) inhabiting the lowland plains of the Argentine Republic and Uruguay; the Caviidæ, the cavies (Cavia) and agouties (Dasyprocta), a family of rodents to which the guinea-pig belongs, and whose members range from Mexico to beyond the forty-eighth parallel of south latitude; the Bradypodidæ, or sloths, confined exclusively to the forest region; the Dasypodidæ, or armadillos, which are found throughout almost the entire region, with one species (Dasypus peba) penetrating as far north as Texas; and the Myrmecophagidæ, or ant-eaters. The Cebidæ, or South American monkeys proper, constitute a very distinct group of quadrumanous animals (Platyrhina), distinguished from the monkeys of the Old World (Catarhina) by several very prominent characters, such as the broad nasal septum (whence the designation of flat-nosed monkeys, Platyrhina), the absence of ischial callosities and cheek-pouches, and the presence of an additional premolar tooth on each side of each jaw, making the total dental formula $\frac{18}{18} = 36$, instead of $\frac{16}{16} = 32$, as we find it in the Old World apes and man. In the smaller group of the marmosets and lion-monkeys (Jacchus, Midas) we find the same number of teeth as in the catarhines, but the relation of the premolars to molars is reversed, i. e., they are arranged according to the formula $\frac{3-3}{3-3}$ pm., $\frac{2-2}{2-2}$ m.; instead of pm. $\frac{2-2}{2-2}$, m. $\frac{3-3}{3-3}$. In none of the American monkeys is the thumb completely opposable to the other fingers, an important distinguishing character; and scarcely less important is the presence, in most cases, of a prehensile tail, which, as such, is developed only in this group of the Quadrumana. The range of the American monkeys, in marked contrast to that of the African, is limited virtually to the forest region, in which alone they find their proper sustenance. Their most southern extension appears to be about the thirtieth parallel of south latitude, and their most northern, the southern portions of Mexico. No representatives of the order are met with in any of the West India Islands, which are also wanting in all Carnivora and Edentata. The Platyrhina comprise, among other groups, the well-known howling-monkeys (Mycetes), spider-monkeys (Ateles), and capuchins. The Neotropical Carnivora embrace a number of larger and minor cats, the most formidable of which are the jaguar and couguar (puma), the former ranging from the pampas of the Argentine Republic to Texas, and the latter, as has already been observed, from Patagonia to about the sixtieth parallel of north latitude in Canada. Among the lesser animals of this family are the jaguarundi, ocelot, and other so-called tiger-cats. Of the weasels, there are no representatives of the genera Mustela or Putorius over the greater part of the region. The Canidæ are represented by various forms of wild-dogs (Icticyon, Lycalopex, Pseudalopex), which are principally confined to the open grass-country; the wolf and fox are both absent, except from certain portions of Mexico, which ought, perhaps, more properly to be relegated to the intermediate tract which separates this from the Holarctic region. The only member of the Ursidæ found in the entire continent of South America (with Central America) is the "spectacled" bear (Ursus ornatus), from the Chilian and Peruvian Andes, which, through certain peculiarities of structure, has been separated by some authors from the true bears (Ursus), and placed in a distinct group, Tremarctos. Among the distinctive rodents, other than the cavies and agouties, are the subungulate capybara and paca, and the beaver-like coypu (Myopotamus). A negative feature is the almost total absence of Insectivora. The hoofed animals (Ungulata) are but very sparingly represented in the Neotropical realm, a circumstance in marked contrast to what is presented by the similarly-situated Ethiopian or African region. The antelopes, so characteristic of the warmer parts of both the African and Asiatic continents, are completely wanting, and there are likewise neither indigenous horses, oxen, sheep, nor goats. A comparatively limited number of species of deer are scattered throughout the region, from Mexico to the Rio Negro in Patagonia. Of other even-toed ungulates (Artiodactyla) we have the peccaries (Dicotyles), the American representatives of the Old World family of swine (Suidæ), whose range extends to the Red River, in Arkansas, and consequently considerably beyond the limits of the region ; and the llama, alpaca, guanaco, and vicuña, together constituting the genus Auchenia, which are the New World representatives of the camel family (Camelidæ). It is a most striking fact in the distribution of this family of ruminants, that the only two genera of which it is composed should, in their habitats, be separated by one-half of the circumference of the globe; and that, further, while the one genus, Camelus, belongs strictly to the Northern Hemisphere, the other, Auchenia, is restricted to the Southern. But these are not the only peculiarities distinguishing this singularly discontinuous family, for, while the Eastern representatives are specially adapted to an existence in the hot and parched surfaces characteristic of desert lands, those of the Western Hemisphere, on the contrary, are habituated to the rugged and snow-covered slopes of the South American Cordilleras.* The tapir, which, with the exception of the peccary, is the only pachydermatous South American mammal, presents us with an example of a discontinuous family no less marked than that of the Camelidæ. Its four to six members are, with one exception, all confined to South and Central America, inhabiting the lofty mountain regions of from eight to twelve thousand feet elevation, as well as the lowland equatorial forests. The only extra-limital representative of the family is the Tapirus Malayanus, or white-banded Malay tapir, whose home, the Malay Peninsula, Sumatra, and Borneo, is separated from that of its American congener by an interval of nearly one-half the earth's equatorial circumference.

The bird-fauna of the Neotropical realm is no less striking by its diversity than the mammalian. It comprises representatives of upwards of six hundred and eighty genera of land-birds, of which some five hundred and seventy, or just five-sixths of the entire number, are peculiar to it.[26] The vernal migration naturally tends to spread many of the South American avian types northward, and thus a large number of even the more strictly Neotropical forms have what might in a measure be considered North American or Holarctic representatives. Of the humming-birds (Trochilidæ), a distinctively South American family, comprising about one hundred and twenty genera, and upwards of four hundred species, no less

* The vicuña is rarely found at a lower level than thirteen thousand feet; the llama descends to three thousand. It has already been remarked, when treating of the influence of climate upon distribution, that, while the camel is more properly an animal of the warm country, it yet winters, with apparent comfort, as far north as the region of Lake Baikal, in latitude 52° to 53°. Again, while most suitably adapted to a desert region, the animal, it appears, can conveniently accommodate itself also to rugged mountain-slopes.

than fifteen species are found within the limits of the Holarctic
realm and the Sonoran transition tract, one species, the ruby-throat
(Trochilus colubris), on the east coast of the continent of North
America extending its range northward beyond the Canadian border, and one (Selasphorus rufus) on the west as far north as Sitka.
So, again, the Conurinæ, or macaws, an equally distinctive Neotropical group, with about eighty species, have a solitary Holarctic
representative in the Carolina parakeet (Conurus Carolinensis),
whose range, at the present time, seems not to extend much farther
than the State of South Carolina, but which, until a comparatively
recent time, penetrated as far north as Nebraska. The Cœrebidæ,
or sugar-birds, whose brilliancy of plumage rivals that of the hummers, have an outlying member in Certhiola Bahamensis, of which
a colony has been established on one of the Florida Keys, or just
beyond the limits of the Neotropical realm.[27]   Other characteristic
families of South American birds are the toucans and araçaries
(Rhamphastidæ), a strictly frugivorous group, recalling by the
structure of the bill the distant Old World horn-bills; the jaca-
mars (Galbulidæ) ; the saw-bills, or motmots (Prionitidæ) ; the
Pipridæ, or manakins ; the Cotingidæ, or chatterers, which include, besides the cotingas and pompadours, the famous cock-of-
the-rock (Rupicola), umbrella-bird (Cephalopterus), and bell-bird
(Chasmorhynchus); the Dendrocolaptidæ, or tree-creepers, with
upwards of two hundred species; the wonderfully variegated tana-
gers (Tanagridæ), with upwards of three hundred species, which
may in a measure be considered to occupy the place of the Temperate Zone finches and sparrows, and of which the common scarlet
tanager (Pyranga rubra) and summer-redbird (Pyranga æstiva)
are familiar North American examples; to the same group belong
the South American spice-birds of the genus Calliste, and the or-
ganist (Euphonia); the Cracidæ, curassows and guans, which are
the largest game-birds of the region, and which take the place of
the Old World grouse and pheasants; and the Tinamidæ, or tina-
mous, a group of birds recalling in their general appearance the
partridges, and possessing certain affinities with the ostriches. The
exquisitely decorated trogons (Trogonidæ) present us with one of
the most remarkable instances of a discontinuous family, whose
representatives are found at opposite points of the earth's equatorial
circumference—the Neotropical and Oriental regions.   The inter-

vening continent of Africa appears to be almost entirely deficient
in the members of this family. The Struthionidæ, or ostriches, are
similarly divided, the South American continent possessing three
(of the genus Rhea) out of the five or six species of which the
family is composed. The noble and ignoble birds of prey are both
well represented in the Neotropical realm, and it is here that the
most powerful and largest bird of flight, the Condor (Sarcorhamphus gryphus), is to be found. True eagles of the sub-family
Aquilinæ are absent from the greater part of the region, and there
are no representatives of the common genera distinctive of the
Temperate Zone—Aquila, Haliaëtus (the bald eagle), or Chrysaëtos
(the golden eagle). The harpy eagle (Thrasaëtus harpya), more
properly a buzzard, penetrates as far north as the Texan frontier.
Among the characteristic families of birds which the Neotropical
region shares with the western division of the Holarctic are the
Tyrannidæ, or tyrant-shrikes, the New World representatives of
the Old World fly-catchers (Muscicapidæ), and the Icteridæ, or
hang-nests, the New World representatives of the Old World
starlings (Sturnidæ), of which the common Baltimore oriole and
the cassique (Cassicus cristatus) are familiar examples. It is a
singular fact that the crows and ravens (Corvus), which are
otherwise nearly cosmopolitan, and which comprise upwards
of fifty species, are completely wanting in the greater part of
the Neotropical realm, no species being found south of Guatemala.

The Neotropical reptile-fauna is scarcely less well-marked than
the mammalian or the avian. It includes the giant boas and anacondas of the genera Boa, Epicrates, and Eunectes, the coral-snake
(Elaps), which has one or two extra-limital representatives in the
United States, the venomous crotaloids, with the true rattlesnakes,
Lachesis, and Craspedocephalus (jararaca), both alligators (cayman)
and crocodiles, and no less than about one hundred and fifty species
of the singular lizards constituting the family Iguanidæ. The ano-
lids and amphisbænians are represented by numerous species. The
tailed amphibians, such as the newts and salamanders, are almost
absent, but in their place there is an unusual development of the
tailless forms, the toads and frogs (horned-frog, Ceratophrys; Hemi-
phractus), especially of the tree-frogs (Hylidæ).—The fresh-water
fishes of the Neotropical realm are specifically more numerous than

those of any other region, with perhaps the exception of the Holarctic. According to Günther they comprise nearly seven hundred
distinct forms, although representing only nine families. About
one-third of the species belong to the family Characinidæ, and
a somewhat larger number to the cat-fishes (Siluridæ). The
toothed-carps (cyprinodonts) are represented by sixty or more
species. Among the distinctive fishes of the region are the
electric eel (Gymnotus electricus), from the equatorial regions,
and the remarkable lung-fish (Lepidosiren paradoxa) of the Amazon.

The Neotropical tract may be conveniently divided into the
following sub-regions: 1, the *Brazilian*, comprising Brazil, Guiana,
Venezuela, Colombia, Ecuador, Paraguay, and the cis-Andean portions of Peru and Bolivia, inclusive of the eastern slope of the
mountain-axis, essentially a region of dense and luxuriant forests;
2, the *Chilian*, principally a region of open plains and pampas,
comprising Chili, Patagonia, the Argentine Republic, Uruguay,
and the remaining parts of Peru and Bolivia, extending to about
the fourth parallel of south latitude; 3, the *Mexican*, including
the Isthmus of Panama, Central America, and Southern Mexico ;
and, 4, the *Antillean*, or the sub-region of the West India Islands. In the first of these sub-regions, the Brazilian, the faunal
facies is essentially that of the Neotropical realm taken as a whole,
inasmuch as there is scarcely a single group of important or typical
South American animals which has not its representatives here.
Furthermore, the majority of these forms have their greatest development in this tract. Among its most distinctive negative
elements may be cited the chinchillas, the spectacled bear, the
llamas (with the alpaca, vicuña, and guanaco), the rheas (South
American ostriches), and the condor—members of the fauna of
the Chilian sub-region—which are either wholly wanting, or but
barely pass beyond the regional confines. Positive distinguishing
characters among the Mammalia may be found in the special development of the quadrumanous and edentate types—among the
former, in addition to the more widely distributed forms, such as
Cebus, Ateles (spider-monkey), and Mycetes (howler), the woolly-
monkeys (Lagothrix), the sakis (Pithecia), the douroucoulis or
night-monkeys (Nyctipithecus), squirrel-monkeys (Chrysothrix), and
some thirty or more species of marmosets, which appear to be con-

fined to the tropical forests; and among the latter, several species of armadillo, the great ant-eater (Myrmecophaga jubata), and the various forms of two-toed (Cholœpus) and three-toed (Bradypus) sloths—the spiny-rats (Echimyidæ), most of whose representatives are confined to this region, and the manatee or vacca marina (Manatus), which ascends the river Amazon. Among the more restricted birds are the capitos, the trumpeter (Psophia), screamer (Palamedea), hoazin (Opisthocomus), pauxi, and boat-bill (Cancroma).

The fauna of the Mexican or Central American sub-region corresponds closely with that of the sub-region just described, from which it differs mainly by the comparative paucity of its developed types, and by the more pronounced infusion of the Holarctic or northern element. As representatives of the latter we have the shrews, the hare (one species also in Brazil), ground-squirrel, fox, and Bassaris. The very limited number of distinctive types include the Central American tapir (Elasmognathus Bairdii), Myxomys among the mice, and Heteromys among the pouched-rats.—The Chilian fauna, some of whose more prominent features have already been indicated, is broadly distinguished from the faunas of the north by its negative characters, as well as by the few distinctive types which more or less belong to it—llama, alpaca, vicuña, guanaco, spectacled bear (Tremarctos ornatus), Patagonian cavy (Dolichotis), coypu (Myopotamus coypu), chinchilla, viscacha (Lagostomus), Chlamydophorus (among the armadillos), and several peculiar genera of mice and the rat-like octodonts. The puma, deer, and skunk extend their range to the Strait of Magellan, and the wolf-like dogs of the genus Pseudalopex, the guanaco, and several mice (Reithrodon, Hesperomys) into Tierra del Fuego. The monkeys, tapirs, peccaries, and sloths are wanting.—With respect to its mammalian-fauna the Antillean sub-region, as might be expected, presents the most negative features. There are neither carnivores, monkeys, nor edentates, the only orders represented being the bats, rodents (Capromys, Hesperomys), and insectivores. The last are represented by two species of the genus Solenodon, whose nearest allies are the Centetidæ of Madagascar. An agouti inhabits some of the islands of the Lesser Antilles (St. Vincent, Sta. Lucia). The resident land-birds are comprised in about one hundred genera and upwards of two hundred species, about one-third of the former and

5

nearly nine-tenths of the latter being peculiar. The remaining species are South American or Central American forms. In addition to this number there are some ninety or more migrants from North America.

THE ETHIOPIAN REALM.

Next in importance to the Neotropical realm in the number, variety, and peculiarity of its animal productions, is the Ethiopian, or African. This region comprises the entire continent of Africa south of the Tropic of Cancer, and likewise that portion of Arabia which lies to the south of the same line; the Island of Madagascar, with some neighbouring groups of smaller islands, is also included. By some naturalists the northern boundary is extended as far north as the Atlas Mountains, thus including the entire Desert of Sahara. With the limitation first assigned almost the entire region lies within the tropics, and is thus the most strictly tropical of the faunal regions. In its physical features it presents several well-marked peculiarities. In the first place, we have the vast expanse of desert, which in the north occupies a transverse band varying in width from about four to nearly ten degrees of latitude. This is succeeded by what may not improperly be termed the open pasture-lands, which, as a narrow belt bounds the Sahara on the south, curves southwards at about the position of Kordofan, and occupies the greater portion of the continent lying east of the thirtieth parallel of east longitude and south of the fifth parallel of south latitude. A very considerable portion of this pasture tract forms a plateau of from four thousand to five thousand feet elevation. Included within it, and bounded on the west by the Atlantic Ocean, is the region of the great equatorial forests, to the present day a *terra incognita* in great part to both geographers and naturalists. That portion of the African continent lying south of the Tropic of Capricorn differs in many respects, both as to its physical configuration and its vegetable products, from the region to the northward, and is characterised by a vegetation which is at the same time one of the richest and most remarkable on the globe. With this marked peculiarity in its vegetable development there is of necessity a certain amount of faunal peculiarity superadded as well, but this is not sufficiently pronounced to permit of a separation of this tract from the tract lying immediately to the north. We have thus on the continent three strictly-defined faunal sub-regions : 1, the pasture-lands already described, constituting the *East-Central African* sub-region, through whose vast expanse there is manifest a strong identity in the character of the animal products, the same or very closely related animal forms being in many instances found at the extreme points of the region; 2, the forest tract, constituting the *West African* sub-region, whose animal products naturally differ very essentially from those of the last; and, 3, the desert or *Saharan* sub-region, containing a comparatively limited fauna, which, with almost insensible gradations, merges into the fauna of the Mediterranean transition tract. To the same division belong in great measure the desert tracts of Arabia, or that portion of the peninsula lying to the south of the Tropic of Cancer. The Island of Madagascar, with Mauritius, the Seychelles, &c., forms an independent sub-region of its own.

Zoological Characters of the Ethiopian Realm.—The mammalian-fauna of the Ethiopian region is characterised no less by the remarkable development of its carnivore and hoofed animals (Ungulata) than by the peculiarities presented in its quadrumanous types. Of the hoofed animals there are two families which are absolutely restricted to the region: the Hippopotamidæ, or hippopotami, and the Camelopardidæ, the giraffes. The former comprise two species, the common hippopotamus (H. amphibius), which is found in nearly all the larger African rivers from the Cape to the Sahara, and from the Zambesi to the Senegal, and the smaller Liberian hippopotamus (Chœropsis Liberiensis), from the river St. Paul on the west coast, characterised by the possession of only one pair of incisors instead of the normal two pairs.* The latter includes but a single species, the well-known giraffe (Camelopardalis giraffa), which ranges throughout the greater portion of the African open country, and to a certain extent also invades the forest region. As to the pigs (Suidæ), a family very closely related to the Hippopotami, the Ethiopian region is deficient in the genus Sus, which

---

* In antiquity the hippopotamus appears to have been very abundant in the waters of the Nile as far down as Lower Egypt. Even as late as 1600 hippopotami were trapped at Damietta, situated at the mouth of one of the arms of the Nile, and in the early part of this century they were still observed by Rüppell in Nubia. At present they are found in the Nile only in its upper course.

comprises the common hog or wild-boar of Eurasia, but its place is taken by the so-called "water-hogs" and "wart-hogs," of the genera Potamochœrus and Phacochœrus. Of the Rhinocerotidæ, a family which this region shares with the Oriental or Indian, there are four or five species or varieties, all of them two-horned. By far the most important of all the African ungulates are the ruminants. We have here an extraordinary development of the antelopes, which in the number of their species far surpass those of all the other regions put together. No less than from eighty to ninety distinct species have already been described, and doubtless many more remain in the districts that have not yet been explored. Among the numerous genera of these animals, which comprise forms ranging in size from the dimensions of a large ox (eland) to those of a rabbit (m'doqua, guevi), there are none that are found in any other faunal region, excepting Gazella, the gazelle, and Oryx, to which the gemsbok belongs, the former represented by a limited number of species in the desert regions of Western and Southwestern Asia (Arabia, Persia), and the latter, by a single species, also from the Arabian desert. The antelopes may be conveniently divided by their habits into four groups: 1. The *desert antelopes*, or such as frequent the desert regions, like the gazelle; 2. The *bush antelopes*, or those which habitually frequent the forest recesses, like the koodoos, water-bucks, and bushboks; 3. The *rock antelopes*, which, like the klipspringer, recalling in aspect and habits the European chamois, frequent the mountain-fastnesses ; and, 4. The antelopes of the open plains — gemsbok, blessbok, hartebeest, gnu, springbok — which comprise the greater number of species, and which, as the springbok, not unfrequently congregate in herds of several hundreds or even thousands. The more familiar forms of ruminants, such as the deer, sheep, and goats, are, with the exception of an ibex found in the Abyssinian highlands, completely absent. The only deer-like animal found on the African continent south of the Sahara is the chevrotain (Hyæmoschus aquaticus), from the region lying between the Senegal and the Gaboon, a small animal closely allied to the Oriental muskdeer, whose nearest representatives, the Traguli, inhabit the southeastern extremity of the continent of Asia, and the adjacent islands of the East Indian Archipelago. The wild-ox (Bos) is also absent, but its place is occupied by the Cape buffalo (Bubalus Caffer), whose domain extends throughout the greater portion of South, Central, and Western Africa. Characteristic non-ruminating ungulates are the zebras and quaggas (Equus zebra, Burchellii, Grevyi, and quagga), and the Abyssinian wild-ass (Equus tæniopus), by many naturalists supposed to be the progenitor of the domestic animal. Among the beasts of prey (Carnivora) there are the lion (possibly two species), leopard, panther, the spotted, striped, and brown hyenas, jackal, and the aard-wolf (Proteles), an animal in many respects intermediate between the dog and hyena, and constituting the type of a distinct family (Protelidæ), which is peculiar to the region. The tiger is absent, as it is, in fact, from the entire continent of Africa. The wolf and fox are also both wanting; but the latter is replaced in the Saharan and adjoining districts by the closely-related fennec (Fennecus). There is a remarkable development of the civets (Viverridæ), with a host of genera that are peculiar to the continent; the best known among these are the civets proper (Viverra), genets (Genetta), and the ichneumon (Herpestes), all of which, however, are found also beyond the limits of the region. Bears are entirely wanting, the only African representative of the Ursidæ (Ursus Crowtheri) being extralimital, a native of the Atlas Mountains. The Ethiopian Quadrumana, or apes, constitute a part of the Old World group of the Catarhina, distinguished from the monkeys of the New World, as has already been stated in treating of the Neotropical realm, by the comparative narrowness of the nasal septum, the presence, in most cases, of ischial callosities and cheek-pouches, the universal absence of a prehensile tail, and the number of teeth, which never exceed the normal number (thirty-two) characteristic of the human species. This group comprises the most perfectly organised, or most hominine of the quadrumanous species, and, at the same time, those in which the fiercest and most savage disposition is combined with a less advanced structural development. As representatives of the former we have the anthropoid or man-like apes, constituting the family Simiadæ, which, in the African continent, comprises the chimpanzees and gorillas, and, in Asia and Malaysia, the gibbons and orang-outangs. The two species of chimpanzee (Troglodytes niger and T. calvus), as well as the gorilla (T. gorilla), are both restricted to the forest region of Equatorial Africa, especially the west coast; but it is still a matter of considerable uncertainty how far inland their range may extend. The researches of von Heuglin,

Schweinfurth, and more recent travellers, seem to show, almost beyond doubt, that the gorilla, which until recently was considered to be limited in its haunts to the west-coast region—the forest tracts lying between about ten degrees of north latitude and the Gaboon River, including the Crystal Mountains—is in reality also an inhabitant of the deep interior of the continent, frequenting the forest recesses which bound the western tributaries of the Nile. Lower in the scale of organisation, but scarcely inferior in size in many cases to the anthropoid apes, are the dog-faced monkeys, constituting the family Cynopithecidæ. These, which embrace many of the most savage forms of all the monkey tribe, inhabit the greater portion of the Ethiopian region, the forests as well as the open plains and rocky fastnesses of mountain solitudes. Among the better known and more formidable members of this extensive family, which is also well represented in the Oriental region (macaques), and less numerously in the Austro-Malaysian (the islands of Batchian and Timor) and Tyrrhenian (the Barbary ape of the Rock of Gibraltar) transition regions, are the baboons, mandrills, chacmas, Diana monkeys, and mangabeys, the first three being characterised by a prolonged snout, similar to that of the dog, at the extremity of which are situated the nostrils; the tail is rudimentary, or almost completely wanting. The Colobi constitute another extensive group of African apes. True monkeys, as well as the more distinctive of the African Mammalia—such as the lion, leopard, hyena, zebra, antelopes, giraffe, hippopotamus, and rhinoceros—are wholly wanting in the Island of Madagascar, whose principal mammalian feature is constituted by the lemurs (Lemuridæ), or half-monkeys, a group of animals usually considered to form a sub-order of the Quadrumana, in certain peculiarities of structure closely approximating the most ancient progenitors of the ungulates. The presence of lemurs on the Island of Madagascar, the continent of Africa, and Southern India (with Ceylon), has led some naturalists to the conclusion that at one time direct land connection existed between the several regions, an assumption that is by some naturalists considered to be further borne out by other equally well-marked faunal characteristics. To this supposed formerly-existing land-mass of the Indian Ocean, which, if it ever existed, may or may not be represented in part by the sunken "Chagos Banks," and the outlying islands, such as the Seychelles,

Laccadives, and Maldives, the name of " Lemuria " has been given.*
A very remarkable quadrumanous animal of the Island of Madagascar, and the only representative of its family, is the aye-aye
(Cheiromys Madagascariensis), formerly described as a squirrel,
which has many points of relationship with the rodents. The only
other orders of Ethiopian Mammalia that need be specially referred
to are the Proboscidea and the Edentata, the former represented
by the African elephant (Loxodon Africanus), and the latter by
the scaly ant-eaters (Manis, Pangolin), and the curious animals
known as aard-varks (Orycteropus). The members of the genus
Hyrax, which includes the shaphan or coney of the Bible, animals
in several characters allied to the rodent on the one side, and the
rhinoceroses, among pachyderms, on the other, constitute, in the
opinion of many naturalists, a distinct order, Hyracoidea, apart
by itself. Several species (Hyrax, Dendrohyrax) are recognised, all
of about the size of the rabbit, and, with one exception, the coney,
which is also found in Syria and Palestine, restricted to the African
continent.

The bird-fauna of the Ethiopian realm is by no means as rich,
either in the actual number of its forms or in those that are peculiar, as the Neotropical. Neither do we find, as a rule, that brilliancy and variety in the plumage which distinguish the birds of the
South American continent, although gaudily-coloured birds are not
exactly rare. Among these are the Irrisoridæ, a group of birds
allied to the hoopoes, and remarkable for their metallic hues; the
Meropidæ, or bee eaters, of which the common bee-eater of Southern Europe (Merops apiaster) is a well-known representative; and
the curious forest-loving birds, known as the turacos and plantain-
eaters (Corythaix, Musophaga), in a measure related to the South
American toucans, constituting the family Musophagidæ. The
Ethiopian region is the home, *par excellence*, of the insectivorous
honey-suckers (Nectarinidæ †), a family of birds bearing a superficial resemblance to the American hummers, which they also, in
many cases, rival in the brilliancy of their plumage. The honey-
guides (Indicatoridæ), formerly classed as cuckoos, and to an extent

---

* It is here that, by some anthropologists, has been located the most ancient abode of man.

† Also abundantly represented in the Australian and Oriental regions.

partaking of their habits, but probably more closely related to the woodpeckers, are found in almost all parts of the region. The fly-catchers, warblers (Sylviadæ), true finches (Fringillidæ), and weaver-birds (Ploceidæ) are numerically well represented, more especially the last, of whose two hundred and fifty, or more, species about two hundred are found within the limits of this region. This family comprises, among other birds, the small speckled and red-billed finches, known as the estrilds and amadines, the tailor-bird (Textor), the true weavers (Ploceus and Symplectes), and the long-tailed whydahs (Vidua), from the west coast. The parrots are but feebly represented in the Ethiopian region, the macaws and cocka-toos (Conurinæ and Cacatuidæ) being wholly wanting. With very few exceptions—Palæornis—all the African parrots belong to the group of the Psittacini, of which the common grey parrot (Psittacus erithacus) is a familiar example. No species is found to the north of the fifteenth parallel of north latitude. Of the gallinaceous birds there is a marked representation of the grouse tribe, especially of the genus Francolinus, and among the pheasants we have all the species of guinea-fowls (Numidinæ), whose nearest allies appear to be the American turkeys. Birds of prey are very abundant, comprising, in addition to the common forms of vultures, eagles, &c., the hawk-like bird known as the "secretary" (Serpentarius), a near ally of the South American cariama. Finally, the Ethiopian region pos-sesses, although not exclusively, the ostriches (Struthio camelus, the common form, and S. molybdophanes, from the Somali ter-ritory), of the family Struthionidæ.

The reptile-fauna is very rich and varied, and comprises a con-siderable number of peculiar forms. Of Ophidia we have a large development of the vipers (Viperidæ), and among these one of the deadliest of venomous serpents, the puff-adder (Clotho). Of the larger constrictors, the rock-snake (Hortulia) and Seba's python represent the Pythonidæ. The lizards comprise, among other sin-gular forms, the Agama, the typical Old World representative of the American iguanas, and the chameleon, with its distinctive changing hues. Crocodiles are met with in nearly all the larger streams.

The fresh-water fishes of the African realm are limited, accord-ing to Günther, to somewhat more than two hundred and fifty species, representing fifteen distinct families or groups. About sixty of these are siluroids (cat-fishes), fifty cyprinoids (carps), and about an equal number members of the family Mormyridæ. Owing to the broad distribution of the different types, which are spread throughout the greater extent of the continent, a division of the region into ichthyic sub-regions is rendered impossible. Of some fifty-six species found in the waters of the Upper Nile, no less than twenty-five are absolutely identical with forms belonging to the West African rivers, and doubtless most of these also occur in the waters of the unexplored tracts of the interior. Greater dissimilarity exists between the northern and western faunas and those of the south, where the relationship has been rendered generic instead of specific. Thus, the fishes of Lake Nyassa and the Zambesi River are specifically distinct from those of the great equatorial lakes, and their outflowing northern and western waters. Africa has representatives of two genera of ganoid proper, Polypterus and Calamoichthys, and one genus of lung-fishes, Protopterus (P. annectens), the last closely related to the Lepidosiren of South America. With the fish-fauna of this region the Ethiopian agrees in the partial possession of the characinids (about thirty-five species) and the chromids, and the genus Pimelodus among the cat-fishes. How the transference of similar or identical types was effected to such widely remote areas, whether through the intermedium of the waters of a continental tract now submerged beneath the Atlantic, or by way of the northern streams, it is impossible to say.

The Ethiopian faunas, taken collectively, exhibit a remarkable homogeneousness throughout, so that the delimitation of even the three greater faunal sub-regions becomes difficult. The East Central African sub-region is, strictly speaking, representative of the entire tract, where the vast majority of all the distinctive types are found. The West African sub-region is more properly the home of the anthropoid apes, the chimpanzee and gorilla, and of the numerous species of Cercopithecus and Colobus. The antelopes are much less abundantly represented than in the plateau districts, although they comprise a number of peculiar types, especially of the bushboks (Cephalophus). Other characteristic ungulates are the zamoose, a species of buffalo (Bubalus brachyceros—possibly also found in Abyssinia), the Hyæmoschus aquaticus, and the Chœropsis Liberiensis (hippopotamus). The insectivores present as a distinctive type the otter-like Potamogale velox ; the rodents the singular

Anomalurus, recalling the flying-squirrels ; and the edentates a distinct species or variety of aard-vark (Orycteropus Senegalensis).

In the Saharan Desert tracts, where the necessary conditions for existence are largely wanting, there is a marked impoverishment of the fauna. The more formidable carnivores, such as the lion and leopard, are absent from most districts, leaving their places to be filled by some minor cats, the hyena, jackal, fox, and fennec. The hoofed animals are represented (in some parts) by the buffalo, and a limited number of antelopes (Gazella, Oryx, Addax). Among rodents the families of rats and jumping-mice (Dipus, Scirtetes) are fairly represented, in addition to which we have the porcupine and hare (Lepus Mediterraneus). The ostrich is sufficiently abundant throughout most of the region. Among the desert reptilian forms may be mentioned the monitors (Varanidæ), scinks, sand-lizards (Sepidæ), and agamas.

The deficiencies in the Madagascar mammalian-fauna have already been indicated. As representative types we have, in addition to the lemurs and aye-aye, several civets (Galidia, Galictis), the singular cat-like carnivore known as Cryptoprocta (C. ferox), a water-hog (Potamochœrus), a sub-fossil species of hippopotamus, and the native hedgehogs (Centetidæ). The bird-fauna is made up largely of Asiatic and African types, although peculiar forms are abundant. Many of the reptilian forms, as the ophidian genera Heterodon, Herpetodryas, Philodryas, have American representatives.

ORIENTAL REALM.

This region comprises all that portion of the Asiatic continent which is not included in the Holarctic and Tyrrhenian tracts (excepting the southern portion of the Peninsula of Arabia, which is Ethiopian), the Island of Ceylon, Formosa, the Philippines, Sumatra, Java, and Borneo, besides some minor island groups. Within its limits are, therefore, included the whole of extra-Himalayan Hindostan, Farther India, the Malay Peninsula, and that portion of China lying south of the Nanling range. A very considerable part of this region is covered with the most luxuriant forest growth, which extends even to an elevation of from eight to ten thousand feet along the slopes of the Himalaya. This forest character more particularly distinguishes the Indo-Chinese and Indo-Malayan sub-regions, the former of which includes Burmah, Siam, Anam, Southern China, the southern Himalaya slopes, and the luxuriant tracts lying along the base of these mountains, known as the " Terai "; and the latter, the Malay Peninsula, with the Indo-Malaysian islands already mentioned. A third sub-region, the Indian, is constituted by the Indian Peninsula, exclusive of the Carnatic. The surface here consists largely of open pasture or grass lands, the fundament being in great part the alluvium of the existing rivers. In the northwestern part, bounded by and partly lying within the valley of the Indus, is the Indian Desert, where we encounter a considerable intermixture of strictly Indian, Holarctic, and Ethiopian animal types. An essentially forest character again distinguishes the southern extremity of the Indian Peninsula—the Carnatic—and the Island of Ceylon, which together form the fourth sub-region, the Cingalese.

Zoological Characters of the Oriental Region. — A cursory examination of the Oriental mammalian fauna shows it to be largely made up of characteristic African forms, for which reason, indeed, some naturalists have been induced to unite this region, either in whole or in great part, with the Ethiopian. We have here the same extraordinary development of the quadrumanous, carnivore, and ungulate types, although in respect of these last very material differences present themselves which are sufficiently distinctive of the two regions. Thus, in the Oriental region there are no representatives of either the Camelopardelidæ or Hippopotamidæ, families peculiar to the African continent; and the only member of the Equidæ—the horses, asses, and zebras—the onager (Equus onager), is found in the debatable land along the Indus, which unites the Oriental and Holarctic tracts. There is also a great falling off in the number of antelopes, of which there are scarcely more than a half-dozen species—comprising among others the gazelle, the true antelope, and nylghau ; but their place is in great measure taken by the solid-horned ruminants, the deer, which, as has been seen, are completely wanting in the Ethiopian region, but have here no less than about twenty species, ranging in size from the diminutive muntjac (Cervulus) to the giant rusa. This is also the home of the beautiful axis. The chevrotains or mouse-deer (Tragulidæ), a small group of diminutive deer-like animals characterised by the presence of tusks in the upper jaw, have but one extra-limital

representative, the Hyæmoschus, from Western Africa, already referred to.   Of the oxen we have the Indian buffalo (Bubalus Indicus
or buffelus), whose range at the present day (as a domestic animal)
comprises, in addition to its native haunts, a considerable part of
Northern Africa and Southern Europe (Hungary, Greece, Italy),
and three or four species of wild cattle—the gaour, gayal (Bibos)—
distributed over the greater portion of the region, from Java to the
Indian Peninsula.   The sacred cow or Brahmin bull, commonly
known as the zebu, is now found only in a domesticated state.
The goats have but a single representative in the entire region—in
the Neilgherry hills; the sheep are completely wanting.   The thick-
skinned ungulates are represented by four or five species of rhinoceros, both one- and two-horned, whose most eastward abode
appears to be the Island of Java; a solitary species of the South
American family of tapirs (Tapirus Malayanus), and about six
species of swine (Suidæ).   Of the Carnivora there is, as in the
Ethiopian region, a large development of the civet-cats (Viverridæ),
most of the genera representing the family in this region being peculiar to it (Viverricula, Paradoxurus, Arctogale, Cynogale).   The
Mustelidæ comprise several well-known Holarctic forms, such
as the true weasel (Mustela), martin (Martes), otter (Lutra), and
badger (Meles), the last found only in Southern China.   The cats
(Felidæ) are represented by many of the more prominent types of
the Ethiopian region—such as the lion, leopard, and panther—in
addition to which we have the ounce and tiger, the latter extending its range as far north as the fifty-third parallel of latitude
(Lake Baikal), and westward to the borders of the Caspian Sea.
It is found in the islands of Java, Sumatra, Bali, and Saghalien,
but is absent from both Ceylon and Borneo.   The dogs (Canidæ)
differ in several respects from the representatives of the same
group of animals entering into the Ethiopian fauna, and are more
nearly allied to the Holarctic forms.   The true wolf is absent; but
its place is filled by several races of closely-allied wild-dogs, which
hunt wolf-like in packs over certain portions of the region, and
the jackal.   The fox is represented by several species.   Only one
of the three recognised species of hyena, the striped hyena (Hyæna
striata), whose range embraces the entire northern part of the continent of Africa and a considerable portion of Western Asia, enters into the region.   One primary distinguishing feature separating

the carnivore-fauna of the Oriental region from that of the Ethiopian is the presence in the former of bears, which comprise here not only the singular forms known as the sun-bears (Helarctos), confined to the Indo-Malayan sub-region, and the honey-bears (Melursus), whose range extends from the Ganges to Ceylon, but also the true bears of the genus Ursus. The Indian elephant inhabits nearly all the wooded tracts from the Himalaya slopes to Ceylon, and eastward to Borneo and Sumatra.* The Oriental region is, *par excellence*, the headquarters of the true mice and squirrels, there being no less than about fifty specific representatives to each of the genera Mus and Sciurus, or about one-half the number of all the forms that have been ascribed to these genera. Both the round- and flat-tailed flying squirrels (Pteromys, Sciuropterus) are distributed throughout the region, the former, with the exception of three or four extra-limital species found in Japan, being restricted to it, and the latter distributed throughout a considerable portion of the Holarctic region, and on both sides of the Atlantic. The marmot is found in the debatable lands of the Western Himalayas, at heights exceeding eight thousand feet. The bats have a much more extended development in this region than in any other, except the Neotropical, there being upwards of one hundred distinct specific representatives. These include members of all the generally recognised families except the Phyllostomidæ, or simple leaf-nosed bats (to which the South American vampyres belong), and consequently embrace the short- and long-eared bats, the horse-shoe bats, and the fruit-eating bats (Pteropidæ), commonly known as flying-foxes. The most important of the Oriental Quadrumana are the macaques (Macacus), to which the Barbary ape of the Rock of Gibraltar belongs, and which inhabit the entire region, and the long-tailed Semnopithecus, which has nearly the same range, and several of whose representatives inhabit the more elevated forests, even during the winter, at heights exceeding eleven thousand feet. To this genus belongs the peculiar Bornean "nose-monkey" (S. nasalis). The anthropoid apes are confined to the southeastern portion of the region, and principally to the larger islands of the Malay

---

* The Javanese elephants do not appear to be indigenous to the Island of Java. The Sumatra elephant, which was considered by Professor Schlegel to represent a distinct species apart from the Indian (Elephas Sumatrensis), is now usually referred to the latter species.

Archipelago. They comprise one or two species of Orang (Borneo and Sumatra), the long-armed gibbons (Hylobates), whose range extends from Java to Assam and China, and the siamang, a native of Sumatra and the Malay Peninsula. The half-monkeys, or lemurs (Nycticebus, Stenops), which were found to be so characteristic of the Madagascar fauna, show a very great diminution in numbers in the Oriental region. Their most distinctive or anomalous type is the tarsier or spectrum lemur (Tarsius spectrum), an inhabitant of Sumatra, Borneo, and Celebes (Austro-Malaysian), which of itself constitutes a distinct family (Tarsiidæ). In the structure of its feet—the extraordinary and unequal development of the toes—and in several other peculiarities it is closely related to the aye-aye, from which to the true lemurs it appears to form a passage. Until recently classed with the lemurs, but now considered as representing the type of a distinct family of insectivorous animals, are the cat-monkeys or flying-lemurs (Galeopithecus), which inhabit the larger islands of the Malay Archipelago and the Philippines.

The bird fauna of the Oriental region is exceedingly rich and varied, and, as might have been expected from its position, comprises in its assemblage a very large proportion of Holarctic, Ethiopian, and Australian forms. But the number of distinctive forms are sufficiently numerous, and eminently serve to characterise the region. Among these are the laughing or babbling thrushes (Timalidæ), a family which, though not absolutely restricted to the region, has its headquarters there. Nearly two hundred of the two hundred and fifty described species occur here, and are found in nearly all parts of the region. Less numerous in species, but scarcely less distinctive, are the Leiotrichidæ, or hill-tits, found in all parts of the Himalaya Mountains; the Pycnodontidæ, or bulbuls; Phyllornithidæ, or green bulbuls; and the minivets of the genus Pericrocotus, one of the group of caterpillar-eaters. Of the warblers (Sylviadæ), the peculiar tailor-bird (Orthotomus) has about thirteen to fifteen species. The starlings are represented, among other forms, by the sacred mynah (Eulabes) and the roseate pastor (Pastor roseus); the bee-eaters by Merops and the resplendent Nyctiornis; the sun-birds (Nectarinidæ) by the Nectarophila and Arechnothera. The flower-peckers (Dicæidæ), a group of small, gaily-coloured birds representing the South American sugar-birds, are tolerably abundant throughout the greater portion of the region. The South American cotingas likewise appear to have their representatives in the large and gaudy broad-bills (Eurylæmidæ). Trogons, scarcely inferior in the beauty of their plumage to their American congeners, are sufficiently abundant in the forest districts, where, also, we meet with a multitude of the remarkable horn-bills. Contrary to what might have been expected, seeing their abundance in the Australian region (and partially so also in the Ethiopian), and the apparently favourable conditions for their existence here, the parrots are but feebly represented, and belong, with one exception—a cockatoo from the Philippine Islands, the only one found within the limits of the region—to the family of the ringed-parakeets (Palæornithidæ). Of the gallinaceous birds the pheasants (Phasianidæ) are largely represented. This is the home of the jungle-fowl (Gallus), from one of whose species, the G. Bankiva, inhabiting the region from the Himalayas through Central India eastward to the islands of Java and Timor, is in all probability descended the greater portion of our domestic poultry. Distinct species of the jungle-fowl are found in Southern Hindostan, Ceylon, and Java, and, possibly, some of the domestic varieties may have been produced as the result of interbreeding between these various forms and the G. Bankiva. The peacock is found throughout a considerable portion of the region, to which it is indigenous, from Ceylon to the Himalayas, and eastward to China; belonging to the same family, and scarcely less resplendent in their plumage, are the argus, impeyan, tragopan, and fire-backed pheasants. The remarkable group of Australian birds known as "mound-builders," megapods, or brush-turkeys, constituting the family Megapodidæ, which would seem to be closely related to the South American curassows, have two representatives (of the genus Megapodius), one, possibly introduced, in the Nicobar Islands, and the other in the Philippines and Borneo.

The reptilian-fauna comprises, among other serpents, the giant python (of the family Pythonidæ), the cobra-di-capello (of the family Elapidæ, or coral-snakes), and one-half of all the pit-vipers, or members of the family of rattlesnakes (Crotalidæ), although the rattlesnake proper is absent; among lizards, the Varanidæ, or water-lizards, to which the monitor belongs, the geckoes, the agamas, or eastern iguanas, the last embracing no less than eighteen

species of so-called flying-lizards (Draco), animals provided with
a lateral expanded tegumentary membrane ; and of crocodilians,
the true crocodile (Crocodilus) and the gavial (Gavialis), the latter
restricted to the rivers of the Indian Peninsula.

The several faunal sub-regions, especially those of the continental tract, are most intimately related to one another, and do not
admit of sharp delimitation.   Yet a number of forms, whether
negatively or positively, may be said to define each.   Thus, among
the Mammalia, the lion, hyena, fox, lynx, mellivore, buffalo, nylghau, gazelle, and true antelope, may be said to represent distinctive types of the first, or Indian sub-region, being absolutely restricted to it, or just passing beyond its confines.   The Cingalese sub-
region is characterised, among other forms, by the loris, which it
alone possesses; by a peculiar genus of civet-cats (Onychogale), and
Platacamthomys among the mice.   The fauna, especially of the
island of Ceylon itself, is related on the one side to that of the
Himalayas, and on the other to that of the Indo-Malayan sub-region,
differing broadly from the fauna of the central and northern portions of the Indian Peninsula.   The relationship with the Malayan
fauna is especially marked in the case of the Lepidoptera and Coleoptera among insects, many of the more distinctive or abundantly
represented types belonging exclusively to the two faunas under
consideration.   The individuality of the Cingalese reptilian and
amphibian faunas is well marked through the number and variety
of peculiar genera, which, perhaps, more than any other animals,
serve to characterise the sub-region.   All the members of the
Uropeltidæ, or rough-tailed burrowing-snakes, appear to be confined to this tract.   The poisonous snakes of the entire peninsula
of India are, according to Fayrer, comprised in eleven genera, representing three families, Elapidæ, Viperidæ, and Crotalidæ (pit-
vipers), and some twenty-five species.   Among the more venomous
forms are Naja (the cobra, which ascends the Himalayas to a height
of eight thousand feet), Ophiophagus, and Bungarus, of the Elapidæ; Daboia and Echis representing the vipers; and Trimeresurus
belonging to the pit-vipers.   In addition to the terrestrial Thanatophidia the marine-snakes (Hydrophidæ), which inhabit brackish
estuaries and tide-water streams, furnish an additional contingent
of thirty-five or more species.

The two remaining sub-regions, the Indo-Chinese and the Indo-

Malayan, are most intimately related to each other in the general characters of their faunas, as they are also to the Indian; the former, however, incorporates a more decidedly Holarctic element, while the latter is almost strictly tropical. Among the more distinctive mammalian types of the Indo-Chinese tract, or such as do not enter into the composition of the Malayan fauna, are the true bears (Ursus), panda (Ailurus), fox, badger, Arctonyx (Melidæ), and a peculiar genus of civet-cats, Urva. On the other hand, this tract is wanting in the anthropoid apes of the genera Simia (orang—Borneo and Sumatra) and Siamanga (Malacca, Sumatra), the spectre-lemurs (Tarsius — Sumatra, Borneo), flying-lemur (Galeopithecus), tapir (Tapirus Malayanus—Malay Peninsula, Sumatra, Borneo), and sun-bear (Helarctos), which belong to the Malayan fauna. The Himalayan districts (with parts of China) are preeminently the home of the pheasants, which, in addition to several genera peculiar to the Oriental region (Pavo, Argusianus, Polyplectron, Euplocamus), comprise a number of forms, the impeyan, tragopan, &c., held in common with the Holarctic; the peacock ranges from Ceylon and the Himalayas to Java, and the argus from Siam to Borneo.

THE AUSTRALIAN REALM.

This region, as usually defined, embraces, in addition to the continent of Australia and the Island of Tasmania, all the Austro-Malaysian islands lying to the east of Borneo and Bali—i. e., beginning with Celebes and Lombok—the vast Polynesian Archipelago, and New Zealand, with its accompanying islets. This last has by some naturalists, and more especially by Professor Huxley, been recognised as a distinct region, although by the majority of zoogeographers it is usually regarded as a sub-region of the Australian. In this place only Australia proper (with Tasmania), Papua (New Guinea, and the minor Papuan islands), and New Zealand are considered to enter into the formation of the Australian region.

The greater portion of the Australian mainland consists of a table-land of moderate elevation, characterised by a harsh and dry climate, and a general absence of water. Hence the surface of the country bears a more or less barren aspect, supporting but a scant vegetable growth, which is parched throughout the greater part of the year, and consequently rendered unfit for promoting a vigorous development of animal life. A considerable part of the tract passes off into an almost uninhabitable desert. Along the coast-line, and more especially on the eastern border, where vapour-condensing surfaces present themselves in the form of elevated mountain-crests, there is an abundance of aqueous precipitation, to which a luxurious forest-growth responds. Less than one-half of the continent lies within the tropics. Papua, which constitutes the second sub-region, exhibits in great part a mountainous character, and, from the abundance of aqueous precipitation distinctive of the equatorial portions of the earth's surface, a vegetation of truly tropical luxuriance. A dense forest growth likewise covers the greater portion of New Zealand (third sub-region), where also we have the most elevated mountain-summits in the region.

Zoological Characters of the Australian Realm.—The Australian region is, by both positive and negative characters, the most marked of any on the earth, and, indeed, so remarkable are its faunal peculiarities that it has been thought by some to constitute properly in itself a main zoogeographical division, as opposed to the rest of the world. The most striking feature of this fauna is the general absence, among the Mammalia, of such forms as, under favourable conditions, are to be met with in all other portions of the earth's surface. In the whole of this region all the terrestrial Mammalia that are to be found in the Old World are absent, except a solitary (possibly two) species of hog (Sus), found in New Guinea, the bats, and rodents, the former being represented by about ten genera, and the latter by the single family of the mice (Muridæ). of which the true mice (Mus) comprise more than one-third of all the various forms. The edentates, insectivores, carnivores, and monkeys, except such as have been introduced through the agency of man, and possibly the " dingo," or Australian wild-dog, which may prove to be indigenous, are wholly wanting, their place being filled by a wonderful variety of the implacental mammals—the marsupials and monotremes—the lowest of the entire mammalian series. No implacental mammals occur at the present time in any portion of the Old World outside of the limits of the Australian and the connecting Austro-Malaysian regions, and their only representatives in the Western Hemisphere are the opossums (Didelphyidæ), upwards of twenty species of which range throughout the forest districts of the Neotropical realm (with two species in North America). The Australian marsupials fall into six distinct families: The Macropodidæ, kangaroos and kangaroo-rats, are numerically the most important and the most broadly distributed of the several types, and comprise a diversity of forms, which are variously adapted to living in the scrub, in the desert, along rocky and precipitous mountain summits (Petrogale), or on trees. The arboreal genus Dendrolagus is thus far known only from New Guinea and Queensland. The great kangaroo (Macropus giganteus), inhabiting the southern half of the continent of Australia and Tasmania, attains in the male a length, from the tip of the nose to the root of the tail, of upwards of five feet. The kangaroo-rats, rat-kangaroos, or potoroos (Hypsiprymnus), are diminutive forms, of about the size of a hare, with rounded instead of elongated ears, and a general rat-like appearance. The second family is that of the Dasyuridæ, or native cats, a group of carnivorous and insectivorous marsupials, which range in size from the dimensions of a mouse to those of a wolf. The most distinctive forms belong to the genus Antechinus, a group of insectivorous animals, which, in outward appearance, are but barely distinguishable from the ordinary mice. The most formidable member of the family is the Tasmanian "tiger," or "hyena" (Thylacinus cynocephalus), a striped carnivorous marsupial, having the general aspect of a wolf or dog, which it also fully equals in size, measuring as much as five feet in length. The third family is that of the Myrmecobiidæ, which includes but a single species, the native anteater, or striped myrmecobius (M. fasciatus), a small squirrel-like animal, of about the size of the common squirrel, inhabiting Southern and Western Australia. The Paramelidæ comprise the bandicoots (Perameles), small kangaroo-like animals of about the size of the rabbit, and the singular pig-footed marsupial known as the Chœropus, a graceful animal somewhat recalling in appearance the mouse-deer of the Oriental region. The fifth family, the Phalangistidæ, exhibits in its individual components a greater diversity of form, and greater specialisation of structure, than is to be found in any other mammalian family. As described by Wallace, we find represented in this family the tailless koala, or native sloth (Phascolarctos); the prehensile-tailed opossum-like phalangers (Phalangista); the beautiful flying-opossums (Petaurista, Belideus, Acrobata), so closely resembling in form the flying-squirrels of North America and India, but often no larger than a mouse; the dormouse-like Dromicia, one species of which does not equal in size the ordinary harvest-mouse; and, finally, the little Tarsipes, a true honey-sucker, provided with an extensile tongue, and of the size of a mouse. All of these various forms are more or less adapted to an arboreal existence. The remaining family of the marsupials is that of the wombats (Phascolomyidæ), nocturnal burrowing animals of about the size of the badger, and somewhat of the appearance of a bear, subsisting chiefly on roots and grasses. The most anomalous and remarkable of all the Australian mammals, indeed of all Mammalia, are the oviparous monotremes — the duck-bill (Ornithorhynchus) and native hedge-hog (Echidna)—strictly speaking, burrowing edentate animals, having certain points of affinity with both birds and reptiles. With few exceptions (Cuscus, Belideus), all the marsupial genera of the Australian region are confined to the continent of Australia, Tasmania, and New Guinea (and the Aru Islands), but a very insignificant fraction of the entire number of marsupial species being represented in any of the Austro-Malaysian islands (Mysol, Celebes). Even in New Guinea, as compared with Australia, the number of such representatives is rather limited; but we here meet with at least one type of placental mammal, the hog, which is not met with on the continent of Australia. On the other hand, the (in Australia) fairly well represented family of rats and mice exhibits in the Papuan sub-region (Aru Islands) but a solitary example, the Uromys.

In respect of its bird-fauna, the Australian region presents us with peculiarities that are scarcely less marked than those which distinguish the Mammalia. In the number and beauty of its forms this region is only second to the Neotropical, while in such as show the most marked peculiarities of structure it is unsurpassed. Many of the most familiar types of Old World birds are represented, and in sufficient number—such as the warblers, thrushes, fly-catchers, shrikes, and crows; but, on the other hand, some of the most broadly diffused families are wholly wanting. Thus, the true finches (Fringillidæ), which have otherwise a universal distribution, appear to be wanting in all parts of the region, being replaced by the weaver-finches (Ploceidæ). The vultures are also completely absent, as are likewise the woodpeckers (Picidæ) and pheasants (Phasianidæ). The true paradise-birds (Paradiseinæ), whose special development in New Guinea and the other Papuan Islands forms such a marked feature in the avian-fauna of that sub-region, have but two representatives (Manucodia) on the Australian mainland; the bower-birds (Tectonarchinæ), on the other hand—members of the same family—have here their greatest development (seven species), and are associated with two species of rifle-birds (Ptiloris). The Australian region is distinguished by being the sole possessor of the families of lyre-birds (Menuridæ) and apteryxes (Apterygidæ), the latter one of the most remarkable groups of existing birds (natives of New Zealand), whose exact relationships have not yet been satisfactorily determined. The struthious birds of the genus Dromæus, the emu (two species), are confined to the continent of Australia, as is also one of the genera of brush-turkeys (Leipoa). Talegallus, the true brush-turkey, and Megapodius, the megapod or mound-builder, have both representatives in the Papuan Islands and Australia, the former being restricted to the region under consideration. The greater number of the species of Mega-podidæ, some twenty or more, are distributed throughout the Austro-Malaysian islands and Polynesia, with two outlying species in the Philippines, Borneo, and the Nicobar Islands. The parrots have an extensive development, and represent in the main forms that are not known beyond the limits of this and the transition region. "No group of birds gives to Australia so tropical and foreign an air as the numerous species of this great family, by which it is tenanted, each and all of which are individually very abundant. Immense flocks of white cockatoos are sometimes seen perched among the green foliage of the loftiest trees; the brilliant scarlet breasts of the rose-hills blaze forth from the yellow flowering Acaciæ; the Trichoglossi or honey-eating parakeets enliven the flowering branches of the larger Eucalypti with their beauty and their lively actions; the little grass parakeets rise from the plains of the interior and render these solitary spots a world of animation; nay, the very towns, particularly Hobart Town and Adelaide, are constantly visited by flights of this beautiful tribe of birds, which traverse the streets with arrow-like swiftness, and chase each other precisely after the manner the Cypseli are seen to do in our own islands. In Tasmania I have seen flocks of from fifty to a hundred of the Platycercus flaviventris, like tame pigeons, at the barn-doors in the farm-yards of the settlers, to which they descend for the refuse grain thrown out with the straw by the threshers.[28]" About sixty species of the order are known from the mainland of

Australia alone. The Cacatuidæ, which are found only in this and the adjoining transition region, with a solitary species, Cacatua hæmaturopygia, in the Philippines, comprise, among other forms, the commoner species of the genus Cacatua, as the sulphur-crested and rose-breasted cockatoos, and the black cockatoos of the genus Calyptorhynchus, which last are restricted to the continent of Australia and Tasmania. An aberrant group of the parrots are the New Zealand nestors (Nestoridæ), some of whose members appear to be addicted to carnivorous habits,[20] and which would seem to have certain points of relationship with the South American macaws. In New Zealand, likewise, are found the owl-like nocturnal parrots of the family Strigopidæ. One of the most distinctive groups of birds of the Australian region are the honey-suckers (Meliphagidæ), whose representatives are scattered all over the Austro-Malaysian and Polynesian groups of islands, from Celebes to the Marquesas Islands, and from Tasmania to Hawaii. Of some two hundred or more species of this family but a single one, Ptilotis limbata, enters the Oriental region (the Island of Bali). The nearly related honey-suckers (Nectarinidæ) are represented by several forms more or less distinctive of the region (Arachnecthra, Arachnothera). Australia with New Guinea, and the adjacent islands, may be considered to be *par excellence* the natural home of the pigeons (Columbæ), nearly one-half of the total number of genera (forty or more) being represented here, and by types many of which are found nowhere else. They include the most gaudily ornamental representatives of the order (Ptilopus), which in the brilliancy of their plumage yield but little to the parrots ; various forms of ground-pigeons (Geophaps, &c.) ; and the beautiful crested-pigeon known as the goura, from New Guinea and some of the adjacent islands. The turtle-dove (Turtur) is found in New Guinea, but the nearly universally distributed Columba, to which the ordinary rock- or wild-pigeon (C. livia) belongs, and which represents the ancestral form of most of our domestic breeds, is wanting.

The reptile-fauna of this region is much less distinctive than either the mammalian or avian. Snakes, amphibians (but only the tailless forms), and lizards are abundant, the great bulk of the last being constituted by the scinks and geckoes. Of the amphibians the true toads (Bufonidæ) are represented by a limited number of species in Australia, although the genus Bufo itself is wanting;

Rana, the frog, has a single species on the peninsula of York (R. Papua), but is more abundant in New Guinea. Tree-frogs (Hyla) are numerous. A peculiarity distinguishing the continental ophidian fauna is the great preponderance of venomous over non-venomous serpents. The proportion of the former to the latter is in some sections, as in South Australia, as six to one. Two-thirds of all the species belong to the family Elapidæ, to which the American coral-snakes and the Indian cobra also belong. One species of crocodile is found in some of the Australian waters.

The fish-fauna is very limited, less than forty species being known from the entire realm. Two species of lung-fishes, the "barramundas" (Ceratodus Forsteri and C. miolepis), inhabit the waters of Queensland. The paucity of forms is, doubtless, in part attributable to the limited number of fresh-water streams.

THE POLYNESIAN REALM.

The scattered island groups of the Pacific Ocean are so deficient in the faunal elements that distinguish the main zoogeographical divisions of the earth that they may be said to constitute a region framed more by negative than by positive characters. For this reason they have by many naturalists been relegated to the rank of a mere sub-region. Yet, when we compare the Polynesian fauna with the faunas of the other recognised regions, it becomes not a little difficult to determine just where it should be placed, although there would seem to be hardly a question as to the preponderating relationship being with the Australian fauna, with which it has generally been united. The addition, however, of so enormous an annex to a region which combines in its faunal elements such a remarkable individuality, and with which, after all, there is not very much in common, does not appear natural; and the less so, when we recognise the full importance of the characters derived by the Polynesian tract from regions other than the Australian, and its own special peculiarities. For these reasons it seems advisable to consider the region as one apart by itself.

Zoological Characters of the Polynesian Realm.—This tract is at once distinguished from the Australian, as well as from all others, by the abscence of all Mammalia excepting bats. Of these last there are representatives of two families, the Oriental fruit-eating bats, or flying-foxes (Pteropidæ), and the cosmopolitan Vespertilionidæ, distributed almost everywhere throughout the region. The flying-foxes are, however, absent from both New Zealand and the Sandwich Islands. The more important families of birds are mainly such as have an extensive representation in the Australian realm, or are held in common by this and the Oriental or Ethiopian realm. Among these are the caterpillar-eaters (Campephagidæ), flower-peckers (Dicæidæ), weavers (Ploceidæ), and swallow-shrikes (Artamidæ). With insignificant exceptions all the families of birds that are wanting in the Australian region are likewise wanting here; in addition to which, several of the more representative families of Australian birds, as the birds-of-paradise, bower-birds, lyre-birds, cassowaries, cockatoos, and apteryxes, are also wanting. On the other hand, the region contains three families which are absolutely confined to it ; these are the dodo-pigeons (Didunculidæ), from the Samoan Islands, the Drepanidæ, from the Sandwich Islands, and the heron-like Rhinochetidæ, from New Caledonia. The mound-builders (Megapodidæ) and honey-suckers (Meliphagidæ) have a very extensive distribution. The more nearly cosmopolitan families include among others the thrushes, warblers, crows, cuckoos, king-fishers, swallows, goat-suckers, swifts, pigeons, falcons, owls, and herons. Most of the genera of Sandwich Island birds are peculiar; hence it might be considered doubtful whether the tract inhabited by them should properly be considered to constitute a part of the region under consideration.

The reptile-fauna is feebly developed. Lizards, principally geckoes and scinks, range throughout the greater part of the region, as do likewise a very limited number of serpents, which are, however, absent from the more distant islands. The appearance of a member of the American family of Iguanidæ (Brachylophus) in the Feejee Islands is not a little surprising. Three species of Cornufer (Ranidæ) inhabit the Feejee Islands, and Bufo dialophus apparently the Sandwich Islands, but otherwise the Amphibia are almost wholly wanting. The fresh-water fishes are very limited in number, and exhibit a remarkable sameness throughout, comprising principally such forms—eels, atherines, mullets, gobies—as can readily accommodate themselves to brackish-water conditions. A siluroid (Arius) is found in the Sandwich Islands.

THE TYRRHENIAN TRANSITION REGION.

The fauna of this region is an association of elements derived from the faunas of the Holarctic, Ethiopian, and Oriental realms, with a preponderance on the north side of the Mediterranean of the Holarctic element, and on the south side probably of the Ethiopian. The number of absolutely peculiar forms, or of forms which barely pass beyond the confines of this tract, is not great. Among the Mammalia we have three such genera : Dama, the fallow-deer, found on both sides of the Mediterranean; Addax, an antelope, confined to North Africa and Syria; and Psammomys, a mouse, restricted to Egypt and Palestine. By far the greater number of the mammalian types occurring on the north side of the Mediterranean are such as might be considered to belong to the European division of the Holarctic tract ; but yet there are a considerable number of both genera and species which are entirely or nearly unknown there, and which either represent peculiarities, or forms belonging to the more tropical regions to the south. Such are the genet, ichneumon (found in Spain), porcupine, jackal (Dalmatia), Corsican deer (Cervus Corsicanus), and moufflon (Corsica, Sardinia, Crete, mountains of Greece). Until quite recently, too, the indigenous animals included also the lion and hyena. The ape of the Rock of Gibraltar (Macacus Inuus), although found on the Barbary coast, is more strictly an Oriental type. The Ethiopian affinities are further established on the African and Asiatic sides by the elephant-shrews (Macroscelididæ), coney (Syria), the antelopes of the genera Oryx, Alcephalus, and Gazella, and several additional members of the larger Carnivora—leopard, serval, and hunting-leopard. In the early part of this century the hippopotamus still inhabited Lower Egypt. The wild-asses which inhabit the desert plains included between the Red Sea and the Indus River may be considered as a link between the Holarctic and Ethiopian faunas.

The bird-fauna is on the whole very much more nearly Holarctic than anything else, a very large proportion of the species being such as inhabit Europe north of the Alps. But this is due in considerable part to the interchanges which are effected through the northerly and southerly migrations. According to Canon Tristram,[30] no less than two hundred and sixty out of three

6

hundred and twenty-two species of birds inhabiting Palestine are European forms, one hundred and thirty-four species (land and water birds) being common to Britain and Palestine. Of the Persian avi-fauna one hundred and twenty-seven species are also found in Europe.[31] Of the Oriental and Ethiopian birds which are not known north of, or barely transgress, the Tyrrhenian tract, may be mentioned the francolin, the quail-like Turnix, pastor, honeysucker (Nectarinea), hoopoe (Upupa), oriole, Ceryle and Halcyon among the kingfishers, the bee-eater (Merops), flamingo, and the genera Gyps, Vultur, and Neophron among the vultures. The ostrich enters the desert regions of Syria.

The reptilian and amphibian faunas contain a very considerable number of forms peculiar to the region. Of some fifty-three species found in Italy, only twenty-six penetrate into the region north of the Alps, and of this number from five to eight also enter the Ethiopian region.[32] According to Böttger, of the forty species inhabiting Morocco, twenty-two also belong to Spain, and but seven of these pass into the northerly Holarctic tract. On the other hand, only eight of the Moroccan species are known to inhabit the Ethiopian region. Algeria, according to the researches of Strauch, is represented by seventy-six species, of which twenty-seven are also Italian, and but ten Holarctic. On the other hand, eleven species are positively known to inhabit the Ethiopian region, and, according to Forsyth Major, not unlikely eighteen others will also be found to do so.[33]

THE SONORAN TRANSITION REGION.

This tract, which, as already stated, comprises the peninsula of Lower California, the State of Sonora in Mexico, New Mexico, Arizona, and parts, not yet absolutely defined, of Nevada, California, Texas, and Florida, is, as far as the Mammalia and birds are concerned, not very clearly differentiated; the intermingling of northern and southern elements, with a decided preponderance in favor of the former, is very great, and the peculiarities insignificant. Two species of Bassaris, a member of the raccoon family, appear to be confined to California, Texas, and the highlands of Mexico. Among the more peculiarly Neotropical forms that enter the tract are the jaguar, the peccary, a solitary species of armadillo, and a bat of the genus Nyctinomus. The reptilian and amphibian faunas are much more distinctly of a southern than of a northern facies. Of some fifty-five or more species of lacertilians, very nearly three-fourths of the number are iguanas (Iguanidæ), and four are geckoes. Fully one-half of the total number of genera represented are not found in other portions of the North American continent. The serpent-fauna comprises about twenty-two genera, one-half of which are peculiar. Eleven of the thirteen species of North American rattle-snake are found here ; the otherwise common coluber is wanting. Among the tailless amphibians, the Bufonidæ (Bufo) have their headquarters here, more than one-half of all the North American species being represented. The tree-frogs (Hylidæ), and frogs proper (Ranidæ), are, on the other hand, both very deficient. The tailed amphibians, of which there are upwards of fifty species in the region to the north, are almost completely wanting.[14]

THE AUSTRO-MALAYSIAN TRANSITION REGION.

This region, which is situated intermediately between the Oriental and Australian realms, naturally partakes of an intermediate position also in respect of its fauna. While the animal types borrowed from the adjoining regions preponderate to a very marked extent, the number of specific forms that are absolutely peculiar, especially among birds, is very remarkable. According to Mr. Wallace nearly three-fourths of some two hundred species of land-birds inhabiting the Moluccas are peculiar to those islands, and very nearly one-half of a hundred and fifty or more species found on the island of Celebes are absolutely confined to it. Yet, of the one hundred and twenty genera here represented (Celebes), only nine are peculiar. With few exceptions, all the families of birds that are represented in either the Australian or the Oriental region are also represented here. As belonging to the former, the birds-of-paradise have but a single species, the standard-wing (Semioptera Wallacei), found in the islands of Gilolo and Batchian ; the honey-suckers appear to be almost entirely absent from Celebes, although they are to be found in several of the other islands. The Oriental babbling-thrushes (Timalidæ) and bulbuls (Pycnonotidæ) barely enter, while the hill-tits (Leiotrichidæ) are completely wanting. It is not a little remarkable that the creepers (Certhiadæ), stone-hatches (Sittidæ), and tits (Paridæ), which have their representatives in the regions on either side of this one, should be entirely wanting here.

The placental Mammalia are represented by about twenty species, exclusive of the Cheiroptera, mainly of Asiatic types. Sixteen of these are found on the island of Celebes, and comprise several rodents (mice and squirrels), a wild-hog, a deer (Cervus hippelaphus), very closely related to, if not identical with, a Javan species; a civet-cat, a spectre-lemur (Tarsius spectrum), and three peculiar forms, unlike anything found either on the continent of Asia or the Malayan islands. These are a black and almost tailless baboon-like ape (Cynopithecus nigrescens), an antelopean buffalo (Anoa depressicornis), and the babyroussa (Babirusa alfurus). Several of these, or closely allied forms, are also found in the Moluccas, and in the group of islands and islets which stretch eastward from Lombok to Timor. We have here, in addition, one of the commonest species of Malay monkey, the Macacus cynomolgus, a Paradoxurus, and a possible second species of deer, Cervus Timoriensis. The implacental mammals of the connecting region comprise only Cuscus and Belideus.

V.

DISTRIBUTION OF MARINE LIFE.

ONE of the most important results of deep-sea explorations is the establishment of the fact that the distribution of oceanic life has no depth-limit. Contrary to the opinion so long entertained by naturalists that this life was confined to a shallow zone, extending but a few hundred feet beneath the water's surface, it is now known that representatives of all the marine invertebrate classes, and probably also fishes, exist at the greatest depths that have been reached by the dredge, and that in all likelihood a fair proportion of these penetrate even to the profounder abysses of four or five miles. The most extensive organic deposits accumulating in the trough of the sea are made by the Radiolaria and Foraminifera, whose world-wide distribution and prodigious development give the determining character to the oceanic floor. The well-known " Globigerina," or " Atlantic " ooze, a composition in principal part of the calcareous tests of four or five genera of Foraminifera (Globigerina, Orbulina, Spheroidina), constitutes the bed of the sea at nearly all points between the depths of four hundred and two thousand, or two thousand five hundred fathoms, except where, as in the immediate neighbourhood of coast-lines, the bottom is formed by the continental debris. Over this vast calcareous area animal life is far more abundant than over the comparatively sterile region where, for any reason, the ooze is wanting, and where, consequently, there is a marked deficiency in the material necessary for the proper development of many of the lower forms of life. Thus, it has been noticed that in such localities chiefly the shell-less orders of animals, as the Holothuroidea and the Annelida, are represented. Beyond a depth of some two thousand five hundred fathoms shell accumulations almost completely cease, their place being taken by a "red-clay" deposit, whose exact nature is, perhaps, not yet clearly understood. The absence of foraminiferal tests from the areas of greatest depth is doubtless due to the dissolution of the calcareous matter of the shell during its descent from the surface to the bottom, most of the pelagic forms, there is good reason to believe, being restricted in their range to a superficial zone of a few hundred fathoms depth.

The sponges, although they attain a maximum development in a zone of five hundred to one thousand fathoms, extend to the greatest measured depths, and are represented in the deepest parts by all the recognised orders, with the exception of the Calcarea (calcareous sponges), which appear to be confined to shallow water. The Hexactinellidæ, among siliceous sponges, to which the "glass-rope sponge" (Hyalonema) and Venus' flower-basket (Euplectella) belong, and whose earliest representatives appear already in the fauna of the Cambrian period, preponderate in the abyssal regions. The ordinary horny sponges (Keratosa), while they possess a very considerable vertical range, have their special development in the coralline zone. In the deepest parts of the sea corals are but sparingly distributed, and by far the greater number of species belong to the type of simple corals, and to the family Turbinolidæ, a large proportion of the genera passing back to the Tertiary period, and a few to the Cretaceous. No true Paleozoic forms have as yet been discovered. But five genera of Madreporaria are known whose range extends to, or exceeds, fifteen hundred fathoms (nine thousand feet), and only a single one, Bathyactis, which transgresses the twenty-five hundred fathom line. The vertical range of some of the species is very extraordinary, most notably so in the case of Bathyactis symmetrica, which is found in all depths between thirty (Bermuda) and twenty-nine hundred fathoms (east of Japan).[b]

Several species of Medusæ have been obtained from depths reported to exceed two thousand fathoms; but it is, perhaps, open to question whether some, or even most, of these apparently deep-sea forms are not in reality inhabitants of a comparatively shallow superficial zone, and have not been simply caught in the hauling of the net. There appears to be strong evidence, however, for concluding that at least a few of the forms are actually inhabitants of deep water.

Among the deep-sea Echinodermata there are representatives of all the modern orders—crinoids, brittle-stars, star-fishes, urchins, and holothurians—but none of the ancient palechinoids, cystids, or blastoids are known. The pear-encrinites (Apiocrinidæ), for a long time supposed to have become extinct with the Mesozoic era, continue their succession in the genera Rhizocrinus, Bathycrinus, and Hyocrinus, forms more strictly abyssal in character than the Pentacrinus, whose greatest development seems to be confined to a zone of a few hundred fathoms. Bathycrinus gracilis has been dredged in water of a depth of two thousand four hundred and thirty - five fathoms.[36]

The Asteroidea (star-fishes) and Ophiuroidæ (brittle-stars) are diffused throughout all the oceanic zones that have thus far been dredged, the former abounding more particularly at moderate depths. The singularly aberrant, and universally distributed, genus of star-fishes, Brisinga, is one of the commonest and most distinctive forms of the abyssal fauna, being found in all depths from four hundred to three thousand fathoms. Of the brittle-stars, of which there are about five hundred species described, more than two hundred are restricted in their range to a zone of thirty fathoms. Despite this apparent localisation of the species to a shallow belt, there are no less than sixty-nine species which descend below one thousand fathoms, and about eighteen below two thousand. None of the genera have been positively identified with fossil forms, although not unlikely the Jurassic Ophioderma may in part belong to Ophiura or Pectinura. The affinities of the Triassic Aspidura are still doubtful.[37] The relationship existing between the modern echinoid fauna and the faunas of past geological periods is much more marked; indeed, this relationship may be considered as one of the most distinctive features of the deep-sea fauna. Not only do a considerable number of the living genera date back to the Cretaceous period, but a fair proportion of those of the families Cidaridæ, Echinidæ, Salenidæ, &c., are already found in the deposits of the Jura, the Lias, and even in the Trias. The Tertiary genera are very largely developed, and the utmost similarity prevails, even among the species. So close is the identity existing between the West Indian urchins and those occurring fossil in many of the European Tertiary beds (older and median Tertiary), that it becomes practically impossible, or nearly so, to distinguish between the species.[38] This correspondence manifests itself among both the regular and irregular forms. In their bathymetrical distribution the echinoids appear to be governed by much the same conditions as have been observed in the case of the brittle-stars. By far the greater number of species—about two hundred—are of a littoral habit, occupying the belt of one hundred to one hundred and fifty fathoms, although not a few of them, like Echinocardium Australe, which descends to two thousand six hundred and seventy-five fathoms, penetrate deep into the abyssal zone. The number of continental species, or such whose normal habitat is included approximately between the one hundred to one hundred and fifty and the five hundred fathom line is forty-six, and a slightly larger number, fifty, may be considered to be strictly abyssal. The species passing below two thousand fathoms are rather limited, and only one is known—Pourtalesia laguncula—whose range embraces the twenty-nine hundred fathom line. This form is also found in the continental belt.[39]

The sea-cucumbers, or holothurians, which are very generally distributed throughout the oceanic abyss, constitute one of the most distinctive elements of the deep-sea fauna. As has already been seen, they, together with certain annelids, form a large part of the fauna of the "red clay," or of the region which lies beyond the reach of foraminiferal shells.

Deep-sea crustaceans are very abundant, and many of them are remarkable for their colossal size, their bizarre forms (Nematocarcinus gracilipes), and brilliant red coloring. Partaking of the first character are the giant blood-red shrimps of the genus Aristæus, and several members of the order Schizopoda. A Gnathophausia was obtained off the Azores, by the officers of the "Talisman," measuring no less than twenty-five centimetres (nearly ten inches) in length. The peneid and caridid shrimps, among the long-tailed decapods, are strikingly numerous, and present many very singular forms. It would seem, from the observations of the "Challenger," that the Brachyura, or crabs, were confined almost entirely to comparatively shallow water, although at depths of one thousand to fifteen hundred fathoms they appear to have yielded a sufficient harvest to the naturalists of the French expedition. Hermit-crabs were collected by the "Talisman" in water of from four thousand to five thousand metres.[40] Many of the pedunculated barnacles are of uncommonly large size, surpassing in this respect the shallow-water forms. As might have been expected from our knowledge of cave-faunas, and the habits of those animals, there are a number of blind crustaceans inhabiting the deep, as Nephropses, Polycheles, &c., the last in a manner representing the Jurassic Eryon. Many of the species, on the other hand, are profusely phosphorescent.

Neither of the three more important orders of mollusks, the Lamellibranchiata, Gasteropoda, or Cephalopoda, enter very largely as components of the deep-sea fauna, although of the first two scattered individuals are not exactly uncommon at nearly the greatest depths. Leda and Arca were obtained from a depth of 16,000 feet. The Cephalopoda are the least numerous, and not unlikely the majority of the apparently deep-water forms represent in reality only captures from shallow water. Wyville Thomson has called attention to the remarkable fact that only on one occasion did the officers of the "Challenger" take the animal of Spirula, "although the delicate little white coiled shell is one of the commonest objects on the beach throughout the tropics—sometimes washed up in a long white line, which can be seen from any distance."[41] The Brachiopoda, while enjoying a very broad geographical distribution, are by no means numerous, either specifically or numerically. Although seemingly on the verge of extinction, it would appear as though the actual specific diminution since the beginning or middle of the Tertiary period has not been very great. Most of the recent species are technically shallow-water forms, by far the greater number being found above the five hundred fathom line. Ten species range to depths of six thousand feet and over, and one, Terebratula Wyvillei, was dredged in twenty-nine hundred fathoms. All depths have furnished specimens of Polyzoa.

Our knowledge respecting the bathymetrical distribution of the deep-sea fishes, owing to the difficulty of determining whether the specimens hauled by the net have been actually taken in the depths indicated by the sounding-line, or have been simply captured during the ascent of the net, is not very precise, and barely sufficient to permit of any general conclusions being drawn from it. That fishes abound at very great depths there can be no question; but whether they are equally distributed in the great zone lying between the surface and bottom waters, may still be considered doubtful. The researches of the "Challenger" would seem to indicate that this intermediate area is largely, if not almost wholly, destitute of such forms, whereas the evidence brought by the "Talisman" tends in just the opposite direction. Thus, in one haul taken off the Cape Verde Islands, in four hundred and fifty metres water, the net brought up no less than one thousand and thirty-one fishes, mostly belonging to the genus Melanocephalus.[42] At depths of one thousand to one thousand five hundred metres in the North African Atlantic, they are stated by Milne-Edwards to abound, and on the bank lying some one hundred and twenty miles off Cape Nun, where in water of from two thousand to two thousand and three hundred metres M. Vaillant obtained the singular Eurypharinx pelecanoides, they are still very varied. Many of the species possess an extraordinary vertical range, accommodating themselves apparently with ease to the most varied conditions of pressure. Alepocephalus rostratus is met with in a zone included between nine hundred and three thousand six hundred and fifty metres, and much the same distribution characterises Scopelus Maderensis. Macrurus affinis is found between five hundred and ninety and two thousand two hundred metres. The greatest depth from which any fish has been obtained is about five thousand metres (Bathyopis ferox).[43]

The deep-sea fishes, although frequently characterised by many very remarkable abnormalities of structure, such as the enormous development of the head or jaw, the ribbon-like body, and the possession of phosphorescent organs, do not belong to any peculiar order, and are in the main simply modified forms of surface types. A large proportion of the species belong to the families Ophidiidæ, Scopelidæ, and Macruridæ.

In summing up the results obtained from a first general survey of the collections obtained by the "Challenger," Sir Wyville Thomson believes that we are warranted in arriving at the following general conclusions :[44]

"1. Animal life is present on the bottom of the ocean at all depths.

"2. Animal life is not nearly so abundant at extreme, as it is at moderate depths; but, as well-developed members of all the marine invertebrate classes occur at all depths, this appears to depend more upon certain causes affecting the composition of the bottom deposits, and of the bottom water involving the supply of oxygen, and of carbonate of lime, phosphate of lime, and other materials necessary for their development, than upon any of the conditions immediately connected with depth.

"3. There is every reason to believe that the fauna of deep water is confined principally to two belts, one at and near the surface, and the other on and near the bottom; leaving an intermediate zone in which the larger animal forms, vertebrate and invertebrate, are nearly or entirely absent.

"4. Although all the principal invertebrate groups are represented in the abyssal fauna, the relative proportion in which they occur is peculiar. Thus Mollusca in all their classes, brachyurous Crustacea, and Annelida, are on the whole scarce; while Echinodermata and Porifera greatly preponderate.

"5. Depths beyond five hundred fathoms are inhabited throughout the world by a fauna which presents generally the same features throughout; deep-sea genera have usually a cosmopolitan extension, while species are either universally distributed, or, if they differ in remote localities, they are markedly representative; that is to say, they bear to one another a close genetic relation.

"6. The abyssal fauna is certainly more nearly related than the fauna of shallower water to the faunæ of the Tertiary and Secondary periods, although this relation is not so close as we were at first inclined to expect, and only a comparatively small number of types supposed to have become extinct have yet been discovered.

"7. The most characteristic abyssal forms, and those which are most nearly related to extinct types, seem to occur in greatest abundance and of largest size in the Southern Ocean; and the general character of the faunæ of the Atlantic and of the Pacific gives the impression that the migration of species has taken place in a northerly direction, that is to say, in a direction corresponding with the movement of the cold undercurrent.

"8. The general character of the abyssal fauna resembles most that of the shallower water of high northern and southern latitudes, no doubt because the conditions of temperature, on which the distribution of animals mainly depends, are nearly similar." ("The Atlantic," II.)

Nature of the Deep-Sea Fauna.—Much diversity of opinion exists among naturalists as to the nature of the deep-sea fauna.

That it is not one governed by conditions of temperature alone, or in principal part, as has very generally been conceived, is made manifest by an examination of the bathymetric distribution which particular animal groups affect. Thus, the reef-building corals, which for their proper development require an average temperature of 70° to 75° Fahr., and a temperature never falling below 68° Fahr., are confined to a superficial zone of twenty fathoms; yet at most parts of the oceanic surface inhabited by these animals a suitable temperature would be found at depths fully five times as great, and in some quarters even very much greater. Over the tropical Pacific, for example, a temperature of 77° Fahr. prevails to a depth of eighty fathoms, and of 70° Fahr. down to one hundred fathoms, so that, as far as temperature alone is concerned, the coral animal might just as well have found a congenial home in those greater depths as in the shallower one of ten to twenty fathoms. Indeed, in the Red Sea the coral isotherm would still be found at the very bottom, or in a depth of water of six hundred fathoms; but here, as elsewhere, the limiting line is found at twenty fathoms. With reference to the vertical distribution of these animals, therefore, the matter of temperature would seem to be but little involved.

What is true of the corals doubtless applies in considerable part to many other animal groups; but it must be confessed that our knowledge respecting the thermal conditions necessary for the existence of most marine organisms is so limited that we can hardly premise at the present day upon any safe deduction being based upon it. Professor Fuchs [45] has quite recently emphasised the fact, however, as tending to prove the non-influence of temperature in determining distribution, that over the entire world almost all the important types of the deep-sea fauna are already represented at the comparatively insignificant depth of ninety to one hundred fathoms, and consequently inhabit a zone the extremes of whose average temperature may be separated by fully thirty to forty degrees. Thus, it is pointed out that on the Pourtales Plateau, off the coast of Florida, which begins at ninety fathoms, and descends to three hundred fathoms without showing any essential modification in its inhabiting fauna, deep-sea forms are very plentiful, especially corals, siliceous sponges, and echinoderms; and the same is the case with the famous Barbadoes grounds. A well-marked deep-sea fauna has long been recognised in the Atlantic and Mediterranean waters of Europe as occupying the one hundred fathom zone; and it has been equally observed along the coast of Brazil, the Philippines, and elsewhere. The fact that almost everywhere this upper limit of faunal distribution should correspond with a line of nearly uniform depth is certainly very remarkable, and one that argues strongly against the notion of thermal influences. For, if the determining factor in vertical distribution were really the matter of temperature, we should naturally expect to find the defining line between the surface and deep-sea faunas to be differently located for different parts of the earth's surface, rising in the polar and high temperate regions, where the surface temperature of the water is itself very low, and falling in the region of the tropics, whereas, as a matter of fact, no such condition obtains. Indeed, on the principle generally entertained, there ought to be in the high northern and southern latitudes no such thing as an abyssal fauna, inasmuch as the thermal conditions requisite for its existence would be those corresponding to the surface fauna as well, a nearly uniform temperature extending through the sea from top to bottom. We should then expect to meet with the characteristic deep-sea forms of corals, brachiopods, vitreous sponges, echinoderms, &c., seemingly indicative of a low temperature, in the littoral region, but, as is well known, they do not occur there, although they are sufficiently abundant in deep water. It is true that certain animals occurring in the deeper parts of the warm seas are known as surface forms only in the Arctic waters; but these are inconsiderable in number, and in the main uncharacteristic, so that they can scarcely be considered as a link uniting the littoral with the deep-sea faunas; in a general way the two are as sharply defined in the Arctic Seas as anywhere else. But, if it is not the matter of temperature that is principally involved in the formation of a deep-sea fauna, what is? The question does not, perhaps, at the present moment admit of a definite solution; but a suggestion thrown out in this direction by Professor Fuchs deserves careful attention. After reviewing the possibilities that may arise from such proximate causes as differences in the chemical characters of the water, the quantity of absorbed air contained in it, and currental motion, all of which must assuredly be of insignificant import, this eminent authority arrives at the conclusion that the only factor which can, in any material way, affect vertical distribution is *light*. It is claimed in confirmation of this view that the limit of light-penetration in the oceanic waters, as fixed by Secchi, Pourtalès, and Bouguer, corresponds closely with the forty to fifty fathom line, marking the upper boundary of the deep-sea fauna, or, more strictly, the line separating the littoral from the deep-sea fauna. Below this line, therefore, the fauna is one of darkness, and above it, except in so far as certain animal groups may be nocturnal in their habits, one of light. In support of this proposition Professor Fuchs emphasises the fact that, " with their character of animals of darkness, numerous peculiarities in the organisation and nature of the deep-sea animals agree. Thus it is known that very many deep-sea animals either have uncommonly large eyes, after the fashion of nocturnal animals, or are completely blind; it is also well known that they are, for the most part, either pale and colourless, or unicolourous, and that varied colouration is exceedingly seldom met with among them; and, finally, it is likewise well known that a very large proportion of deep-sea animals, in many groups, indeed the majority, are vividly luminous. This last peculiarity is of special importance, for it is clear that luminosity can be of consequence only to such animals as are destined to live in darkness, and, in point of fact, scarcely any luminous animals are known to us from the littoral region."

While it may, perhaps, be admitted that temperature is not the only, or even principal, agent in determining distribution, it must, nevertheless, be confessed that certain grave objections present themselves to the theory which looks upon light as the determining factor; indeed, the objections are much of the same kind as those which have been urged against the thermal theory. If the fact is surprising that corals do not descend below the one hundred and twenty foot line, when the temperature for a very considerable distance beyond that point is still above the normal required by them, is it not perhaps equally surprising that they should be limited at this point at all, seeing that the penetration of light extends to fully double or treble the depth, or, as has been more recently shown by MM. Fol and Sarasin in the case of the Mediterranean Sea, to even ten times that depth ? Can it be rationally conceived that such lowly organisms, devoid of special visual organs, can be so affected by the conditions of light and darkness as not to be able to endure that amount of obscurity which distinguishes the zone immediately underlying the twenty fathom line ? This scarcely appears possible. The extended range of a very large proportion of the animal forms entering into the composition of the littoral fauna, and the extreme rarity of instances in which limitation is so marked as to render most effective the difference between light and darkness, argue strongly against the notion of the all-paramount influence of light as affecting distribution. This objection to the views advanced by Professor Fuchs, as opposed to the doctrine of thermal limitation, is further strengthened by the fact which has been noted in the case of many animal groups (*e. g.*, the Brachiopoda) that the vertical range of surface forms is on an average greater in the boreal and hyperboreal regions—*i. e.*, where the temperature of the water is more nearly uniform—than in the more centrally located regions, where a much broader variation in the temperature of the water manifests itself. It is true, as has been urged by Professor Fuchs, that the littoral fauna is largely dependent for its development upon the existence of coral reefs and coast-binding shell-banks; but in how far this association is connected with the presence or absence of light, still remains to be determined. On the whole, while it may be assumed that we are still largely ignorant of the fundamental facts underlying distribution, it appears more than likely that not a single cause, but a combination of causes, is operative in bringing about the general result. That the deep-sea fauna is a fauna of darkness must be admitted; but this is so from the necessity of the case rather than a matter of choice resting with the animals composing it.

A singular correspondence has been noted as existing between the pelagic (surface) oceanic fauna and the fauna of the oceanic bottom (abyssal)—a correspondence that has likewise been attributed by some to a condition of darkness by which the different organisms are supposed to be governed. To what exact degree the members of this animal assemblage are nocturnal in their habits, or constitute true animals of darkness, the observations are not sufficiently far advanced to permit of a general conclusion being arrived at.

Pelagic Fauna.—Under the designation "pelagic" may be included those forms of life which habitually pass their existence on the free expanse of the ocean, and which only on accidental occasions, if at all, visit the continental borders, or descend to the floor of the sea. Such are the radiolarians, certain foraminifers and in-fusorians, the siphonophorous medusæ (Portuguese man-of-war, Physalia; Velella, Porpita), winged-shells (pteropods), a limited number of gasteropods (Atlanta, Ianthina, Glaucus, &c.), cephalo-pods, and tunicates (Salpa, Pyrosoma). The schizopod and ento-mostracous Crustacea are numerously represented, while a genus of hemipterous insect (Halobates) finds a suitable home clinging to the waves at practically all distances from the land. A number of fishes, such as the herrings, mackerel, tunny, swordfish, flying-fish (Exocœtus), flying-gurnard (Dactyloptera), sea-horse, and most of the sharks, some of which approach the shore during the spawning season, might be added to this list, and among the Mammalia the whales and dolphins.

The primary condition governing the existence of a pelagic fauna is manifestly the development over the oceanic expanse of vegetable life. This is found, for the most part, in the microscopic diatoms and the Oscillatoriæ, the former of which abound more particularly in high northern and southern latitudes, frequently, by their vast numbers, rendering the water thick as soup, and impart-ing to it a peculiar brownish or blackish tint, the so-called "black water" of Arctic navigators. In the temperate and warmer seas the diatoms are largely replaced by the oscillatorians, whose profuse development is no less remarkable. In the Arafura Sea, between Australia and New Guinea, the officers of the "Challenger" found the water continuously discoloured during a period of several days' sail, and giving out the odour of a reedy pond; and in the Atlantic they "passed for days through water full of minute algæ (Tricho-desmium), gleaming in the water like particles of mica."[*] It is to a species of Trichodesmium (T. erythræum) that is due the peculiar red colouring frequently seen over stretches of the Red Sea.

Were it not for this profuse vegetable growth the sea would probably be, in great part, an uninhabited waste. The algæ fur-nish the necessary nutriment to the simpler forms of animal life, which in turn yield their substances to those more highly organised. In this manner a true interdependence of conditions, or balance of life, is maintained. Professor Moseley, however, believes that in some parts of the ocean the quantity of freely suspended vegetable

[*] Moseley, "Nature," xxvi., p. 559.

growth is not sufficient to maintain the animal life which appears to be nourished by it, and he suggests that the deficiency may be made good through a peculiar symbiotic relation which appears to exist between certain lowly-organised plants and animals. Thus, many of the radiolarians and comb-bearers (Ctenophora) contain, embedded in their body-substance, a number of yellow starch-cells, which Brandt recognises as unicellular algæ (Zooxanthellæ), and which are supposed to thrive upon the waste products of the animal, and to yield to it in turn the compounds elaborated in the process of its own development. The relation of mutual benefit which is here stated to exist probably requires further investigation before it can be accepted as an absolute fact.*

One of the distinctive characters of the majority of pelagic animals is their transparency, which renders them very nearly invisible on the surface of the water. The nerves, muscles, skin, and organs generally, are alike hyaline, although in many instances the liver has remained unaffected. The protection thus afforded to such animals as the radiolarians, jelly-fishes, tunicates, and many crustaceans, is compensated for in animals less transparent by a colouring which harmonises with that of the open sea. Thus, the predominating colour is either blue or violet, as we find it in the Portuguese man-of-war, the Velella and Porpita, and in Ianthina and Glaucus among the snails. The fishes are principally steel-blue above and lustrous white underneath, a want of correspondence which is also manifest in other animal groups. Glaucus, just mentioned, whose progression is effected in the manner of the common pond-snail, ventral surface uppermost, has this side coloured blue, and the opposite, or dorsal side, silver-white. Exceptional cases are presented of an extreme brilliancy of colour, as in the copepod Sapphirhina, which is said to rival in metallic lustre the humming-birds, and to display the colours of the spectrum with the intensity of the gleam of the diamond.[46]

Most of the pelagic animals, except the lowest, are devoid of a shell, or, when present, the shell is usually very thin and fragile,

* Professor Hensen estimates that in some parts of the Baltic there are upwards of 140,000,000 plants (Rhizosolenia, Chætoceros) in every ten cubic metres of water, and maintains that this prodigious quantity is produced in the course of about two months. "Bulletin of the United States Fish Commission," August, 1885.

as in Argonauta, Cleodora, Atlanta, Carinaria, Ianthina, &c. Abnormalities of structure, especially in the case of the immature forms of littoral species, are frequent, leading to such modification of outline as to obscure in great measure the general parental relationship. The supposed young of the conger-eel develop into small transparent ribbon-shaped fishes, largely devoid of hæmoglobin in their blood, and with an exclusively cartilaginous skeleton; the young of certain flat-fishes (Platessa) die without ever reaching maturity, and before the eyes have become asymmetrically placed; and, in the case of some of the rock-lobsters (Palinurus), the flattened larvæ attain to gigantic proportions. Other instances of such abnormal development might be mentioned. The unusually large size of the eyes in some of the annelids and crustaceans (Alciops, Corycæus) recalls a similar characteristic belonging to many of the deep-sea forms of life; likewise the total absence, or very rudimentary condition (as in pteropods), of these organs. The power of emitting phosphorescent light is another feature held in common by many of the surface forms of life with the fauna of the deep. Pyrocystis and Noctiluca, amœboid bodies on the border-line between the Foraminifera and Infusoria, appear to be the principal contributors to the general oceanic phosphorescence.

A remarkable feature of the pelagic fauna is the vast swarms or schools in which many of the forms are found, association being the rule rather than the exception, and the broad expanse over which the greater number of the types are spread. The genera are of almost universal distribution, and many even of the species, of both the higher forms (fishes) and the lowest, are identical over the most distantly removed quarters of the globe; the polar faunas, however, which are constituted principally by the crustaceans, pteropods, and whales, differ materially from the faunas of the temperate and equatorial belts, lacking largely in the medusæ, the tunicates, and pelagic fishes. The varying salinity of the oceanic waters appears very sensibly to affect this fauna, whose distribution is, accordingly, in a measure governed by it. Thus, the surface-fauna of the Baltic is very meagre, and in the upper part of the basin, where the waters are nearly fresh, it is reduced to little more than a very limited number of crustaceans. Some of the medusoids, however, as Aurelia and Cyanea, appear to be but little affected by a deficiency in the salt-supply, and, indeed, according to Moseley, they would seem to prefer a habitation near the mouths of fresh-water streams, being seen to crowd up towards the heads of fjords and inlets. In the Hawkesbury inlet, New South Wales, the Scyphomedusæ were observed by this naturalist swimming in shoals where the water was so pure as to be quite drinkable.

Much uncertainty still exists as to the relation which the free oceanic fauna bears to the fauna of the deep-sea, an uncertainty due to the difficulty of determining the actual depth whence the different organisms caught in the net were obtained. Alexander Agassiz maintains, as the result of experiments made with the Sigsbee net, the most improved appliance thus far invented for the purposes of deep-sea exploration, that "the surface-fauna of the sea is really limited to a comparatively narrow belt in depth [about fifty fathoms], and that there is no intermediate belt, so to speak, of animal life between those animals living on the bottom or close to it, and the surface pelagic fauna."[47] Beyond a depth of one hundred fathoms nothing was found. On the other hand, the numerous observations made by Mr. Murray on board the "Challenger," with appliances less perfect than those used by Mr. Agassiz, almost conclusively prove that the depth penetration is very much greater than is here indicated, and that possibly a direct continuation exists in the case of certain groups of animals between the pelagic and abyssal faunas. The fact seems to be pretty satisfactorily established, however, that the true zone of free oceanic life, or that which is most numerously inhabited, is a shallow one, and that whatever life extends to great depths is comparatively restricted.

It would appear that a large proportion, if not the greater number, of the pelagic animals are more or less nocturnal in their habits, shunning the glare of daylight, and appearing on the actual surface only during the hours of evening and night. Such are most of the pelagic fishes, crustaceans, pteropods, heteropods, and fora-minifers, which in their hidden depths for a long time eluded the search of naturalists. The radiolarians, jelly-fishes, and certain crustaceans, on the other hand, seem to prefer the open daylight, appearing at all hours on the surface during calms; and the same is the case with a number of fishes, as the flying-fish and dolphin (Coryphæna). There is thus a perpetual oscillation in this upper zone of life, which, as dependent upon an excess or deficiency in illumination, probably does not extend much beyond a depth of fifty fathoms.

There can be but little doubt that a pelagic fauna antedated all the faunas of the globe, and that from it, through a long process of modification and adaptation, have been derived the faunas of the shore, the abyssal deep, the land-surface, and the various fresh-waters. The identity, or close resemblance, existing between the larval forms of many of the most divergent animal groups clearly indicates the lines along which modification has resulted, for it can scarcely be conceived, as Professor Moseley well insists, that this general identity in larval structure could have been brought about, as the result of natural selection after the adult forms had largely diverged from one another. The earliest traces of a pelagic fauna are indicated in the rocks of Cambrian age, where, as representative of it, we find, besides the remains of pteropods, the impressions of jelly-fishes, which were apparently not very far removed from some modern Scyphomedusæ. The marine animals that are deficient or lacking in the composition of the pelagic fauna, and not improbably have always been lacking, are the sponges, alcyonarian corals, sipunculoid worms, brachiopods, lamellibranchs, and echinoderms. The true infusorians (Ciliata) appear to be but very feebly represented, although there is an abundance of the Cilioflagellata.

Nature of the Littoral Fauna.—That the littoral fauna is either wholly or in great part a derivative of the free oceanic or pelagic fauna there is every reason to believe. The supposition that the latter came into existence before the former is at once a natural one, and is supported, apart from general zoogeological considerations, by the character of the mutually related littoral larvæ, whose adaptation to a pelagic existence clearly indicates the nature of their primal condition. There is, further, every reason to believe that the earliest plants were also largely pelagic, and that not until these had firmly established themselves as permanent forms along the sea-border was there developed a shore-fauna. In exposing themselves to the manifold conditions, such as the breaking of the surf, tidal action, shore-wash, attacks of enemies, which a change of abode entailed upon the members of the pelagic fauna, these were by force of adaptation compelled to undergo particular modifications of habit and structure which rendered them better fitted to their new surroundings. This we see in the defensive armour or encasement with which a very large number of the shore animals are provided, a character which eminently distinguishes them from the inhabitants of the open ocean. The shells, which are with the latter in most cases very thin and fragile, as in Atlanta, the argonaut, and the pteropods generally, are in the vast majority of shore animals thick and resisting, and capable of withstanding the numerous strains and impacts to which they are subjected.

But the same causes which have been operative in producing modifications in the pelagic fauna have been influential in bringing about a no less important series of modifications in the littoral fauna as well. The intermingling of fresh and salt waters about the embouchures of rivers, or a deficiency, as in the ice-bound north, in the salinity of the sea itself, will have gradually paved the way for the formation or evolution of animal forms destined to live eventually in fresh water. Hence, the origin in principal part of the fresh-water faunas. Similarly, frequent exposure to the atmosphere beyond the interacting influence of the aqueous medium, as in the region of "between tides," will have developed a method of respiration, or respiratory apparatus, other than that which is dependent for its action upon the presence of water. The remarkable series of modifications which the Amphibia (frogs, toads, salamanders) undergo from their larval condition, when, as inhabitants of the water, they breathe by gills, to their adult stage, when respiration is in most cases effected through the intermedium of lungs alone, most forcibly illustrate the progression which, at least in one division of the animal series, the vertebrates, has led to the formation of the air-breathing or terrestrial fauna. But other instances of adaptation from an aqueous to a terrestrial existence are not wanting. Many fishes have their gills so modified as to permit of a very protracted existence on dry land, while, as is well known, the lung-fishes (Dipnoi) have developed true lungs. Land-crabs are very abundant in the Tropics, roaming about in the interior at very considerable distances from the shore. In Japan they have been observed at an elevation of four thousand feet above the sea. The remarkable cocoanut crab, Birgus latro, is provided with a pair of true lungs, developed on the walls of its gill cavities.

Professor Moseley[48] thus sums up the relations of the littoral fauna: "The fauna of the coast has not only given origin to the terrestrial and fresh-water faunas, it has throughout all time, since life originated, given additions to the pelagic fauna in return for having received from it its starting-point. It has also received some of these pelagic forms back again to assume a fresh littoral existence. The terrestrial fauna has returned some forms to the shores, such as certain shore-birds, seals, and the polar bear; and some of these, such as the whales and a small oceanic insect, Halobates, have returned thence to pelagic life.

"The deep-sea fauna has probably been formed almost entirely from the littoral, not in most remote antiquity, but only after food, derived from the *débris* of the littoral and terrestrial faunas and floras, became abundant in deep water. It was in the littoral region that all the primary branches of the zoological family-tree were formed; all terrestrial and deep-sea forms have passed through a littoral phase, and amongst the representatives of the littoral fauna the recapitulative history, in the form of series of larval conditions, is most completely retained."

Lake Faunas.—It would appear that in all large lakes three distinct faunas can be recognised: 1. The littoral fauna, comprising the animals of the shore-line, which do not habitually descend to a much greater depth than fifteen or twenty feet. 2. The deep fauna, whose representatives live along the floor of the lake, at depths usually exceeding sixty to a hundred feet, a limited number of forms occasionally rising to the surface; and 3. The pelagic fauna, whose members occupy the free surface of the lakes, rarely or never reaching the shore-line or descending to the bottom. The zone inhabited by the last measures from fifty to a hundred metres in depth.

Our general knowledge respecting the pelagic fauna is still very limited, and is based almost exclusively upon observations made upon the European lakes. From these it would seem that the fauna is a very restricted one, consisting, as far as is known, of some twenty-five species of entomostracous crustaceans (ostracods, cladoceres, and copepods), a fresh-water mite (Atax crassipes), about six species of rotifers, and a limited number of infusorians. No lake has thus far yielded all these forms, and the majority of lakes are largely deficient. Between the years 1874 and 1878 Forel [49] obtained in the Lake of Geneva only eight species: Diaptomus castor,

Cyclops (sp. undet.), Daphnia hyalina, D. mucronata, Bosmia longispina, Sida crystallina, Bythotrephes longimanus, Leptodora hyalina. Of the total number of twenty-four species which were obtained by Pavesi [50] from the Italian lakes, belonging almost exclusively to such genera as occur in Lake Geneva, only four (Daphnia hyalina, D. galeata, Bosmia longispina, Leptodora hyalina) are known to inhabit the Lago Maggiore, and an equal number the Lago di Como. The Lago d'Iseo, on the other hand, has ten species, and Orta and Mergozzo eleven each. As far as the Crustacea are concerned, the Swiss lakes appear to be less rich in point of species than the Scandinavian; but they alone, with the adjoining lakes of Annecy and Bourget, in Savoy, have thus far yielded any variety of forms of lower organisation than the articulates. M. Imhof's investigations have brought to light, as constituents of the Swiss pelagic faunas, two species each of flagellate (Dinobryon) and cilioflagellate (Peridinium, Ceratium) infusorians, two species of true infusorians (Epistylis lacustris, Acineta elegans), which live attached on the crustaceans, and the six rotifers already referred to, belonging to the genera Conochilus, Asphanema, Anuræa, Triarthra, and Polyarthra. Doubtless some of these forms will also be found in the more northern and southern lakes.*

The general characters common to the animals of the pelagic region, which are the outcome of their particular mode of life, are thus briefly summarised by Forel: "They must swim incessantly, without ever being able to rest upon a solid body, and, instead of any organ of adhesion, they possess a highly developed natatory apparatus; their specific gravity, which is nearly the same as that of the water, enables them to swim about in the water without any great muscular exertion. They are rather sluggish animals, and escape the enemies that pursue them rather by their transparency than by their activity; they are, indeed (and this is their characteristic peculiarity), perfectly transparent, like crystals; and only their strongly pigmented black, brown, or red eye appears distinctly. This nearly perfect transparency of the pelagic animals may be regarded as a mimicry acquired by natural selection; only the animals

* "Ann. and Mag. Nat. Hist.," December, 1883; January, 1884. Since the above was written Imhof has identified several of the Swiss crustaceans, rotifers, &c., in the lakes of Alsace-Lorraine. "Zoologisher Anzeiger," December, 1885. [50a]

which are as transparent as the medium in which they live have held their own."

The lacustrine pelagic animals perform daily vertical migrations of the same character as has been noted in the case of the oceanic pelagic fauna, descending to the regions of obscurity during the day-time, and ascending by night. The animals appear to shun the light of the sun, and even of the moon, and hence retire to a depth probably not far from the limits of light penetration; the fauna is, therefore, one of darkness. The greatest depth whence specimens were obtained by Forel in Lake Geneva was about one hundred and fifty metres; but at this depth only Diaptomus was found. At a depth of fifty metres, in the Lago d'Orta, Pavesi found a very profuse fauna, represented by seven species; in the Lago d'Iseo, at five, fifteen, and thirty metres, the catch appears to have been exceedingly abundant ("*la pesca fu prodigiosamente abbondante*"); but, at one hundred metres, where the temperature of the water was 19° C., as compared with a surface temperature of 23° C., the fauna was decidedly scanty, although five distinct forms were obtained.

To what extent the downward extension of the pelagic fauna is governed by conditions of temperature, or in how far this limitation is dependent principally upon the presence or absence of light as a determining factor in the evolution of plant life, still remains to be ascertained. Forel, in 1874, found that paper sensitised with chloride of silver was still acted upon by the diffused light of the Lake of Geneva at a depth of about forty-five metres in summer and one hundred metres in winter, while ordinary shining objects disappeared from view at a depth of sixteen to seventeen metres. Asper, in August, 1881, obtained positive results through the use of plates sensitised with an emulsion of bromide of silver at a depth somewhat exceeding ninety metres in the Lake of Zurich; and more recently (1884–'85) Fol, Sarasin, Pictet, and others, have been able to detect the penetration of light in Lake Geneva to a maximum depth (in winter) of two hundred metres. In summer the penetration is considerably less. Fol and Sarasin [51] have also demonstrated that, in the Mediterranean, the solar rays penetrate to a depth nearly double that to which they were found to descend in the Swiss lakes, or to four hundred metres, and that at a depth of three hundred and eighty metres the intensity of light is as great as in Lake

Geneva at one hundred and ninety-two metres. At this depth, however, the impression produced upon the sensitised plates was of no greater value than that which would have been produced, under ordinary conditions, on a clear night, without a moon.

A remarkable feature of the lacustrine fauna is the very broad distribution of most of the species. Not only is there a general resemblance between the pelagic faunas of all the European lakes that have thus far been examined, from Scandinavia to Italy, and from Italy to Bohemia and the Caucasus, but a strict identity, at least as far as the species of Entomostraca are concerned. The species that occur in the one lake are also the species of the other lakes, although the respective littoral and deep faunas may be largely distinct. Further, it would appear that the same species are constituents of the pelagic faunas of American lakes as well, and not improbably make up the greater part of them. Professor S. I. Smith,[52] in his investigations of the fauna of Lake Superior, determined the presence, in the surface waters, of Daphnia galeata and Leptodora hyalina, common forms in the lakes of both Southern and Northern Europe, and of Daphnia pellucida, which was described by Müller as a pelagic inhabitant of some of the Danish waters.

As to the origin of the pelagic fauna little positive is known. That it is not a direct derivative of the different littoral faunas is very nearly certain, for were this the case we should expect to meet with largely differing assemblages of pelagic forms in all lakes where the littoral or deep faunas likewise differ; but, as has been seen, this is not the case. Yet there can be little or no question that it really represents a modification of some primary shore-fauna, whose members, through force of circumstances, were compelled to adapt themselves to new conditions of existence. The supposed method of its differentiation and further distribution is thus indicated by Forel: "I believe we must find the cause of the differentiation of the pelagic fauna in the combination of two different phenomena — namely, the daily migrations of the Entomostraca, and the regular local winds of the great lakes. It is well known that on the borders of great masses of water two regular winds prevail, one of which blows at night from the land towards the water, the other by day from the water to the land. The nocturnal animals of the shore-region, which swim at night at the surface,

7

are at this time driven towards the middle of the lake by the sur-
face-current of the land-winds, sink during the day, being driven
away by the light, into the deep water, and thus escape the surface-
current of the lake-winds, which would otherwise have carried
them again to the shore.   Constantly driven farther every night,
they remain confined to the pelagic region, as they are not carried
back again during the day.   Thus a differentiation takes place by
natural selection, until at last, after a certain number of generations, there remain only the wonderfully transparent and almost
exclusively swimming animals which we know.   When this differentiation has once taken place, the pelagic species is conveyed [in
the condition of resting eggs] by the migratory water-birds from
one country to another, and from one lake into another, where it
reproduces its kind if the conditions of existence of the medium
are favourable.   In this way we may find the pelagic Entomostraca
in lakes which are too small to possess the alternation of winds,
the animals having been differentiated by the action of the winds
in other larger lakes."

It might, however, be asked with Pavesi, if the general uniformity of pelagic faunas has been brought about through a method
of distribution such as is here indicated, how has it happened that
some lakes should be so largely deficient in pelagic forms as compared with other, and nearly contiguous, lakes ?   The lakes of
Northern Italy may be taken in illustration of a condition of this
kind.   Seeing that identical forms have been scattered to such
widely separated quarters of a continent, as Italy, Scandinavia, and
the Caucasus, it certainly appears a little surprising that immediately adjoining districts should have been so irregularly stocked
with the distributed material.   It might, however, be conceived
to be a matter of accident, and, indeed, at first sight the condition
appears to be more in the nature of a support to the theory stated
than as an argument against it.   But if accidental conditions of
this kind have happened, why has it not also accidentally happened
that some of the lakes should have retained a fauna, formed through
modification of their own particular littoral or deep fauna, distinct
from that of any other lake ?   Still, the objection here raised is not
an insuperable one, and offers much less difficulty in the way of the
partial solution of the problem than does the circumstance of the occurrence of identical forms in the lakes of Europe and North America.

Deep Faunas of Lakes.—The most systematic and thorough investigations that have been made into the nature of deep lacustrine faunas are those of Forel upon the fauna of Lake Geneva.[53] As the result of the observations of this naturalist it would appear that the abundant fauna of the floor of this lake comprises representatives of nearly all the primary divisions of fresh-water—inhabiting Invertebrata, and that even a fair proportion of the secondary groups are also represented, although by a very limited number of species in nearly all cases. Included in the lowest forms are several amœbæ, and Epistylis, Opercularia, and Acineta among infusorians. The hydroids are represented by the common brown hydra (Hydra rubra—to one hundred metres), and the rotifers by Floscularia. Three orders of worms are indicated—nematoids, cestoids, and turbellarians—and two of annelids proper, the hirudines and chætopods (Lombriculus, Tubifex, &c.). The turbellarians (Planaria, Mesostomum, Dendrocœlum) have no less than eleven species, one of which, Vortex Lemani, is found at all depths between fifteen and three hundred metres. A cestoid was dredged from a depth of two hundred and fifty-eight metres. The crustaceans are represented by a limited number of species belonging to the amphipods (Gammarus cæcus), isopods (Asellus cæcus), cladoceres (Lynceus), ostracods (Cypris, Candona), and copepods (Cyclops, Canthocamptus). Other articulates are four or five species, and as many genera, of arachnids (Arctiscus, Hydrachnella, &c.) and the larvæ of some tipuliform insects. The limited number of mollusks inhabiting the depth is not a little remarkable; of the lamellibranchs there is the single genus Pisidium, with about three species, and of the gasteropods only the genera Limnæa and Valvata. Although the Unionidæ (Anodon) are very abundant in the littoral fauna, they are completely absent below. One species of Limnæa (L. abyssicola) was found to be sufficiently abundant at a depth of two hundred and fifty metres, a circumstance to which Forel calls attention as indicating the readiness with which an air-breathing mollusk can accommodate itself to conditions largely at variance with those which are considered necessary to conform to certain structural peculiarities.*

---

* When brought to what might be considered its proper position, the surface of the water, the mollusk almost immediately adapted itself to the new conditions of existence, apparently without undergoing any inconvenience.

Although the fauna, taken as a whole, may be said to possess certain special characters, yet, broadly considered, it is only the representative, by slight modification, of the fauna of the littoral zone. It possesses no really well-defined or abnormal features of its own. Most of the forms are of small size, and a number of them, whose surface representatives are active and good swimmers, appear to have taken to sluggish habits; neither Cyclops nor Lynceus would rise when placed in an aquarium. Blindness is exceptional, and it is a surprising fact that the animals suffering from this defect (Gammarus cæcus, Asellus cæcus) are comparatively shallow-water forms (thirty metres), whereas those living at the greater depths, down to three hundred fathoms, are well provided with visual organs. The faunas of different zones of depth do not appear to differ sensibly from one another, except in the elimination or excess of a number of species. The greater number of these would seem to be distinct from their analogues of the littoral zone.*

Many of the species found in Lake Geneva are identical with forms found in the other Swiss lakes, and in the lakes of Savoy, as identified by Imhof, and there is good reason for supposing that a general analogy, if not absolute identity, unites the different deep lacustrine faunas of the same region. Professor Smith obtained from deep water (exceeding fifteen fathoms) in Lake Superior[54] Hydra carnea (from eight to one hundred and forty-eight fathoms), a Pisidium (from four to one hundred and fifty-nine fathoms), several species of worms (Sænuris, Nephelis, Tubifex, &c.), the larvæ of various tipulids and ephemerids, and among crustaceans Mysis relicta and Pontoporeia affinis (from shallow water to one hundred and fifty-nine fathoms). The last two, which were also found by Stimpson in Lake Michigan, are forms belonging to Lake Wetter in Sweden, supposed by Lovén to have been derived by modification from marine species.

* So stated by Forel in his report of 1876, although in 1874 he and Plessis appear to have maintained the opposite view.

PART II.

GEOLOGICAL DISTRIBUTION.

I.

The succession of life.—Faunas of the different geological periods.

THERE is no fact more patent in the history of the organic
world than that there has been from first to last a progressive evolution from lower to higher forms in the chain of beings that successively peopled the earth's surface. Casting our eye back over
the vast series of rock deposits which together constitute the fossiliferous scale of geologists, from the Cambrian to the Post-Pliocene,
and which together have a maximum development of probably
not less than two hundred thousand feet (or forty miles), we remark along the most ancient horizon the traces of animals which
bespeak the organisation of some of the lowest forms of life with
which we are at present acquainted; in the middle distance we
note the appearance of forms whose organisation marks a decided
advance upon that of their predecessors; and, finally, in the fore-
ground, we are brought upon the threshold of those highly complicated forms which to-day people the surface of the earth. The
simpler forms of life came into existence first; the most complex
last. It must not be implied, however, that with the progressive
and steady evolution of higher forms there has been an equally
progressive destruction or elimination of the forms of lower organisation; both have kept pace with each other, so that, at the
present day, although innumerable groups have completely disappeared, the lowest is found flourishing side by side with the highest.
This inter-association of lower and higher forms has manifested it-

self in all the geological formations that are known to us, from the Cambrian period to the present day, and there can be no doubt that, were a fossiliferous formation discovered of older date than the Cambrian, or immediately underlying it, we should find precisely the same juxtaposition, although to a more limited extent, of organisms of higher and lower development. Only then when we could fathom the first-born deposit, or trench upon the period when life first came into existence, would we, in all probability, be circumscribed in our survey to animals exhibiting a nearly uniform low grade of organisation. Such a point has probably not yet been reached, or, if reached, its existence can only be indicated with doubt, since the oldest rock deposits (the Laurentian) into whose composition an organic element unquestionably largely entered have lost all or nearly all traces of their primary fossiliferous character. With this wholesale obliteration have, consequently, disappeared the traces of the earliest and most primitive types of life-forms.

If Eozoon and Archæospherina, from the Laurentian limestones, be considered actually to represent organic forms, as is maintained by many prominent geologists and naturalists, then, indeed, are we presented with a large series of deposits in which apparently all the organic elements belong to one uniformly low type—the type of the Foraminifera—not yet the lowest, but very nearly it. But even granting the animal nature of the two structures here indicated, it would yet be very unsafe to affirm that they represent the only forms of life that tenanted the earliest seas; multitudes of other forms may have flourished and perished, and left no traces behind them, or had their traces completely obliterated at some remote subsequent period. We should then be no wiser for their existence. The succeeding Cambrian period ushers in with it such a host of multiform beings—beings of comparatively high organisation—that it becomes almost impossible to conceive that their ancestry should date back only to a period so little removed from the Cambrian as the Laurentian, unless, indeed, the hiatus separating the Laurentian from the Cambrian is very much greater than is indicated by its stratigraphical position. But if the ancestral forms of the Cambrian stock already existed in the Laurentian seas, what has become of their remains ? Why is it that in these oldest so-called fossiliferous rocks we meet with only Eozoon and

Archæospherina ? Surely it could not have been that there was such a disposition of the remains as to leave nothing but these two forms belonging to the lowest type. It appears far more plausible to assume with those who uphold the mineral nature of Eozoon and Archæospherina that we have no traces of this ancient pre-Cambrian fauna remaining, and that, consequently, the destruction was complete. How far back beyond the Laurentian the root of the present existing chain of organisms may have extended it is impossible even to conjecture.

Cambrian Fauna.—It is certainly a surprising fact, whichever way it be considered, that, with the formation bringing the first unequivocal evidences of organic life, we should meet with that multiplicity and variety which characterise the faunal assemblage of the Cambrian period. Most of the greater divisions of the animal kingdom, possibly not even excepting the vertebrates, were there represented, and most of these already in the lowest or oldest deposit — protozoans, cœlenterates, echinoderms, worms, articulates, and mollusks. And more than this, some of these groups were already represented by a full, or nearly full, complement of the orders that have been assigned to them by naturalists, and which include all the various forms that have thus far been discovered as belonging to the groups. Thus the Cambrian echinoderms are represented by forms belonging to three out of the six usually recognised orders—the Cystidea, Crinoidea (ocean-lilies), and Asteroidea (star-fishes). The last two have representatives living at the present day, whereas the former is entirely extinct. We have here, then, the most ancient ocean-lily and star-fish (Palasterina), and it is interesting to note what distinct relations these two forms hold to their modern representatives. While the Crinoidea attained their maximum development in the seas of the Paleozoic period—Silurian, Devonian, and Carboniferous—since which time they have been pretty steadily declining, until at the present moment they are represented by scarcely more than a half-dozen distinct generic types, the Asteroidea have been just as steadily increasing, and, indeed, attain their maximum development in the modern seas.

It may appear at first sight anomalous how two groups, so widely dissimilar from each other, and having such varying developments, should have appeared simultaneously in the same period of the earth's geological history, the Cambrian. But it must be borne in mind that in the Cambrian formation we have only what is seemingly the oldest fossiliferous formation, and that the ancestral forms of both ocean-lilies and star-fishes lie buried in rock deposits of undeterminably older age. If, on the hypothesis of evolution, we uphold the inter-derivation, or derivation from one another, of these two forms, then it is but fair to assume that the crinoid, which is structurally the lowest, appeared at a period considerably anterior to the star-fish, which must have required for its specialisation a no inconsiderable lapse of time. And it is a singular fact, and one strikingly confirmatory of this view of relationship, that we have in both these forms certain peculiarities of structure which effect a sort of transition from the one to the other. Thus, in some of the fixed crinoids the plume or tuft separates from the column after a certain period of existence, and then leads an independent existence, to all appearance a stellarid (the Comatula). Conversely, the officers of the late "Travailleur" deep-sea dredging expedition obtained off the coast of Spain, and from depths respectively of nineteen hundred and sixty and twenty-six hundred and fifty metres, two individuals of a new genus of star-fish (since named by Perrier Caulaster peduncularis), which exhibited on the dorsal surface a true peduncle, demonstrated to be absolutely homologous with the stalk of the crinoid. Yet, despite this obvious relationship, it is not a little surprising that no pedunculated star-fish has thus far been found fossil, nor any comatulid crinoid to antedate the Jurassic period (Antedon).

The Cambrian Mollusca comprise representatives of five of the six classes that now inhabit the seas, namely, the Brachiopoda, Acephala, Pteropoda, Gasteropoda, and Cephalopoda. Here again, therefore, we have an apparent simultaneous appearance of lower and higher forms; but, as before, we must look to a much earlier period for the ancestral traces, if any have been preserved, of the first or most primitive type. The genetic relationships of these various molluscan groups cannot, in the present state of the science, be determined with any degree of certainty; but, if a low degree of organisation indicates antiquity, which certainly appears to be the case with many groups of animals, then it may be fairly assumed that the Brachiopoda were the first to appear. It is a surprising fact in the history of these animals, and one which is, perhaps, not repeated to the same extent in any other group, that while hosts of genera, and even complete families of this order, which flourished in the seas intermediate in time between the Cambrian period and our own day, should have successively disappeared, a few individual types seem to have survived from first to last, without having undergone any essential modification of structure. Thus, the Lingula, or Lingulella, of the Cambrian rocks is but very little, if at all, different from the existing Lingula, and it has indeed been considered doubtful by some authors whether even specific characters could be assigned to distinguish some of the earlier from the later forms, separated by an interval of millions of years. The same persistence of type is represented in the genus Discina.

Side by side with these lower molluscan types, but appearing at a somewhat later period, the Upper Cambrian, we find, as has already been stated, forms belonging to the highest order, the Cephalopoda (Orthoceras, Cyrtoceras), another apparent contradiction to the doctrine of progressive higher development. Considering the group of the cephalopods by itself, however, we observe that its earliest types belonged to the lower of the two divisions into which the cuttle-fishes have been divided—the tetrabranchiate, or four-gilled order—a division to which the somewhat later appearing, and now probably disappearing, Nautilus also belongs. These primitive cephalopods were succeeded in time by other members of the same order—Gyroceras, Nautilus, Goniatites—until the Triassic period was reached, when the first dibranchiate form, Belemnites, appears. From this period down to the close of the Mesozoic era both the two-gilled and the four-gilled forms occur in such abundance that it would be almost impossible to state to which group belonged the preeminence. But in the meanwhile a general alteration and succession in the representative cephalopod type had been taking place. The early forms already mentioned, Orthoceras, Gyroceras, Cyrtoceras, and their allies, belonging to the family of the Nautilidæ, are succeeded in the Triassic period, where their last traces (excepting Nautilus) are to be met with, by the members of the more complicated group of the Ammonitidæ, whose earliest precursors (three or more species from the Carboniferous formations of India, and a solitary species from the Carboniferous of Texas) would seem to have been foreshadowed by Goniatites, a type structurally intermediate between the Nautilidæ and the Ammonitidæ.

Similarly, in the case of the Cephalopoda dibranchiata, the Belemnitidæ are succeeded by forms more nearly resembling the calamaries, or cuttle-fishes (Teuthidæ), of to-day, whose remains are found already in the Jurassic deposits (Onychoteuthis, Teuthopsis, Belemnosepia). With the beginning of the Tertiary period * we note the final disappearance of the varied group of the Ammonitidæ, and with them the last traces of all but one of the lower or four-gilled order of cephalopods. The single exception is the Nautilus, which, as a persistent type, almost unaltered from the Silurian to the present period, alone survives to contest the seas with the members of the higher or dibranchiate order.

The predominant Mollusca of our modern seas, the Gasteropoda and the Acephala, were but feebly represented in the seas of the Cambrian period; but it seems not improbable that some of the earliest forms—*e. g.*, Capulus, Pleurotomaria, among the snails— belonged to types absolutely identical with those living at the present time. It is not until we have completely passed over the Paleozoic era, which, so far as the Mollusca are concerned, may be said to constitute the age of the Brachiopoda, that these two orders of shell-fish attain any special significance. From the beginning of the Mesozoic era onward they steadily crowd the deposits with their remains, until, finally, with the Tertiary formations, and the formations succeeding these, they constitute the most characteristic and most important invertebrate landmarks to the geologist and paleontologist.

It has been contended, and with apparent force, that the irregular appearance in time of the Mollusca—*i. e.*, the almost simultaneous introduction of forms belonging to both the lowest and the highest orders, and the final supremacy in the existing seas of the type of the Acephala, a group of mollusks inferior in organisation to the Cephalopoda, the Gasteropoda, and the Pteropoda—is incompatible with the doctrine of evolution, which, as argued, requires for its confirmation the introduction first of the lower forms, the development from these of the more advanced, and, ultimately, the appearance of those that are most perfect or specialised in structure. It must be recollected, however, that, as far as the almost simultaneous introduction of lower and higher forms is con-

* A species from the Lower Tertiary of California.

cerned, the obstacle is more apparent than real, for, as has already
been insisted upon, it is impossible to determine how far back beyond the Cambrian, or first unequivocally fossiliferous formation,
life may have already existed, and, consequently, to what very
ancient period the ancestry of the molluscan type may extend. As
a matter of fact, the most ancient mollusk, the Lingula, or Lingulella, is almost the lowest in structure of any with which we are
acquainted, and if in the rock deposits we meet with its remains
but barely antedating those of the very much more highly organised Orthoceras, we have yet strong grounds for concluding that its
first appearance was very much earlier, only that, through the general obliteration of all remains in the preceding geological period,
direct evidence to that effect has been lost. As to the other objection, that the predominant forms persisting at any given epoch
should be those whose structure manifests the highest development,
it may be remarked that the evolutionary force requires no such
result as the outcome of its operative action. It is among such
forms as, in their mutual relations to their surroundings, whatever
these may be, are best adapted or fitted for combatting the numerous elements that constantly interpose themselves in the path
of existence, that we must look for examples of greatest persistence
and development—for the survivors in the struggle for existence.
Hence, while the highest developed forms in any given series of
animals will present themselves in or about the period most removed from the birth of that series, yet it need not follow that
the higher series will ultimately outlive, or even predominate over,
the representatives of a lower parallel series of the same class of
animals, whose fitness for struggling in the battle for existence is
not infrequently vastly superior to that of the higher class. We
need not be, therefore, surprised at finding, in a given class of animals, some of the more perfect forms disappearing from the world's
horizon before the less perfect, and these last, consequently, the
survivors in the general battle for life. But, while no general law
can be formulated regarding the *disappearance*, as conditioned by
the degree of perfection, of the various series of a given class of
animals, or, regarding their relative development in any one period
of the earth's history, the law of *appearance* or succession already
stated—*i. e.*, the introduction of lower forms before those of a
higher order—can very generally be maintained.

An objection to the evolutionary doctrine, similar to that which has been drawn from the distribution of the Mollusca, is also furnished by the articulated animals, or, more particularly, by the class of the Crustacea. The members of this class boast of a lineage, as far as has yet been determined, very nearly, if not fully, as ancient as that of the Mollusca, one extending back to the earliest Cambrian period. But, while the most ancient mollusks with which we are acquainted belong in great part to orders, families, and even genera, whose representatives still flourish in the existing seas, the most ancient crustaceans, or at least the majority of them, the Trilobita, have long since become totally extinct; hence the impossibility of determining their true relationships. However uncertain or obscure this relationship may be, whether it is with the Phyllopods, as claimed by some, or, what is much more likely, with the Xiphosura (king-crabs) and arachnids, as argued by others, there can be no doubt, if homologies of structure can be relied upon, that the members of this group of animals represent a high grade of structural organisation, and especially if the period of their appearance is taken into consideration. But here, just as in the case of the Mollusca, we have the strongest evidence for concluding that their earliest appearance dates far beyond the Cambrian period, as is proved almost conclusively by the simultaneous appearance in the oldest Cambrian strata of some of the simplest and most complicated trilobitic forms, Agnostus and Paradoxides, which are at the same time also among the smallest and the largest forms of the entire order. A further evidence of the pre-Cambrian antiquity of this group is furnished by the circumstance of the abundance in which the earliest remains are found, an abundance which, though perhaps not equal to that characteristic of the succeeding Silurian and Devonian trilobitic faunas, is yet sufficient to impress a distinct individuality upon the fauna of the period. While with the Cambrian trilobites we find associated other forms of crustacean animals, such as the phyllopods (Hymenocaris) and ostracods (Primitia, Leperditia), the highest members of the class, the Decapoda (crabs and lobsters), appear not to have been as yet evolved. Indeed, it is not until the entire Silurian period and a considerable portion of the Devonian are passed that we meet with an example of the ten-legged order of crustaceans. Barely had these higher forms asserted themselves on the field of life ere a decline in the supremacy of their predecessors is made manifest. With the middle of the Devonian period the beginning of the trilobitic decay becomes apparent, and, after the close of that period, *i. e.*, in the Carboniferous, less than a half-dozen types remain, and even these are of comparatively rare occurrence. At the close of this lastnamed period the trilobites disappear totally and forever from the scene.*

Broadly looking over the Cambrian fauna, we find it to be distinguished by two important features. One of these is the fact that it is entirely destitute of both land and fresh-water forms, or such as are strictly adapted to breathing directly the oxygen of the atmosphere or that of fresh water. All the forms thus far encountered are, as far as we know, of a strictly marine nature. The absence of land animals will scarcely appear surprising in view of the complete, or nearly complete, absence of a land vegetation, and the correlative want of the nourishing material requisite for that character of organisms. The absence of fresh-water forms is not so readily accounted for, unless it be that there were formed at that time no fluviatile or lacustrine accumulations of sufficient magnitude to have left their traces behind them. It is not impossible, however, that some, or even many, of the recognised marine fossils, or such as have a marine habit, of the Cambrian formation, are in reality estuarine or brackish forms, as it can scarcely be conceived that all the deposits that were formed at the mouths of the ancient rivers should have been so totally destroyed or covered over as to have left absolutely no vestiges of their former existence. Doubtless, some of these have been preserved, along with their contained fossils, although the exact nature of such deposits may be disguised from us by reason of our imperfect knowledge concerning the true habits of their representative organisms. Nor would it be absolutely safe to affirm that some of these organisms, undistinguishable from what at the present day are indisputably marine types, may not in reality have been of a purely fresh-water habit in those early days.

The other distinguishing feature of the Cambrian fauna is the

---

* Shumard has described representatives of the genera Phillipsia and Proetus in deposits of the Sierra Madre, of the Southern United States, claimed to belong to the Permian period ; the determination of age may be considered to be very doubtful, however.

absence of positive indications of the existence of vertebrated animals. The only objects that have thus far been described as pertaining, with any show of probability, to the members of this highest division of the animal kingdom, are the singular bodies known as conodonts, which, in the opinion of their discoverer,.Pander, and of some other naturalists, represent the teeth of fishes belonging to the order of the myxinoids (hags and lampreys), with the exception of the lancelet (Amphioxus) the lowest of the entire class of Pisces. The weight of opinion, however, seems to relegate these problematical bodies to the Invertebrata, and not improbably, as has been urged for some of these forms, they represent the jaw-teeth of certain annelids.

Silurian Fauna.—The fauna of the Silurian period marks a decided advance upon its predecessor. The chain of organisms which, with the exception of the somewhat doubtful conodonts, was hitherto constituted exclusively by the members of the invertebrate series—sponges, echinoderms, mollusks, articulates—exhibits here for the first time indisputable representatives of the more highly organised group of the vertebrates; but not until the Upper Silurian deposits are reached. We here meet with the remains of two distinct orders of fishes, the sharks or dog-fishes (Elasmobranchii), as represented by Onchus and Thelodus, and the bucklered Ganoidei—Pteraspis—the former still very abundant in the modern seas; the latter, which include, among other forms, the sturgeon and alligator gar, probably nearly verging on extinction.* In both these orders the osseous framework or skeleton is frequently in a more or less imperfect condition—complete ossification being the exception rather than the rule—and hence, in so far, these primitive vertebrates exemplify a low grade of organisation compared with those—like the bony fishes, and most of the animals above them—in which the vertebral column is completely ossified, or reaches its furthest development. Nor are other characters wanting proving inferiority of organisation. We have here, therefore, another illustration of the very important fact—a fact sustaining the inference of the progressive evolution of higher from lower

---

* The oldest fishes were, until recently, supposed to belong to the British Ludlow beds ; but the discovery, by Professor Claypole, of ichthyic fragments in deposits below the " Water-Lime " of Pennsylvania would seem to remove them still farther back in the geological scale.

forms of life—that the representatives of each class of animals were first ushered in in their simplest or most embryonic forms, and that not until these had attained a considerable development was there a noticeable appearance of the more highly constituted forms. It is a significant (even if not a very remarkable) fact that, prior to the first introduction of this lowest class of the Vertebrata, all the larger divisions of the Invertebrata, as now recognised by naturalists, had already come into existence. Of these, the diversity of form in the Silurian deposits, no less than the numerical development, is very great, and equally so in almost all the classes represented.

The most marked feature of the Silurian invertebrate fauna, as contrasted with the Cambrian, is furnished by the corals, which, barring a few forms doubtfully belonging to the Cambrian of Sweden, have here their earliest representatives. These primitive types of the Actinozoa, as well as nearly all others of the Paleozoic series of deposits, have generally been recognised by naturalists to constitute two well-defined groups, the Tabulata (Favosites, Halysites, Heliolites, Alveolites, &c.) and the Rugosa, or cup-corals (Cyathophyllum, Streptelasma, Omphyma, Zaphrentis, &c.), in both of which the calyces are divided up into superimposed chambers by transverse plates or tabulæ—the former with very rudimentary septa, the latter with the septa well developed, and the outer calicular wall greatly thickened. In the majority of these cup-corals the septa are disposed in multiples of four (Tetracoralla), whereas in nearly all recent Madreporaria this disposition is effected in multiples of six (Hexacoralla). Our recently acquired knowledge of the deep-sea fauna, and a more intimate acquaintance with the anatomy of some of the more aberrant species of coral, tend to show that the supposed sharp delimitation of the Paleozoic actinozoan fauna does not in reality exist. The tabulate corals, for example, whose final extinction with the Paleozoic era has generally been insisted upon as one of the most decisive of geological landmarks, would seem to hold a number of forms more or less closely related to types living in the modern seas, which in themselves combine most diverse features in their organisation. The genera Halysites and Syringopora appear to be not distantly removed from the recent organ-pipe (Tubipora); Favosites is placed among the Poritidæ; and Heliolites not impossibly represents an ancestral form of the group to which the modern Heliopora, recently claimed to be an alcyonarian, also belongs.* The rugose-corals, apart from the very limited number of forms which occur fossil in deposits newer than the Paleozoic—? Holocystis (Cretaceous), Conosonilia (Tertiary)— have apparently two living representative types in Guynia and Haplophyllia. On the other hand, the modern star-corals, if we exclude from this group the Favositidæ, have but a feeble development in the earlier deposits, although several recent families are represented (Poritidæ, Eupsammidæ, Astræidæ). Of the genera belonging to this group, Protaræa, Stylaræa, Prisciturben, and Calostylis date back to the Silurian period.

The remarkable development of corals in the Silurian seas makes it not a little difficult to account for the total, or almost total, absence of their remains in deposits of the preceding Cambrian age. It is scarcely credible that the animals of this class should not have already then existed; but if so, what has become of them ? In explanation of this anomaly some geologists have urged that the strata of Cambrian age which contain recognisable fossils are all of a deep-sea origin, and that in the shallow and littoral deposits, where we might be expected to look for the traces of the organisms in question, organic remains have been completely obliterated through rock-metamorphism of one kind or another. The evidence supporting this hypothesis is, however, far from satisfactory, and the problem must be considered as one still awaiting solution.

The graptolites, a group of organisms whose earliest remains are found in the transition rocks which unite the Cambrian and Silurian formations, and whose organisation appears to be most nearly reflected in that of the recent sertularians or sea-firs of the class Hydrozoa, constitute an important element in the Silurian

---

* The reference of Favosites to the Poritidæ, it must be confessed, is based upon rather slender evidence, and perhaps scarcely less so the placing of Heliolites among the Alcyonaria. Hörnes ("Elemente der Palæontologie," 1884) justly emphasises the artificiality of a classification in which the number and disposition of the septa and tentacles are made the basis for a division into primary groups, and in which other equally important characters are completely lost sight of. The relationship existing between past and recent forms, taken in conjunction with the general homogeneousness of character exhibited by the Tabulata, would seem to imply that the classification of the recent Actinozoa requires serious emendation.

fauna, becoming practically extinct with the close of that period.*
It is a noteworthy circumstance in connection with the history of
this family that the more complicated or double-stemmed forms, such
as Diplograptus, Didymograptus, Phyllograptus, and Dichograptus,
preceded, in the order of appearance, the simple-stemmed forms, like
Monograptus and Rastrites, proving, contrary to what might have
been naturally supposed, that the latter were not the ancestral types
of the family. On any evolutionary hypothesis the simpler forms
appear to have been brought about as the result of degeneration.
In modern type hydrozoans the Silurian, as all other Paleozoic, deposits are very deficient, a circumstance, doubtless, due in considerable part to the perishable nature of the organisms belonging
to this class. The impressions of jelly-fishes have, however, been
indicated in both the Cambrian and Silurian rocks of Sweden.
Stromatopora, a very broadly distributed genus, whose affinities
are now generally conceded to be with the Milleporida, passes into
the Devonian formation.

Of other Invertebrata, such as the echinoderms, mollusks, and
articulates, there is a vast profusion of forms, which, apart from
the mere matter of numbers, are in many respects sharply contrasted
with their predecessors of the Cambrian period. The brachiopod
mollusks, the predominant forms of which, as Spirifer, Atrypa,
Athyris, Strophomena, Rhynchonella, and Pentamerus, belong to
the group of the Brachiopoda articulata, are structurally considerably in advance of the inarticulate genera Lingula, Lingulella,
Discina, and Obolus, which make up almost the whole of the
corresponding Cambrian fauna; the latter, as far as is known, contains but a single precursor of the articulate division, Orthis. The
prodigious development of the Silurian Cephalopoda would of itself
be sufficient to distinguish the period from the period preceding.
While up to the present time only two species of this class, an
Orthoceras and a Cyrtoceras, are positively known from Cambrian
deposits, no less than eleven hundred species, referable to a very
considerable number of genera of Nautilidæ—Orthoceras, Cyrtoceras, Gyroceras, Endoceras, Gomphoceras, Phragmoceras, Lituites,
Nautilus, &c.—have been described from the Silurian basin of Bohemia alone. The total number of species of this period may be

* The somewhat problematical Dictyonema passes into the Devonian:
Triplograptus, if a true graptolite, is Devonian.

estimated at between two and three thousand.   The genus Goniatites, which effects a partial transition between the Nautilidæ and
the Mesozoic ammonites, appears for the first time in the later
Silurian deposits.

It is a surprising fact, considering the remarkable development
of the Cambrian trilobitic fauna, that not only are none of the
earlier species represented in the Silurian deposits, but that by far
the greater number of generic types, and more particularly those
which by a special individual or specific development are rendered
most important, as Paradoxides, Dikelocephalus, Olenus, Sao, and
Conocephalus, should be also wanting.   Only seven out of some
twenty-seven genera of the primordial zone connect the formations of
the two periods.   The Silurian Trilobita comprise probably in the
neighbourhood of fifteen hundred species, referable to some fifty or
more genera; yet of this vast number there are barely a half-dozen
species which transgress the boundaries of the formation, passing
into the Devonian.   Among the more abundantly represented genera are Phacops, Dalmania, Calymene, Asaphus, Trinucleus, Aci-
daspis, and Cheirurus.

Viewed irrespective of numerical development, the most significant feature connected with the Silurian invertebrate fauna is the
introduction of the earliest "air-breathers."   Until recently these
were supposed to belong to the period following, the Devonian, but
the discovery of a true scorpioid (Palæophoneus) in the Upper Silurian deposits of both Sweden and Scotland, and of an apparent
orthopteroid (Palæoblattina) in the nearly equivalent deposits of
Calvados, France, proves conclusively that a very considerable
differentiation among the air-breathing arthropods had already
taken place, and points to a period very much more ancient for the
first origination of the group.*

Devonian Fauna.—The primitive air-breathing arthropod fauna
just referred to finds a somewhat larger extension in the rocks of
Devonian age, where fragments belonging to some five or six species of insect, possibly representing as many genera—Platephemera,
Gerephemera, Lithentomum, Homothetis, Xenoneura, Discrytus—
have been discovered.   These appear to belong to the modern

---

* The scorpion described by Professor Whitfield ("Science," July 31,
1885) from the Upper Silurian rocks of New York may, as suggested by Mr.
Pohlman ("Science," September 4, 1885), prove to be a young eurypteroid.

group of the netted-veins (Pseudoneuroptera and Neuroptera), although by some authors they, as well as all other Paleozoic insects, are considered to represent a distinct, and now wholly extinct, type of Insecta, the Palæodictyoptera. In whichever way their relationship be viewed, there can be little doubt that they represent very nearly the lowest structural type of their class. It is a very remarkable fact that the wing venation of these primitive insect forms is practically identical with that which characterises the modern insects belonging to the same group or order; the vast lapse of ages between the Devonian period and our own day appears to have effected no essential modification of structure in this particular direction. Besides these neuropterous forms, certain wing fragments have been referred to the members of the higher order Orthoptera, to which the modern grasshopper and cockroach belong. But there seems to be considerable doubt as to the claims of the so called Devonian cockroach,* and it would, perhaps, be as well to consider its position as still a matter of uncertainty. The marked differentiation exhibited by the Devonian insects indicates that they were far more numerous than would appear from the paucity of their remains, and the inference drawn as to their great antiquity has been confirmed by the discovery of the Upper Silurian form already referred to. Coincidently with the appearance of these early inhabitants of the land surface, we remark the first considerable development of a land vegetation, whose earliest traces are to be met with in the Silurian period. With but very few exceptions (certain forms, as Prototaxites, Ormoxylon and Dadoxylon, considered by some authorities to represent true conifers, the first of their kind) all the Devonian plants belong to the lower or non-flowering division, the Cryptogamia, comprising a multitude of ferns, tree-ferns, giant representatives, like Sigillaria and Lepidodendron, of the modern club-mosses (Lycopodiaceæ), and scarcely less gigantic forms (Calamites, Calamodendron, with Annularia, Asterophyllites, Sphenophyllum) belonging to the group of the horse-tails (Equisetaceæ).

Of Invertebrata other than insects the Devonian fauna is very rich in forms; but these show a marked similarity to those of the

---

* Referred by Hagen to the Neuroptera. The same authority considers the age of the rock formation in which the other insect remains have been found as more likely Carboniferous than Devonian.

preceding period. Approximately the same types are represented, and while in the case of certain families a diminution in the number and variety of their representatives is noticeable, in others there is a corresponding increase. The corals, echinoderms, cephalopods, and brachiopods have approximately the same value as in the Silurian period, and, indeed, several of the specific forms that are to be met with in the one formation are also seen in the other. Among the Devonian brachiopods we have the first appearance of the genus Terebratula, a form which has continued to flourish, although in constantly diminishing numbers, from that period down to the present time. With it are associated a number of other genera— Strophalosia, Productus, Uncites—which likewise appear here for the first time. The gasteropods are all of the holostomatous or round-mouthed type — Pleurotomaria, Murchisonia, Euomphalus, Loxonema, Holopea, Platyceras, &c.—while the lamellibranchs, which show a marked increase, both in numbers and variety of form, over their Silurian predecessors, belong, as far as it has been possible to ascertain, exclusively to the Integropalliata, or such as are devoid of a sinual inflection to the pallial line. The families represented are either Heteromyaria (Aviculidæ, Mytilidæ) or Dimyaria (Nuculidæ, Arcadæ, Astartidæ, Cardiidæ), no true Monomyarian being as yet known. We meet here with the first pulmonate, Strophites, a member of the modern family of snails (Helicidæ), and likewise with what appears to be the earliest unequivocal fresh-water invertebrate, a mussel of the genus Anodonta, or one very closely allied to it. It is here, therefore, that we have the earliest undoubted traces of a fresh-water formation.

The trilobites among Crustacea manifest a very rapid decline, and, indeed, in some regions they appear to have completely died out with the close of this period. The giant eurypterids—Eurypterus, Pterygotus, Slimonia — the most formidable of all known living and extinct crustaceans—which first appeared in the Upper Silurian formation, linger on into the succeeding Carboniferous period, when they forever disappear. The Devonian deposits, as has already been stated, contain the earliest remains of the highest order of crustaceans, the Decapoda, or ten-footers, which comprise the modern lobster and crab; the form in question (Palæopalæmon) belongs to the macrurous, or long-tailed division, and is allied to the shrimps.

The vertebrate life of this period exhibits a remarkable development as compared with that of the period preceding. But, as in the latter, all the remains belong to the class of fishes, and indeed principally, or one might say almost exclusively, to the same two orders, the elasmobranchs and ganoids. No animal of a grade higher than fishes had as yet appeared upon the scene, or, if possibly it had, no traces of it have thus far been discovered to indicate its existence there. To such an extent was the fish-fauna developed that the term " age of fishes " has not inappropriately been applied by paleontologists to designate this epoch of geological time. The preponderating types are the ganoids, which appear not only in forms that may be considered more or less remotely related to the type of the modern sturgeon (Macropetalichthys), or to the fringe-finned Polypteri of Africa (Holoptychius, Glyptolepis, Dipterus, Osteolepis), and the American alligator-gars (Chirolepis), but in such as have no representatives in any of the succeeding formations. These are the so-called "bucklered ganoids," which, in addition to the enamelled plates characteristic of this group of fishes, had the head and the anterior portion of the body encased in bony plates, more or less firmly united to each other, and serving as a protective armour. To this group, among others, belong Pteraspis, Cephalaspis, Pterichthys, and Coccosteus, forms which had their forerunners already in the Upper Silurian deposits. Generally placed among the ganoids, and closely related to Coccosteus, are the giant Dinichthys and Titanichthys, which appear to have attained a length of from twenty to thirty feet, and whose dental apparatus closely approximates that of the modern Lepidosiren, one of the lung-fishes (Dipnoi), a group of animals which effect a transition between the true fishes and the amphibians. If this relationship with Lepidosiren be absolutely established, as is claimed to be the case by many of the more prominent anatomists, then it is certainly significant that the advent of the Amphibia (the class of animals immediately above the fishes), in the succeeding Carboniferous period, is preceded by just that group which, in accordance with the principles and workings of evolution, we should expect to find interposed—the group which, on the one hand, combines some of the characters of the Amphibia, and, on the other, those of the ichthyic fishes. But, whatever the exact relationship of Dinichthys may be, there can be little question as to its representing a dipnoan type, or at least a transition form between the true ganoids and the lung-fishes. A similar position is occupied by some of the other crossopterygian fishes of the period, as Dipterus.

Carboniferous Fauna.—The life of the Carboniferous period is marked by two important features: 1. The introduction for the first time of vertebrate animal forms higher in the scale of organisation than the fishes, *i. e.*, the amphibians; and 2. The great development of strictly air-breathing or terrestrial animals. Of these last we have at least four distinct types indicated—the Gasteropoda, Insecta, Arachnida, and Myriapoda. Of the first, which have a solitary forerunner in the Devonian formation, we are acquainted with a comparatively limited number of forms (Pupa, Anthraco-pupa, Dawsonella, Zonites), all of them more or less closely related to forms still living at the present day. The insects comprise not only members of the low order of netted-veins, which are the only forms known to be represented in the Devonian deposits, but those of the more highly organised Orthoptera, and not improbably also Coleoptera (beetles), although most of the remains referred to the latter order are now positively known not to belong there. The Orthoptera comprise, among other forms, some sixty or more species of primitive cockroach, the Palæoblattariæ, which may be considered to represent the ancestral type of the modern social pest (Blatta), whose earliest appearance dates from the Triassic period. To the same order belong the giant walking-sticks recently brought to light from the coal-measures of France, the Titanophasma Fayollei, which measure in length (in one specimen) upwards of twelve inches, and are, therefore, by linear measure, very nearly the largest of recent as well as fossil insects.

This extraordinary development of a form, which may be taken to represent the extreme term of specialisation in an insect, in a period so early as the Carboniferous, is certainly not a little remarkable, and argues very strongly for the great antiquity beyond its own period of the origin of this class of animals. It is also not a little surprising that no representatives of the family of walking-sticks (Phasmida), other than those found in the Carboniferous deposits of France—Titanophasma and Protophasma—have as yet been found in a fossil condition, except such as may have been preserved in amber. Of the Neuroptera, the Haplophlebium Barnesii, from Nova Scotia, attained an expanse of wing of seven inches, nearly equal to the expanse of the largest of the living dragon-flies, to which it appears to have been related. The Carboniferous Arachnida comprise representatives of true spiders (Protolycosa, Anthracomartus), scorpions (Eoscorpius, Cyclophthalmus), and pseudo-scorpions (Microlabis). The scorpions appear to have attained a degree of specialisation very little below that of their modern representatives; but the true arachnids have all, or nearly all, segmented abdomens, and may be considered to mark a transition between the arthrogastric and anarthrogastric forms. The Myriapoda, which have a solitary forerunner in the Devonian rocks of Scotland (Forfarshire), are represented by both the cheilognathous and cheilopodous types, although on account of certain structural peculiarities the greater number of these earlier forms (Euphoberia, Xylobius, Trichiulus) have been constituted into a special order, the Archipolypoda.

Of the remaining invertebrate fauna of the Carboniferous period little need be said. The various groups of the tabulate and rugose corals (Lithostrotion, Syringopora, Cyathophyllum, Amplexus, Zaphrentis), the brachiopods, pteropods, lamellibranchs, gasteropods, and cephalopods, among the mollusks, and the crinoids and blastoids (Actinocrinus, Platycrinus, Cyathocrinus, Dorycrinus, Battocrinus, Pentremites, Granatocrinus) of the Echinodermata, have, as in the Devonian formation, abundant representatives; but they belong in considerable part to genera which now appear for the first time, or to such as had but a feeble development heretofore. The widely distributed group of trilobites, which, as has already been seen, played such an important part in the faunas of the Cambrian and Silurian periods, has here barely four generic representatives, Phillipsia, Proetus, Griffithides, and Brachymetopus, whose species occur in the main part in the deposits situated below the true coal.* With these forms the trilobites disappear forever from the scene. While the deposits of the preceding Silurian and Devonian formations have shown a fair representation of at least two of the primary groups of the Echinodermata, the Asteroidea and Crinoidea, especially of the latter, it is not until the present period that the urchins themselves (Echinoidea) acquire any significance (Archæocidaris, Palechinus, Melonites);

* Professor Claypole has latterly announced the discovery of Dalmania in the "Waverly group" (Lower Carboniferous) of Ohio.

and here, also, for the first time, if we except the Laurentian rocks, with the hypothetical Eozoon, do the Foraminifera appear to enter largely as rock constituents.   The genus Fusulina is developed to an extraordinary extent, and its distribution appears to be but little, if at all, less universal than that of the genera Nummulites and Orbitoides of the Eocene period.   A solitary forerunner of the Nummulites has been discovered in the Carboniferous rocks of Belgium.

In the remarkable development of the elasmobranch (shark) type of fishes, and in the absence of the bucklered ganoids, the Carboniferous ichthyic fauna is sharply defined from that of the Devonian.   With the exception of a considerable number of fin-spines or ichthyodorulites, referred to such genera as Ctenacanthus, Gyracanthus, Oracanthus, &c., whose position is still very doubtful, and which may in part belong to the order of ganoids, all the remains of the former appear to have been more or less nearly related to the modern Port Jackson sharks.   These remains are in the main in the form of teeth—Psammodus, Helodus, Orodus, Chomatodus, Petalodus, Cochliodus—whose (somewhat distant) resemblance to the pavement teeth of the cestracionts has led to their reference to members of that group ; not impossibly, however, they represent a very distinct type.*   The ganoids comprise, in addition to polypteroid forms—Cœlacanthus, Rhizodus, Megalichthys—representatives of the rhomb-plated Lepidosteidei, which include the American alligator-gar.   The most widely distributed and most abundantly represented genera are Palæoniscus and Amblypterus, the former of which is also one of the most abundant fishes of the succeeding Permian period.

The only vertebrates other than fishes which appear in the Carboniferous period, and now appear for the first time, are the Amphibia, that group of animals whose members stand immediately next above the fishes in the scale of organisation, and whose embryonic forms are so clearly ichthyic as to have necessitated the union of the two classes into the one comprehensive division of the Ichthyopsida.   It is not a little significant that the appearance of

---

* Mr. Garman has recently described a species of shark from the Japanese seas, Chlamydoselachus anguineus, which appears to be generically most intimately related to the Carboniferous Didymodus, and which, accordingly, represents about the most ancient type among living vertebrates.

these animals should have been foreshadowed in the Devonian dipteroid ganoids, which, leading up to the lung-fishes on one side,
and not impossibly directly to the amphibians on the other, effect
a transition to the higher class from the side of the fishes. This
succession of higher upon lower types is not a matter of accident,
but a direct outcome of the inevitable laws of evolution. Through
the application of no other law would the numerous accidental or
coincidental occurrences of direct succession, which present themselves throughout the entire geological series, receive an intelligent
explanation. All the Carboniferous amphibians belong to the extinct order of the Labyrinthodontia (Stegocephala), salamandroids
of both minute and gigantic frame, whose members were distinguished by the possession of a dermal (cephalo-dorsal and ventral)
armour of sculptured plates, and in many cases by a peculiar labyrinthine infolding of the enamel of the teeth, a structure unknown
among modern amphibians, but which is in great part shared by
certain members of the ganoid fishes, as the modern alligator-gars
(Lepidosteus) and the genus Rhizodus (Carboniferous). Among
the genera are Anthracosaurus, Hylerpeton, Dendrerpeton, Batrachiderpeton, and the cæcilian-like Dolichosoma and Ophiderpeton.
No other vertebrate higher in the scale of organisation than these
labyrinthodonts is as yet apparent, unless, possibly, the very doubtful Eosaurus be proved to be a true reptile.

The flora of this period partakes essentially of the character of
that of the period preceding, the Devonian. We have here the
same ancient representatives of the modern club-mosses and horse-
tails, the Lepidodendra and their allies,* and the calamites, the
ferns—Neuropteris, Pecopteris, Alethopteris, Sphenopteris, Cyclopteris—giant tree-ferns, and forms that have been referred to the
group of the cycads, an order of plants to which the sago-palms
belong, and which appear to be not distantly removed from the
conifers. No positive indications of the existence of any true
flower-bearing herbaceous plants are yet manifest, and with their
absence the total absence of flower-frequenting or nectar-sucking

---

* The recent anatomical investigations of Renault and Saporta have led
these authorities to consider Sigillaria, at least in some of its recognised forms,
to be much more closely related to the gymnospermous phanerogams than to
the club-mosses; but Professor Williamson has pretty definitely shown that
such a relationship does not exist.

8

insects, the Lepidoptera and Hymenoptera, the two most highly organised orders of insects, is noticeable. The only true trees, or such as are made up principally of woody tissue, of the Carboniferous deposits belonged to the coniferous series, the order of plants which embraces the modern pine and its allies. These ancient evergreens were represented by several distinct genera— Dadoxylon, Palæoxylon, Pinites—which, if the fossil fruit associated with their remains, and known as Lepidostrobus, be justly attributed to them, had their nearest allies among their modern congeners in the berry-bearing yews. No deciduous leaf-bearing trees, such as the oak, beech, or maple, which make up the great mass of our forest growths, can be positively shown to have existed in these early days.

Permian Fauna.—In the formations of the period succeeding the Carboniferous, the Permian, a considerable advance in the structural type is indicated by the animal remains. While the predominant forms of life of the period preceding pass, although in most cases with very diminished numbers, into the present one —in fact, to such an extent as to have induced many geologists to unite the formations of the two periods into a common whole—we meet here with a class of animals whose representatives had not hitherto been detected. These are the true reptiles, most of whose members belonged to the order Theromorpha (Pelycosauria), reptilian forms which in several important characters—the structure of the pectoral and pelvic girdles, humerus, and tarsus—show strong affinities to the lower orders of mammals, the Monotremata and Edentata, of which, not impossibly, they may prove to be the early progenitors. A further approximation to mammalian structure is found in the character of the dentition, which in many forms exhibits a distinct differentiation into incisor and canine teeth. The deposits of the Southern and Western United States, especially of the State of Texas, have yielded a wealth of species and genera belonging to this order (Theropleura, Dimetrodon, Diadectes, Empedocles, Clepsydrops), representative of several distinct families. The modern type of lizards had their nearest analogues in the monitor-like Proterosaurus (Germany, England), whose dentition, however, was of the crocodilian type (thecodont). These early reptiles, while exhibiting many points of structure indicative of comparatively high specialisation, yet clearly proclaim a primitive type of organisation in the rudimentary or embryonic condition of the vertebral column, which is in most cases only partially ossified.

The Amphibia of the Permian period are by most authors placed in the single group of the Labyrinthodontia, although in certain structural departures from the normal type, as in the very rudimentary condition of the vertebral column, and in the absence of the peculiar labyrinthine infolding of the enamel of the teeth, some of the forms may have to be separated from this order. Most of the species were provided with a tail of greater or less length, and the general resemblance to living amphibians appears to have been mainly with the salamandoids, although in several points of structure they more closely approximate the tailless frogs and toads. The relationship with the plated ganoids is well pronounced, and not improbably some of these, as dipneusts, or double breathers, may have been their true ancestors (as well as of the lung-fishes proper, Dipnoi). Among the more prominent genera are Branchiosaurus, Melanerpeton, Urocordylus, Archegosaurus, Eryops, Palæosiren, and Ophiderpeton, the last two apparently apodal, and recalling the cœcilians in outline. In Eryops megacephalus, the largest of American amphibians, from the Permian of Texas, the skull measures eighteen inches in length and twelve inches in breadth.

The fish-fauna of this period partakes essentially of the character of the fauna of the period preceding, from which it has borrowed most of its types. We have here, however, the first unequivocal remains of the genus Ceratodus (Bohemia and Texas), which represents the most ancient generic type of all existing Vertebrata.

Regarding the invertebrate fauna of the period, it may be remarked that a deficiency in the number of forms is noticeable in nearly all the localities where the Permian deposits are developed, a circumstance due to the peculiar physical conditions under which the deposits were formed, and the subsequent alteration, resulting in the obliteration of the contained organic remains, to which, in many places, the rock-masses were subjected. A very large proportion of the known fossils are, as has already been intimated, of clearly Carboniferous types, more especially in the case of the Mollusca. The trilobites, so characteristic of the earlier deposits of the Paleozoic era, are wholly wanting, not a single individual, apparently, of this very numerous order having survived the Carboniferous period. Here, also, we have the almost final disappearance of the two great groups of the rugose and tabulate corals, which by their numbers so eminently characterise the limestone deposits of the Paleozoic series, from the Silurian to its close. The Permian flora is essentially that of the Carboniferous period, and requires no special consideration.

Paleozoic Faunas.—Briefly reviewing the more salient features of the Paleozoic faunas, we find, as far as the invertebrate series is concerned, that with few exceptions all of its recognised classes have their representatives, or, at least, there are representatives of nearly all those classes whose members could reasonably be expected to have been preserved in a fossil state. Thus, of the Protozoa we have the Foraminifera and Spongida; of the Cœlenterata, the Actinozoa and Hydrozoa; of the Echinodermata, the Echinoidea, Asteroidea, Ophiuroidea, Crinoidea, Cystidea, and Blastoidea ; of the Mollusca (and Molluscoida), the Polyzoa, Brachiopoda, Acephala, Pteropoda, Gasteropoda, and Cephalopoda; and of the Articulata, the Crustacea, Arachnida, Myriapoda, and Insecta. Of the classes here enumeratedthere are wanting in the Cambrian the Actinozoa,* and possibly also the Hydrozoa; the Echinoidea, Blastoidea, and Ophiuroidea, among the echinoderms ; and the Arachnida, Myriapoda, and Insecta, among the articulates. In the Silurian the number of missing classes is reduced by six, since we have here representatives of both corals and hydroids, blastoids and brittletars, insects and arachnids; but the last two are represented almost by single individuals. In the Devonian the number is further reduced by one, the class of the Myriapoda, likewise (as is also the case with the insects) represented in almost solitary individuals; only with the Carboniferous period do all the classes acquire for the first time any marked development. We thus cannot fail to remark the progressive evolution of new forms correlatively with the advance of time. Of the vertebrate series the Paleozoic deposits contain the remains of only three of the five recognised classes, the fishes, amphibians, and reptiles, which appear serially in the order of their progressive organisation, the lowest, or fishes, in the Silurian (or, if the conodonts be fishes, in the Cambrian), the amphibians in the Carboniferous, and the highest, or true reptiles,

* Some forms have been doubtfully referred to this period in Scandinavia.

in the Permian. These ancient deposits have as yet yielded no traces of either birds or mammals.

Triassic Fauna.—In the first of the Mesozoic series of formations, the Triassic, we enter, as it were, upon an entirely new phase of organic development. Many of the more characteristic groups of organisms of the preceding era have now either completely disappeared, or only survived in such diminished numbers as to constitute but a very insignificant element in the new fauna. Thus, of the class Brachiopoda but comparatively few of the older generic types are represented; and the same may be said of the other classes of mollusks, and more especially of the Cephalopoda. Of all the various forms of Paleozoic tetrabranchiates there are barely more than a half-dozen surviving types, and of these one, Orthoceras, itself becomes extinct in this period. But, in the place of these ancient types, we have others of the same class which are no less conspicuous for their numbers than for the complexity of form which they subsequently attain, and some of which exhibit a marked advance upon their predecessors in the scale of organisation. The ammonites, whose advent appears to have been foreshadowed in the goniatite and the Devonian Clymenia, now for the first time acquire any importance, and, indeed, if we except certain forms from the Carboniferous deposits of India and Texas—Arcestes, Xenodiscus, Sageceras, Medlicottia—now for the first time appear altogether. The numerous species which in some districts, more especially in the region of the Alps, crowd the deposits of this age, belong in principal part to the families Arcestidæ and Pinacoceratidæ, as representatives of the leiostracous, or smooth-shelled division, and the Tropitidæ, Ceratitidæ, and Clydonitidæ (with the somewhat aberrant genera Cochloceras, Rhabdoceras, Choristoceras, and Clydonites), of the Trachyostraca, or forms with strongly sculptured shells.

In these deposits, also, we meet in the Belemnitidæ with the first unequivocal traces of the dibranchiate or two-gilled, cephalopods, which, if we except the Nautilus, alone of this class of mollusks inhabit the seas of the present day.* The rugose and tabulate

---

* The view entertained by several eminent paleontologists, that the ammonites themselves represent dibranchiate forms, requires further support before it can be fully accepted. The evidence at the present time appears to be fully as much, if not more, opposed to this notion as it is in favor of it.

corals have been succeeded by the modern type of the star-corals, Zoantharia perforata and aporosa (Montlivaltia, Thecosmilia, Isastræa, Thamnastræa, &c.), whose fragmentary remains build up giant reefs (Alps); and, similarly, the more distinctive ancient group of the Echinodermata, the crinoids, whose most characteristic representatives at this period are Encrinus and Pentacrinus, find their successors largely in the more modern Echinoidea, or true urchins (Cidaris, Hemicidaris, Hypodiadema).

It is, however, in the vertebrate fauna that we find the most prominent feature separating the life of this period from that of any of the periods preceding. Not only do we meet here with the remains of fishes, amphibians, and reptiles, but with those of mammals, and not improbably also with the impressions or tracks of birds. Granting these last, which are, however, a little uncertain, it may be assumed that all the classes of the animal kingdom, as now recognised by naturalists, had their representatives. The fishes are still principally referable to the predominant type of the periods preceding, the ganoids, which also in a measure retain the embryonic heterocercal tail, although a tendency towards homocercality is observable in some of the genera, as in Semionotus. The more numerously represented forms—Ischypterus, Catopterus, Semionotus —belong to the group typified in the American gar-fishes, and may be looked upon as the direct descendants of the Carboniferous and Permian Palæonisci. The lung-fishes find an abundant representation in the teeth of Ceratodus, which, as has already been seen, dates from the Permian, and possibly from a still older period. This animal furnishes us with one of those rare instances where a genus of living vertebrates has been founded upon the fossil remains.

The amphibians of the Triassic period show but little advance over the type of their predecessors, all the forms still belonging to the single order Labyrinthodontia, some of whose members attained to prodigious dimensions (Mastodonsaurus, Labyrinthodon). To this group are probably referable the singular handshaped impressions of the animal known as Cheirotherium, or " hand-beast," originally supposed to have been an animal of the frog-type, but now assumed to have been a salamandroid, or animal allied to the newts, and, like them, provided with a tail, although possessing in the structure of the skull certain features belonging to the tailless amphibians—frogs and toads. The true reptiles exhibit a remarkable variety of form, and, as in the succeeding Jurassic and Cretaceous periods, constitute the most marked feature of the faunal remains that have been left to us. Hence, by some geologists the collective era of these three periods—the Triassic, Jurassic, and Cretaceous—or what is generally known as the Mesozoic, has been designated the "era" or "age of reptiles." The modern lizards and crocodiles had both their ancient representatives, the former as indicated by the genera Telerpeton, Hyperodapedon, and Rhynchosaurus, and the latter by Stagonolepis, Belodon, and Parasuchus. But besides these there flourished a multitude of reptiles belonging to several distinct and very widely removed orders, which have left, to our knowledge, no traces whatever of their existence in the present seas. Such are the South African Anomodontia, some of whose members, as Oudenodon, were totally destitute of teeth, and had their beaks encased in horn, after the fashion of the modern turtles (of which they may have been in part the progenitors); while others, as Dicynodon, possessed the horny mandibular apparatus of the former, but were provided, in addition, with a pair of huge and powerful teeth in the upper jaw; * the Theriodontia (as represented by the South African Galesaurus), reptiles whose dentition partook of the character of that of the ordinary Carnivora, and whose earliest types had already appeared in the deposits of the Permian period; and the Plesiosauria, a group of essentially sea-inhabiting reptiles, which acquired a very considerable development in the Jurassic seas, and whose best known exponent is the Plesiosaurus. In the animals of this order the extremities of the limbs, both anterior and posterior, were encased in integument, and thus converted into flippers, very much like those of the whale, and admirably adapted for propulsion through the water. The most characteristic genera of this period were Nothosaurus and Simosaurus. In Placodus the dental armature consisted in principal part of flattened plates, resembling the teeth of the pycnodont fishes, with which animals these reptiles were first confounded.

* Professor Judd has quite recently ("Nature," October 15, 1885), announced the discovery of dicynodont remains in the Elgin Trias of Scotland; the group of animals had hitherto been known only from Africa, India and the Ural Mountains.

The most remarkable of all the Triassic reptiles are the Dinosauria, a group of the greatest importance when viewed from a teleological standpoint by reason of the many structural characters which separate them from the typical reptiles, and approximate them to birds. These avian characters are indicated principally in the structure of the powerful pelvic girdle and hind limbs, which depart very broadly from the normal type of reptilian structure. Thus, the pubic bones, in many cases, instead of projecting forwards as in other reptiles, are directed backwards, more nearly parallel with the ischium, both bones therefore taking a position directed towards the posterior portion of the body, a feature characteristic of birds. In the hind limb, again, the ornithic characters are seen in the great cnemial ridge which is developed on the tibia, the gradual diminution of the fibula towards the distal extremity, the structure of the astragalus, and in the disposition of the digits, three or more, and their accompanying phalanges. The inner and outer digits are shorter than the rest, or quite rudimentary, and the third toe, as in birds in general, is the longest. There are good grounds for concluding that the bones of the limbs, and, doubtless, if this was the case, of some of the other portions of the trunk, were permeated with air-passages, as in birds. The structure of the fore limbs is still only imperfectly understood, but there is no doubt that in many cases they were but very feebly developed, being very much shorter than the hind limbs, and that progression was, either habitually, or at least at times, effected by means of the posterior appendages alone. The remains of these earliest dinosaurs are indicated both by the actual parts pertaining to the skeleton (Zanclodon, Thecodontosaurus, Amphisaurus, Clepsysaurus, Bathygnathus), and by the foot-prints, many of them three-toed, that have been left implanted in the rock-masses. Some of these, which measure fully a foot, or even considerably more, in length, were originally supposed to represent the imprints made by the feet of giant birds, a suspicion strengthened by the serial arrangement in twos in which the tracks are disposed; but now that the structure of the dinosaurs has been more accurately determined, and their ornithic characters and mode of progression recognised, there can be little doubt that they represent the imprints of the reptiles belonging to this order. This view is further strengthened by the circumstance that no actual remains of birds have as yet been discovered in deposits of Triassic age.* But it is not impossible, or even improbable, that some of the smaller foot-prints that are scattered about the larger ones, and which in some instances are disposed in a single series one in advance of the other, indicating a method of progression adopted by certain wading birds, may actually be of an ornithic nature. However this may be, it is certainly a significant fact bearing upon the doctrine of evolution, that no unequivocal traces of birds have thus far been discovered in deposits antedating those which contain the remains of reptiles, which in their several characters most approximate the birds, and in reality effect a transition to them. The progressive evolution of advanced or most specialised types is here clearly indicated.

The Mammalia, the highest class of vertebrates, appear for the first time in the deposits of this age. They are indicated by the teeth and fragments of jaws pertaining to two or three genera, Dromatherium, Microlestes, and Hypsiprymnopsis, forms, as nearly as can be determined, belonging to the low type of the Marsupialia, and, probably, more or less closely allied to the modern banded ant-eater (Myrmecobius) and kangaroo-rats (Hypsiprymnus) of Australia.

Jurassic Fauna.—The life-history of the Jurassic period, while combining certain prominent features not hitherto recognised, presents to us primarily an expansion of those characters with which we have just become acquainted. The remarkable group of the dinosaurian reptiles, whose development in the Triassic period had but barely passed beyond its own beginnings, acquires here renewed importance, apart from the mere matter of numbers, from the circumstance of the gigantic and diverse forms which it includes. Four distinct types of this order are recognised, all of which had representatives in the Jurassic period: 1. The Sauropoda, lizard-footed vegetable-eaters, in which the anterior and posterior pairs of limbs were of nearly equal length, and whose progression was effected on all fours. Among the more important genera of this period belonging to the group are Atlantosaurus, Brontosaurus, Morosaurus, and Cetiosaurus, the first, from the deposits of the

* Since the above was written announcement has been made of the discovery of the skeletal remains of a track-making dinosaur of the Connecticut Valley. Trans., New York Ac. Sciences, Oct. 26, 1885.

Rocky Mountains, measuring from eighty to one hundred feet in length—the largest land animal with which we are acquainted. 2. The Stegosauria (Stegosaurus, Scelidosaurus), armoured vegetable-feeding dinosaurs, some of them of gigantic frame, whose progression, owing to the feeble development of the anterior pair of limbs, appears to have been in great part effected by means of the hinder extremities alone. In Scelidosaurus Harrisoni, from the Lias of Dorsetshire, the hind foot measured three feet and a half in length. 3. The Ornithopoda, bird-footed herbivores, with a very unequal development of the anterior and posterior appendages, the latter closely approximating the structure found in birds. There can be but little question as to the habitually erect posture assumed by such forms as Camptonotus, Laosaurus, and Iguanodon. Iguanodon Mantelli, a Cretaceous species, measured about thirty feet in length from the tip of the nose to the extremity of the tail. No member of this genus is known from the American deposits. 4. The Theropoda, carnivore forms, whose progression was largely erect, and assisted in many cases, probably, by the greatly developed tail acting as a fulcrum, in the manner of that organ among the kangaroos. This type, which is almost alone represented in the Triassic deposits, includes the most formidable members of the order—Megalosaurus, Allosaurus, Dakosaurus, the former apparently attaining a length of fifty feet. The genus Compsognathus, represented by a single species (C. longipes) from the Upper Oolite, possesses probably the greatest number of avian characters of the entire order, and is considered to stand in the direct line of the descent of birds.

Monsters parallel to those of the land-surface inhabited the oceanic waters, such as the finned Plesiosaurus and Pliosaurus, and the not distantly removed Ichthyosaurus and its American toothless ally, Sauranodon. The geographical distribution of Ichthyosaurus is a very remarkable one. While apparently the genus is completely wanting in the deposits of the New World, its range in the Eastern Hemisphere embraced very nearly its whole north and south extent, from Spitzbergen (Ichthyosaurus polaris, Triassic) to Australia (I. australis, Cretaceous). Its greatest development appears to have been in the early part of this period (Lias). From the discovery of fragmentary parts of young individuals within the bodies of more fully developed ones, it has been conjectured that the animal was viviparous, a supposition in a measure strengthened by the ill-adaptation of its structure to breeding on the land-surface. Not impossibly, however, these animals may have been in the habit of devouring their young, or the young of allied species, as it seems many species of snake do at the present day.

In the Jurassic rocks we meet with the first traces of that extraordinary group of reptiles, the Pterosauria, which, in the possession of a tegumentary membrane stretched between the greatly elongated outer digit of the anterior limbs and the bases of the hinder extremities, resembling in many respects the flying-apparatus of bats, were enabled to navigate the air in the manner of birds. To these last, the pterodactyls, as the members of this order are familiarly designated, were closely related in the general conformation of the skull, the pneumaticity of the bones, and the presence of a well developed keel to the sternum or breast-plate, a character among recent animals found only in birds and bats. But while possessing these and other avian features, the pterodactyls depart in many important particulars from the bird type, and notably in the presence of true teeth implanted in sockets, as in the Crocodilia, the structure of the manus, the absence of a feathery integument—the animal having been apparently provided with a naked skin—and the possession of a tail composed of distinct vertebræ.[*] Despite these important differences, however, it may, perhaps, be deemed doubtful whether the animals in question have not as much right to be considered birds as reptiles, the more so as the one great feature separating them from modern birds, the presence of alveolar teeth, has recently been shown to be characteristic of some, if not of most, of the ancient birds. While, therefore, it may not be possible to decide upon the exact position occupied by these singular organisms, there can be but little doubt that they, or possibly some closely-allied predecessors with which we are not as yet acquainted, represent the primitive stock whence the type of the modern flying or carinate bird has been evolved. The birds would then have a double line of ancestry, the one here indicated, and another, culminating in the struthious or non-carinate

---

[*] Professor Marsh has shown that at least in some forms of pterodactyls (Rhamphorhynchus) the extremity of the tail was provided with a tegumentary expansion, or vertical rudder, by means of which the animal doubtless guided its flight.

birds, having its origin in the dinosaurian reptiles, and in a form possibly not distantly removed from the Jurassic Compsognathus. The more important Jurassic genera of Pterosauria are Pterodactylus, Rhamphorhynchus, and Dimorphodon, which differ from each other mainly in the character of the dentition and in the relative development of the tail.

The deposits of this age have yielded, in addition to the reptilian forms mentioned, the remains of true lizards, crocodiles, and turtles. The first are but sparingly represented, and in the main, or wholly, belong to the acrodont division (Geosaurus, Acrosaurus, Homœosaurus, the last closely related to the modern Lacerta). The crocodiles, with some partial exceptions (Streptospondylus, Theriosuchus), belong to the primitive amphicœlous division (Teleosaurus, Mystriosaurus), or those in which there is a retention of the ichthyic character of bi-concave vertebræ, as distinguished from the more modern forms dating from the Cretaceous period, with procœlous (concavo-convex) vertebræ. The turtles, which appear here for the first time, exhibit a remarkable differentiation, and in their diverse forms comprise representatives of several of the more important modern groups, as the Chelydæ (Plesiochelys, Craspedochelys, Pleurosternon), Emydæ (Thalassemys), and Chelydridæ. The recent genus Chelone is found in the Purbeck beds.

The most interesting addition to the fauna of this period is furnished by the bird remains, whose earliest unequivocal traces are found in the famous Archæopteryx of the Solenhofen slates of South Germany (Bavaria), and in Laopteryx, from the deposits of Wyoming Territory—the last a bird probably of about the stature of a crane, but with uncertain affinities. The Archæopteryx, which is known by two more or less well-preserved specimens, and a feather pertaining to a third individual, combines in a most extraordinary manner what have generally been considered distinctively avian and reptilian characters, and, indeed, the animal may be regarded as a type intermediate between the two classes. Thus, in many points of structure of the skull and trunk, no less than in the structure of the tail, which was greatly elongated and made up of numerous distinct vertebræ, it is decidedly reptilian, and this relationship to the class of animals next lower in the scale of organisation is borne out by the discovery, recently made by Professor Dames, that the extremities of both jaws were provided with a number of diminutive teeth implanted in sockets. But, on the other hand, the animal, which was of about the size of a rook, was provided with powerful wings, and these wings no longer consisted simply of a tegumentary membrane, as in the case of the bats and the extinct pterodactyls, but were made up of feathers as in living birds, a character indicating that the animal provided with them was warm-blooded. Feathers were also developed in pairs on either side of the tail; but the rest of the body, according to Vogt, appears to have been completely naked.

The mammalian remains of the Jurassic period consist principally of teeth and jaws, in a more or less complete state of preservation, whose characters indicate animals of diminutive size, pertaining wholly or in principal part to the order of the Marsupialia, or pouched animals. Both the insectivorous and the herbivorous types had their representatives, the former in such genera as Amphilestes, Phascolotherium, and Amphitherium, and the latter in Plagiaulax; and it is not impossible that the more strictly carnivorous type of marsupial also then existed. Stereognathus, further, presents us with the type of a hoofed herbivore, and points to a possible origin of the modern placental ungulata. Thus, the marsupials had, as early as this period, attained a considerable degree of differentiation, though apparently less considerable than that exhibited by them at the present time. Latterly, some naturalists, and notably Professor Marsh, have attempted to show that these most ancient mammals of the Triassic and Jurassic periods were not true marsupials, as these are now recognised, but that they constitute distinct orders, Allotheria and Pantotheria, apart by themselves; there do not appear to be sufficient grounds, however, for the separation here proposed.

In the invertebrate fauna of this period we see, even more than in the vertebrates, the reflection of the fauna of the period preceding, but with the predominant features very largely extended. These are in the main constituted by the mollusks, and more particularly by the cephalopods, lamellibranchs, and gasteropods, the brachiopods (Rhynchonella, Spiriferina, Terebratula, Terebratella), although still sufficiently abundant, no longer having that paramount importance which distinguished them as perhaps the most distinctive type of the Paleozoic faunas. The cephalopods still belong, in the main, to the types of the nautilus, ammonite (Amaltheus, Arietites, Harpoceras, Ægoceras, Stephanoceras, Lytoceras, Phylloceras), and belemnite ; but representatives of groups that appear to have been closely related to the modern calamary (Belo- teuthis, Belemnosepia, Teuthopsis) are not exactly wanting.   The lamellibranchs and gasteropods comprise a most varied assemblage of forms, many of them but barely distinguishable from individual forms living at the present day, and by their great numerical de- velopment give a generally modern aspect to the fauna.   Of the former the modern families Ostreidæ, Limidæ, Mytilidæ, Astartidæ, Lucinidæ, and Cardiidæ are remarkable for their profuse develop- ment, and scarcely less so the now nearly extinct Trigoniadæ and Pholadomyidæ; of the latter, the more important families are still the non-siphonated ones (Pleurotomariidæ, Naticidæ, Trochidæ, Actæonidæ); but a no inconsiderable representation of the Siphon- ata (Cerithiidæ, Aporrhaidæ, Strombidæ, Purpuridæ) is also inter- spersed.   The earliest fresh-water univalves belong to this period (Paludina, Melania, Neritina, Planorbis).

The corals, which are of the type of existing star-corals, may be considered next in importance to the Mollusca, and, indeed, in some instances, as in the Coralline Oolite, they constitute by their own vast numbers the greater portion of the solid rock, not im- probably the vestiges of ancient reefs.   Somewhat less important, but yet very abundant in certain localities, are the fragments of the Crinoidea, which are most distinctively represented by the genera Apiocrinus and Extracrinus, the latter having its modern analogue in the Pentacrinus of the Carribean Sea; but this comparatively little specialised group of the Echinodermata has, ever since the close of the Carboniferous period, been on its decline, and has left its place to be filled by the true urchins (Cidaris, Hemicidaris, Holectypus, Echinobrissus, Clypeus, Collyrites) and brittle-stars (Ophioderma, Ophiurella, Ophioglypha), both of which, but more particularly the former, now for the first time acquire any special importance.   The star-fishes are represented, among other forms, by the type of the modern Uraster.   Of the articulates there is a considerable development of the Crustacea—crabs, lobsters, and their allies—and in the fine-grained rocks the remains of centipedes, spiders, and true insects are not uncommon, the last comprising representatives of all the recognised modern orders.   To this period belong the earliest Diptera, Hymenoptera, and Lepidoptera.

Turning to the flora, we find it to be sharply defined from that of any of the Paleozoic periods, although in the abundance of ferns, many of them of ancient type, and in the absence of the higher forms of plants, it shows an interesting correspondence. Its most marked feature is furnished by the group of the Cycads, of which there are numerous genera recognised, and the pines, whose nearest allies appear to be the southern araucarias. The earliest undoubted representatives of endogens are found in the deposits of this age, some of them clearly indicating a close relationship with the Australian screw-pines (Pandanus). No positive traces of exogenous plants other than conifers have as yet been determined, but it is by no means improbable that they already existed.

Comparing the fauna and flora of the Jurassic period with the existing fauna and flora of any portion of the earth's surface, we remark a striking similarity to the conditions presented on the Australian continent. Here, at the present day, is the home of the marsupials, of the Port-Jackson shark, which had its Jurassic representatives in genera like Acrodus, Hybodus, and Strophodus, and of Ceratodus among the lung-fishes, a form which, though of more ancient date, also had its habitat in the seas of the Jurassic period. Only along the Australian coast do we meet at the present time with the lamellibranchiate genus Trigonia, one of the most characteristic and abundant of the Jurassic mollusks. As regards the flora, a no less striking correspondence is apparent. On the Australian land-surface flourish a considerable variety of ferns, tree-ferns, and cycadaceous plants ; likewise, the Araucaria type of Conifer; and here, principally, do we find the singular plants already referred to as screw-pines (Pandani). Australia is, in fact, that portion of the earth's surface which, as far as its faunal and floral characteristics are concerned, has undergone the least modification since the Jurassic period, and, indeed, it may be said that the present fauna and flora of the continent became differentiated during the interval between the Triassic and Jurassic periods, although, as has already been seen, some of the distinctive types date from a more ancient epoch. The retention of an ancient type of fauna and flora clearly indicates that the continent had retained its isolated position through a period probably extending as far back as the Mesozoic era ; otherwise, if connection with some other continental land-mass had existed at some subsequent period, it

would be barely possible that an interchange between the faunal
and floral characters of that land mass and Australia should not
have been effected to a greater extent than is indicated by the
isolated position, especially of the fauna, now existing.

Cretaceous Fauna.—The Cretaceous period presents in many
respects a marked contrast in its faunal characters to the period
preceding. While many of the most important or characteristic
of the Jurassic invertebrate types still persist, in many cases with
undiminished or even increased force, as, for example, the different classes of the Mollusca, we meet here with a development
of other animal groups which in most of the periods preceding
were of comparatively insignificant import. Such are the Foraminifera, which in their various forms (Globigerina, Rotalia, Textularia, Cristellaria) build up by their remains the great mass of
the chalk rocks, whose enormous extension is one of the most imposing monuments presented to the geologist. The sponges (Si-
phonia, Jerea, Thecosiphonia, Ventriculites) here likewise find their
greatest development, some of the forms having their analogues in
the types that still inhabit the oceanic depths; and the same has
been shown to be the case with the Cretaceous urchins (Echinoidea),
which are represented in great multitude and variety—Cidaris,
Ananchytes, Galerites, Micraster, Discoidea. The corals are in
comparison feebly developed, and can by no means claim that importance which they obtained in the Jurassic period. The Belemnitidæ (Belemnites, Belemnitella) and Ammonitidæ still constitute
the most important of the cephalopod types, the latter especially
presenting a very considerable number of characteristic forms, the
so-called unrolled ammonites—Crioceras (with Ancyloceras and
Toxoceras), Hamites (Ptychoceras), Scaphites, Turrilites, Helico-
ceras, and Baculites.

The bivalve and univalve faunas, while largely made up of
Jurassic types, show a marked advance over the corresponding
faunas of the period preceding in the much greater development of
the siphonate forms. The Sinuata among the former, which, if we
except the very abundant family of the Pholadomyidæ, had hitherto
but scattered representatives, now acquire considerable importance,
especially in the families Veneridæ, Tellinidæ, Glycimeridæ, Ana-
tinidæ, Mactridæ, and Myidæ. Among the non-sinuate forms
the members of the oyster family (Ostrea, Exogyra, Gryphæa) and

the scallops (Pecten), and the genus Inoceramus among the Heteromyaria, are distinguished by their numbers; but the most characteristic elements of the lamellibranch fauna are furnished by two families of very inequivalve-shelled mollusks, the Chamidæ, with the genera Requienia, Monopleura, Caprina, and Caprotina, and the so long misunderstood Rudistæ (Sphærulites, Radiolites, and Hippurites), whose forms so eminently characterise the southern belt of European and American Cretaceous deposits, and which appear and disappear with this period. The siphonate univalves have an almost exclusively modern aspect, and comprise among others representatives of the families Fusidæ, Strombidæ, Muricidæ, Tritonidæ, Buccinidæ, Cancellariidæ, Pleurotomidæ, Conidæ, Olividæ, and Cypræidæ.

Turning to the vertebrates, we find in the lowest class, Pisces, the introduction for the first time of teleosts, or true bony fishes, that ichthyic group which at the present day surpasses, both in individual members and variety, all the other orders of fishes put together. These earliest teleosts, although not very abundant, comprise a considerable number of modern types (Clupea, Esox, Osmerus, Beryx); but it is not till the Tertiary period that they acquire any well-marked development. No amphibian remains have been detected in any Cretaceous deposit. Reptiles, on the other hand, are exceedingly abundant, and comprise most of the types whose existence has been indicated in the Jurassic seas. Thus, of the modern groups, we have turtles, lizards, and crocodiles (of both the amphicœlous—Hyposaurus—and procœlous types —Holops, Gavialis), and, in addition, the first true serpent (Simoliophis). The extinct orders Ichthyosauria and Plesiosauria are still represented, and in Elasmosaurus, belonging to the latter, we meet with one of the most formidable types of the finned Reptilia. Here, also, are found some of the most gigantic of the Dinosauria — Iguanodon, Megalosaurus, Hadrosaurus, Camarasaurus—and the remarkable group of the Pythonomorpha, or "sea-serpents"—Mosasaurus, Leiodon, Clidastes—which in several respects united the characters of both serpents and lizards. The largest of the pterodactyls, or flying reptiles, having an expanse of wing of from twenty to twenty-five feet, or even more, occur in deposits of this period, and are represented by the normal-toothed types, and by such, as the American Pteranodon, in which the jaws appear to have been encased in horn, and to have been entirely edentulous.

Bird remains are sufficiently abundant in certain localities, many of them belonging to forms seemingly not very far removed from some of our modern groups. But, in addition to these ordinary forms, we have some of the most extraordinary of any that have ever been described, and which, from the presence of true teeth in their jaws, have received the name of Odontornithes (toothed-birds). In the genus Ichthyornis, as exemplified in I. dispar, which was of about the size of a pigeon, in addition to the peculiarity of alveolar teeth that of biconcave vertebræ is presented, a structure of the vertebral column characteristic of fishes and many of the extinct reptiles, but not known in modern birds. The wings appear to have been well developed, and in this, and all other respects beyond those just mentioned, the animal conformed strictly to the modern type of bird structure. In the still more remarkable Hesperornis, which in the species H. regalis attained a height of five or six feet, the teeth, instead of being implanted in distinct sockets, were placed in a continuous groove; the extremity of the upper jaw appears to have been bent down in the form of a beak, and to have been edentulous. The breastplate was entirely destitute of a keel or ridge for the attachment of the powerful muscles required for the motion of the wings, so that the bird was doubtless completely denied the power of flight. The presence, in the same geological period and the same geographical area (Kansas), of two birds so closely related to each other in the presence of jaw-teeth, and yet so distantly removed from each other by other peculiarities of structure, argues strongly for the antiquity of this class of animals, and, though the earliest unequivocal traces of birds have thus far been met with in the deposits of the Jurassic period, it is more than probable that their first origin is considerably more ancient.

No traces of any mammalian have thus far been discovered in any indisputably Cretaceous deposit, a circumstance in great part attributable to the particular conditions under which most of the deposits of this period, as known to us, were laid down, namely, their marine origin. But there can be no doubt that at some future day such remains will be found, and, indeed, if the deposits of the Laramie age be conceded to be absolutely Cretaceous, as is claimed (although on most contradictory evidence) by many geologists, then the first of such remains, the Meniscoessus, has quite recently been discovered. As now generally recognised, the Laramie deposits constitute a series intermediate between the Cretaceous and the Tertiary, the faunal characters, as are principally indicated by the abundant remains of dinosaurian reptiles, pertaining to the former, while the plants point directly to the latter. The angiospermous exogens, whose earliest undoubted remains occur in the Upper Cretaceous deposits, here undergo a very considerable development, and may, indeed, be said to represent the stock whence the floras of the subsequent Tertiary and existing periods have been derived. We find here many of our most common modern types, such as the oak, beech, poplar, tulip-tree, magnolia, alder, and plane.

Tertiary Faunas.—With the close of the Cretaceous period and the beginning of the Tertiary, we note the most marked of all the organic changes that characterise the different geological epochs. Whole series of animals, from the lowest almost to their highest divisions, suddenly become extinct, or so nearly verge on extinction as to constitute but a very insignificant element in the succeeding fauna; on the other hand, groups of equal or greater importance, and which had hitherto no (or but very scanty) predecessors, just as suddenly make their appearance. It would seem as though a fresh start had been taken in the peopling of the earth's surface, so different in many respects are the faunas of the Cretaceous and Tertiary periods. But this difference, as it now presents itself, must not be taken to indicate that it in fact even existed as such. The gaps that now separate the one fauna from the other were undoubtedly filled by animal types of intermediate grade, of whose existence we shall only be made cognisant when the hiatus which here breaks into the continuity of the geological system will be more completely filled in. It is illogical, and directly opposed to the workings of evolutionary force, to conceive of a wide-spread group of animals suddenly appearing and springing into prominence ; and no less illogical to conceive of an equally sudden extermination. Hence, where vast differences in the faunas of any two succeeding geological periods present themselves, we have reasonable grounds for concluding that a long lapse of time has intervened between the close of one period and the commencement of the period (as represented) next succeeding—in other words, that there is here a geological break.   Only there where the continuity of the geological system is complete, or where the imperfection of the record is reduced to insignificance, can we hope to meet with an organic chain whose continuity is likewise complete.   No such complete record, or anything approaching it, has as yet been discovered, nor is it at all likely that one will ever be discovered.   But the gaps in the record that occur in one locality or country may be wanting in another, those present here be absent in the third, and so on; hence, by a series of comparisons made between several localities, we can in a measure realise a comparatively perfect record, or at any rate one in which the breaks have been materially narrowed, and with it also a comparatively perfect organic chain. Except possibly in one or two regions of the earth's surface, New Zealand and California, nothing that may be said absolutely to link together the Cretaceous and Tertiary deposits, at least those of the marine series, has as yet come to light; the faunas are largely distinct, and their distinctness is the index of the interval that separates the outgoing of the one and the incoming of the other.

The Tertiary fauna presents to us a clearly modern aspect, and one that characterises all the animal groups represented, from the lowest to the highest.   And the farther we advance in this period the more modern becomes the general faunal facies, so that in the Pliocene, or uppermost division, not only are the genera largely identical with existing ones, but (if we exclude the vertebrates) also the species, notably among the mollusks.   It may be stated in a general way that all the more comprehensive of the animal groups now existing are represented in the Tertiary deposits, and the majority of these date from the Eocene, or earliest division.   We have no longer representatives of those wonderful reptilian orders, the Ichthyosauria, Dinosauria, Pythonomorpha, and Pterosauria, which characterised the greater portion of the Mesozoic era, and continued to its termination; nor do we find any vestiges of the scarcely less wonderful birds of the odontornithic group,* or of the type represented by Archæopteryx.   Both reptiles and birds belong to

---

* An exception may, perhaps, be made in favour of the Odontopteryx, described by Professor Owen, which has the substance of both jaws developed into well-pronounced serrations (or false teeth), an exaggeration of the character exhibited by ducks and geese, to which the bird appears to have been related.

the type of existing orders, and the same may be said of the fishes, principally teleosts. The change in the character of the inverte-brate fauna is somewhat less marked than in the case of the verte-brates; but yet certain important differences present themselves. Thus, among the acephalous and gasteropod mollusks, by far the greater number, in fact nearly all the types, are referable to exist-ing families, and even in the oldest division, the Eocene, to exist-ing genera, or to such as are very closely allied to them. Such characteristic families as the Hippuritidæ and Caprotinidæ, among the bivalves, have completely disappeared, and, if we except some half-dozen or more species found in Australian Tertiary deposits, the same may be said of the Trigoniadæ, as well as of the Am-monitidæ * and Belemnitidæ among cephalopods, about the most distinctive of the invertebrate forms of the entire Mesozoic series. Among the Tertiary invertebrates must be noted the extraordinary development of the foraminiferal forms Nummulites and Orbitoides, which, by their prodigious numbers, make up some of the most stupendous deposits known to us. But that feature of the Ter-tiary fauna which above all others arrests attention is constituted by the class Mammalia.

The most striking fact that presents itself in connection with the history of these animals is their very sudden introduction, both as to individual numbers and diversity of form, almost with the beginning of the period, a circumstance of no little significance when it is remembered that, in the period preceding, if we except the doubtfully placed Meniscoessus, not even a trace of their existence has been detected, and that all such forms as have been found in the earlier Jurassic and Triassic deposits belong, as far as we are able to determine, to the single order of the Marsupialia. In the earliest division of the Tertiary, the Eocene, on the other hand, we meet with the remains of individuals belonging to at least one-half of all the recognised orders of the present day.† Thus, we have marsupials of the opossum type (Didelphis), insectivores, rodents (as represented by the Sciuridæ, or squirrels), cetaceans

* A few ammonitic fragments have been found in the Tertiary deposits of the Tejon group of California, and a Tertiary belemnite is claimed for Aus-tralia.

† The Ornithodelphia, Edentata, Proboscidea, Hyracoidea, and possibly also the Sirenia and true Carnivora, are still unknown.

(Zeuglodon), ungulates, both odd-toed and even-toed (among the former the tapiroid Lophiodon and Palæotherium, and other such forms as Eohippus and Hyracotherium, which, through a series of modified but closely-related types in the Miocene and Pliocene periods—Anchitherium, Hipparion—can be traced genetically to the modern horse; and among the latter the possible ancestors of some of the modern deer, Xiphodon and Anoplotherium, and the suilline Anthracotherium and Palæochœrus), bats, even of existing genera (Vespertilio, Vesperugo), lemurs, or lemuriform insectivores (Adapis, Necrolemur), and not impossibly also the true monkeys. But while most of the forms found in these earlier Tertiary deposits are referable to modern orders, there are others which would appear to have no place in the classification laid down for living forms, and which combine, in many respects, the characters of two or more orders. Thus, it has been convenient to designate an order Amblypoda for a line of animals which, at the one extremity, stand nearest in their relationship to the Proboscidea, or elephants, and at another to the odd-toed ungulates. In it are comprised the Uintatheria, ponderous tusked-animals, rivalling or exceeding in size the modern elephant, and the coryphodons, considerably smaller animals of a generalised type, the probable progenitors of the last. A still earlier type is embodied in the Condylarthra (Phenacodus), from the very base of the Eocene, which represent the most primitive type of known ungulate animals, and which not impossibly are derivatives of some preceding hoofed marsupial. Another order, the Tillodontia, has been established for certain animal forms which, in several respects, combine the characters of the insectivores, rodents, and edentates; and, again, a fourth order, the Creodonta, for forms that seem to hold a position intermediate between the insectivores and carnivores, and not unlikely represent the ancestral line of the latter.

From the researches of paleontologists it would appear that the primitive type of placental mammal is the insectivore, and that from this original type have descended, by gradual modification, most of the varied forms that now people the surface of the earth, and those whose remains lie buried in the deposits of the Tertiary period. At what precise period in the earth's history the Insectivora first appeared it is impossible to say, for, although no remains occur in any deposits antedating the Eocene, there can be little or no doubt, seeing what modifications of insectivore structure are presented in the earliest deposits containing their remains, that they appeared at a very much earlier epoch. The Tillodontia, Creodonta, and Insectivora appear, as it were, simultaneously in the Lower Eocene deposits, and if, therefore, they represent merely modifications of one and the same structure, as is maintained by Professor Cope, who has united the three groups into the one comprehensive order of the Bunotheria, then they must point to a common progenitor (foreshadowed in the Jurassic insectivorous marsupials) removed far beyond the limits of the Tertiary period. Of the three insectivore types here indicated, the true Insectivora, which may be considered as the main or axial stem, have alone survived to the present time. The Tillodontia and Creodonta both became extinct before the middle of the Tertiary period, the latter, however, by gradual modification passing off into the Carnivora, whose earliest undoubted remains are to be found in the deposits of Oligocene, or Miocene age. From the same group of the Insectivora, although apparently at a somewhat later date than the Creodonta and Tillodontia, appear to have been descended the so-called Prosimia, or primitive monkeys, the lemurs, whose earliest remains occur in deposits of both Lower and Upper Eocene age; and to these last, again, is doubtless to be traced the direct line of ancestry of the various types of true monkeys that at the present day inhabit the earth's surface, and whose unquestionable traces are first met with in deposits of Miocene age. The most important non-insectivore type of Lower Eocene mammalian is the ungulate, whose remains, belonging to both the odd-toed and even-toed sub-orders, occur in astonishing abundance, and argue very strongly in favor of a very remote ancestory, one that may not impossibly carry us as far back as the middle of the Mesozoic era.

The progressive modifications of structure which can be traced through the more generalised of the Eocene mammalian groups results in greater and greater specialisation the further we advance in the course of time, and hence, in the Miocene period we meet with more of distinctly specialised (or isolated) groups than in the period preceding. In addition to the recent orders that have been enumerated as belonging to the Eocene period we have the Edentata (represented by such gigantic forms as Macrotherium and Ancylotherium, whose nearest relationship appears to have been with the aard-vark), the true Carnivora, Sirenia,* Proboscidea, and Quadrumana. *Per contra*, most of the older forms have now completely disappeared, and, in fact, no mammalian order, with the possible exception of the Creodonta (Hyænodon), is indicated which has not its living representatives at the present day. Most of the

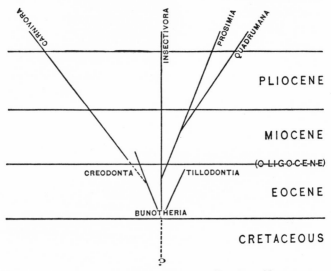

DIAGRAM ILLUSTRATING RELATIONSHIP OF TERTIARY MAMMALIA.

families are such as still exist, and even many of the genera are identical, so that on the whole the mammalian fauna has a decidedly modern aspect. The Miocene Insectivora comprise, among other forms, representatives of the families of hedgehogs, shrew-mice, and moles; the Rodentia, porcupines, mice, squirrels, rabbits, beavers, &c.; the Cetacea are represented by true whales and dolphins; the odd-toed Ungulata by the tapir and a number of allied tapiroids, and the singular giant forms that have been referred to the not distantly removed family of the Menodontidæ (Symborodon, Titanotherium); by true rhinoceroses and other forms (Hyracodon, Aceratherium) closely allied to them; and by Equidæ—Hipparion, Miohippus (Anchitherium)—which differed from the true horses

---

* Eotherium Egyptiacum, from the Mokattam nummulitic limestone, is referred to this group by Professor Owen.

principally in size and the polydactyl character of the feet. Among the even-toed ungulates we find the hippopotamus, the true swine, deer, giraffe, and musk-deer, and, of the hollow-horned ruminants, the antelopes—the sheep, goats,* and oxen being still absent, although some of the antelopine forms would seem to effect a transition between the true antelopes and goats. The Proboscidea comprise, in addition to the true elephant (which, however, as a Miocene animal is known only from the deposits of the Siwalik Hills of India, now frequently referred to the Lower Pliocene, or Mio-Pliocene), the Mastodon, and the aberrant Dinotherium, which was provided in its lower jaw with two prominent recurved tusks. The Carnivora have yielded representatives of the cats—the true Felis and the related sabre-tooths (Machairodus, Dinictis), the most formidable of all known recent and extinct Carnivora—the weasels, civets, hyenas (with the true hyena), and seals. The dogs are represented by the genus Canis itself, and the more primitive Amphicyon, through which a transition is effected to the ursine Hyænarctos, and to the Pliocene true bears. Finally, the Primates have yielded several genera, as Semnopithecus, Pliopithecus, and Dryopithecus, the last referable to the group of the anthropoid or highest apes, and fully equalling in size the human species.

Passing on to the Pliocene period, the mammalian fauna makes a still further approximation to that of the present day in the introduction of a number of modern types that had not hitherto made their appearance, or only just appeared. Thus, we have here the camel (in India), the ox, true bear (in Europe), and horse, in addition to most of the types that have been enumerated as belonging to the Miocene period, and it may be broadly stated that the majority of the genera of this period are such as still exist, although the species are in most cases distinct. The regions where Tertiary mammals have been studied are principally the United States, Europe, and India, between whose faunas there is a well-marked correspondence. While certain of the animal groups referred to are found in the one region and not in the other, and are therefore specially restricted, it may be said that approximately the same groups are represented throughout, although the date of their appearance, or of that of their individual components, may be different for the different countries. Thus, in the Upper Miocene, or older

* Capra Perimensis, from the island of Perim, is possibly a Miocene form.

9

Pliocene, deposits of the Siwalik Hills of India we have the true hippopotamus, bison, bear, and elephant, forms which do not make their appearance until a somewhat later date in Western Europe, Pliocene, or Post-Pliocene. Similarly, the genera Cervus, Hystrix, Felis, Hipparion, and Mastodon, which appear in Western Europe in the Miocene period, are still wanting in the American continent, making their first appearance in the Pliocene. And, likewise, the true bears and oxen in Europe antedate the American forms by one period. It may be stated, as a general rule, that where identical genera of living forms occur in the deposits of both the Old and the New World, those of the Old World are the more ancient; and the same probably holds good, although to a less extent, with families. From these differences in the dates of appearance of certain animal groups, their presence or absence, we are led to discuss the probable origin of our existing faunas, or portions of them.

The existence in Western Europe in Miocene, and especially in Pliocene, times of a fauna consisting of forms which still inhabit the region, and of others as are only to be found at the present time on the continents of Africa and Asia, may appear at first sight somewhat singular. But when we reflect that the climate, during the whole or the greater portion of this period, was probably very much more uniform and warmer than it is at the present time, and possibly not very different from what it now is in the region of the Tropics, the apparent singularity in great measure disappears. Some of the tropical forms, as the giraffe and rhinoceros, may have been indigenous to the region, while others, whose development in South-Central Asia appears to have taken place at an earlier period, not improbably represent immigrants from the heart of that continent. It is practically certain, moreover, that direct land communication existed during a considerable portion of this period with the continent of Africa, with which, consequently, there would have been effected a general interchange of forms. Indeed, it is much more singular that Europe no longer retains its more characteristic African forms; but it must be recollected that, with the advent of the Glacial period, an era of cold set in, and that with this inclement climate a general retreat southward took place, the more tropically constituted animals passing over into the continent of Africa, or suffering extermination by the cold. The most northerly animals, passing southward, occupied the region now more or less vacated by the tropical forms, and hence, in the deposits of the Glacial and Post-Glacial periods, we meet with the remains of the elk, reindeer, hyena, lion, giraffe, elephant, rhinoceros, and hippopotamus mixed together. At about the same epoch it would appear that the present rupture existing between the African and European continents was effected, a separation which precluded the possibility of a return migration from Africa when the more tolerable climate succeeding the closé of the Glacial period set in. Hence the survival only of the more temperate forms of the European fauna. While Africa, therefore, retains its strictly African mammalian forms, it may be considered questionable whether these are not direct importations, in great part, from the region lying to the north. As far as the North American continent is concerned, it appears not improbable, from what has already been said with reference to the earlier appearance in Europe of equivalent mammalian types, that a not inconsiderable portion of its later (fossil) fauna was derived from the Old World. Thus, it appears likely that the bears, swine, oxen, sheep, antelopes, and elephants originated in the Old World, whence they were transplanted by way of some land connection existing in the north into the New World. The true dogs, on the other hand, seem not unlikely to have been developed first on the American continent; and it is also not improbable that the ancestral line of the camel is to be traced to the Western Hemisphere. Our paleontological knowledge of the different countries, even of those more thoroughly explored, is, however, far too insignificant to permit of a definite solution to the problem of the origination of the various mammalian groups, and in the case of most continental faunas but little can be said with positiveness concerning their formation. The North American continent lacks in its Tertiary fauna the giraffe, hyena, and hippopotamus; nor do we find any traces of the group of the Old World monkeys, or, in the greater portion of the region, of the Edentata, whose various forms are now so abundantly represented in the South American fauna.

With the Post-Pliocene period the correspondence existing between the fossil and recent faunas of the several geographical regions is in most cases further increased, and not only by the introduction of many new modern genera, but by the presence, in considerable number, of identical specific types. The modern fauna may now

be said to be broadly marked out. Thus, in the Australian Pleis-
tocene marsupials, Diprotodon, Nototherium, Thylacoleo, and their
allies, we have the forerunners of the various marsupial forms that
now characterise the continental fauna; in the giant birds Palapteryx, Dinornis, Mionornis, &c., from New Zealand, and Æpyornis
from Madagascar, the forerunners of the wingless apteryx and the
struthious birds from the same or neighbouring regions; and in
the giant South American edentates, Glyptodon, Megatherium, Mylodon,* and their allies, the representative, if not the ancestral,
forms of the existing sloth, armadillo, and ant-eater. It is remarkable that these last forms also occur in the Post-Pliocene deposits of the United States; but there can be but little doubt that
their presence there was the result of a temporary migration from
the south, since their remains are only exceptionally found in any
of the preceding Tertiary formations.

We have noted in the Eocene period the presence of a certain
generalised group of mammals, from which, by gradual modifications of structure, the more specialised groups of subsequent periods
have sprung. The demonstration of this successive evolution of
forms is not, however, restricted solely to groups of animals, but it
can be indicated with no less positiveness in the case of certain individual members of a group. The most notable instance of evolutionary modification in a given line is afforded by the horse. Thus,
from the modern horse we can trace downward in the geological
scale a gradual series of modifications in the structure of the teeth
and limbs which, at the further end of the line, characterise an
animal so far removed in general structure from the existing form
that, were not the intermediate forms known, or were it to be considered by itself, it would be recognised not only as the type of a
distinct family, but of a distinct sub-order. From the solidungu-
late type represented in the existing form we reach, by sensible
gradations, an animal of the polydactyl type, or one having several
toes to each foot. A phylogenetic line, but little less complete, can
also be traced in connection with the families Camelidæ, Tapiridæ,
and Felidæ, and others, doubtless, will be discerned with the further progress of paleontological investigation.

* The deposits containing these remains have been very generally considered to be Post-Pliocene; Ameghino and Cope, however, probably correctly
refer them either in whole, or in part, to the Pliocene.

II.

Appearance and disappearance of species.—Reappearance.—Extinction.—Persistence of type-structures.—Variation.—Geographical distribution.—Climatic zones.—Synchronism of geological formations.

It is assumed by all, or nearly all, geologists, that every species of animal, broadly speaking, had a definite belonging in the geological scale; in other words, that its existence was coincident with a certain period in the development of the earth, and with no other. Thus, the well-known and largely-represented brachiopod, Spirifer disjuncta (Verneuilii), whose occurrence has been noted in North America, throughout the greater extent of continental and insular Europe, in Asia Minor, China, and New South Wales, is everywhere restricted to the Devonian formation, and is, therefore, distinctive of that period. Similarly, the no less widely disseminated Productus semireticulatus, a member of the same group of animals, is restricted to the Carboniferous formation. So limited, indeed, appears to have been, in most instances, the range in time of a given species, that the inspection of a single well-determined form will frequently fix, not only approximately but absolutely, the horizon of the deposits whence it was obtained. Belemnitella mucronata, one of the squids, characterises a definite horizon of the Upper Cretaceous; and among the Ammonites we have numerous instances of specific restriction to special " zones " of even the minor divisions of a formation. Limitation of range appears to pertain more strictly to the members of the higher groups of animals than to the lower, or to such forms whose complexity of organisation might be supposed to interfere with a ready accommodation to changing physical conditions of the surroundings. Not one of about one hundred and twenty-five species of fish described from the Old Red Sandstone of Great Britain appears to have lived on into the succeeding Carboniferous period, whereas, of the Invertebrata, comprising the corals, annelids, echinoderms, crustaceans, and the several classes of the Mollusca, the survivors of the Devonian period number no less than twelve per cent. An equally striking case is presented by the Mammalia, where, of the very numerous forms that have been referred to the European and North American Tertiaries, not one is positively known to have passed either from the Eocene division to the Miocene or from the Miocene to the Pliocene; and but a mere handful, if that, from the Pliocene to the period next succeeding, the Quaternary or Post-Pliocene. And yet we are aware that, in certain Eocene localities, no less than five per cent. of the molluscan fauna has survived into the present epoch, and as much as thirty to forty per cent. from the Miocene ! Turning to the Invertebrata themselves, we find among the different classes no less striking confirmation of the law of the persistence of the less highly organised specific types over those more highly organised. Taking the sub-kingdom Mollusca, for example, we find, from an examination of the carefully prepared tables of Mr. Etheridge [55] on British Paleozoic fossils, that, of the three great classes whose members are not free-swimmers, and who would be consequently most likely to fall under the influence of special physical conditions, the order of persistence is, in most cases, Brachiopoda, Lamellibranchiata, and Gasteropoda—*i. e.*, the lowest first and highest last. Species of brachiopods range from every one of the major formations to the formation next succeeding, as from the Cambrian to the Silurian, from the Silurian to the Devonian, from the Devonian to the Carboniferous, and from the Carboniferous to the Permian. The Lamellibranchiata and Gasteropoda, on the other hand, each pass but once from the Devonian to the Carboniferous. Looking at the numerical development of the transgressional forms, we find that fourteen per cent. of the Devonian brachiopod fauna reappears in the Carboniferous, thirteen per cent. of the lamellibranch, and eleven per cent. of the gasteropod. Of the equivalent polyzoan fauna the transgressional forms constitute thirty-seven per cent.

Numerous instances of transgressional forms could be cited from the other branches of the animal kingdom, but to no special purpose. As a rule, the number of such connecting forms is very

limited, and only in the comparatively very recent Tertiary epoch
do they acquire any marked significance. But, if the number of
survivors of any one epoch is very limited, it will naturally follow
that this number will be still further reduced if a question of two
or more epochs is involved. It was, indeed, for a long time maintained by geologists that no species of animal, no matter of what
form of organisation, could possibly have entered into the formation of three successive [epoch] faunas; for instance, that no species
appearing in the Silurian could live completely through the Devonian, and then continue into the Carboniferous. And this notion
is still largely entertained by the geologists and paleontologists of
the present day. However much of truth there may be in such a
doctrine, it must be confessed that its acceptance or rejection will
depend, at least in some part, upon the standpoint from which the
investigator views the nature of species. There is, perhaps, nothing
that has more taxed the ingenuity of the naturalist than to determine just exactly what a species is, what constitutes its absolute
boundaries or limitations. The amount of disagreement upon this
point is so very great, even among naturalists holding approximately
the same views on genesis or evolution, that one might fairly despair
of arriving at anything like a just solution of the problem. During
the period when the doctrine of the immutability of species was a
common faith, there was, indeed, but little difficulty in the matter,
since every reasonably differing form was immediately constituted into a distinct species; and it is in this period that many, if
not most, of our existing paleontological notions have their roots.
Now, however, when the doctrine of descent by modification is
universally acknowledged as one of the great truths of nature, much
greater latitude is permitted to the definition of the word species,
and, in fact, what might at one time be considered as a good species
can, in the light of a newly-discovered chain of " intermediate "
forms, be readily degraded to the rank of a variety. Such has, in
truth, been the history of a great many so-called species. The
difficulty of determining specific longevity thus becomes apparent,
for who can state what will be the fate of forms that now stand
apparently far apart? However divergent may be the views of
authors on the matter of relationship, it is practically certain that
numerous forms of life, exhibiting no distinctive characters of their
own, are constituted into distinct species for no other reason than

that they occur in formations widely separated from those holding their nearest of kin. Whether these be really good species or not may be a matter for further consideration; but it cannot be denied that their primary recognition as such is based upon the assumption that no species, after it once became extinct, ever again came into existence. While there is much that speaks in favor of this doctrine, it may, nevertheless, safely be asked, in what lies the proof of its correctness? Surely we possess no knowledge which will permit us to state just when a given species became extinct. Disappearance from one locality is in itself no indication of absolute extinction, any more than appearance is an indication of primary origination. How, then, can we ascertain, when a given form supposed to be extinct reappears after an interval of a formation, whether that form in reality became extinct or not? It is usually assumed that it did not, and its range in time is correspondingly extended; or, the reappearing form, with no distinctive characters of its own, is elevated to the rank of a new species, and the extinction of the first species insisted upon. But it is evident that this method of forcing the point is in the nature of an *argumentum in circo*, and leaves the question of extinction and reappearance in the condition of "not proven."

While, apart from the proof that is lacking in the matter, the doctrine of non-reappearance seems to commend itself by a certain "plausibility," it may still be doubted whether this supposed plausibility is not more a matter of preconception than of actual fact. If evolution is true, and there are few among scientists who would deny that it is so, can it not readily be conceived that, as the result of the interaction of the physical and organic forces, identical forms may have been evolved as the heads of very distinct lines of descent? And, if so, may not this process have operated through distinct periods of time? Mr. Darwin, in attacking the problem, thus states the case ("Origin of Species," p. 379, ed. 1866): "We can clearly understand why a species, when once lost, should never reappear, even if the very same conditions of life, organic and inorganic, should recur. For though the offspring of one species might be adapted (and no doubt this has occurred in innumerable instances) to fill the exact place of another species in the economy of nature, and thus supplant it, yet the two forms—the old and the new—would not be identically the same, for both would almost certainly inherit different characters from their distinct progenitors. For instance, it is just possible, if our fantail pigeons were all destroyed, that fanciers, by striving during long ages for the same object, might make a new breed hardly distinguishable from our present fantail; but, if the parent rock-pigeon were also destroyed, and in nature we have every reason to believe that the parent-form will generally be supplanted and exterminated by its improved offspring, it is quite incredible that a fantail, identical with the existing breed, could be raised from any other species of pigeon, or even from the other well-established races of the domestic pigeon, for the newly-formed fantail would be almost sure to inherit from its new progenitor some slight characteristic differences." That the reasons here given satisfactorily explain why, in the *vast majority* of instances, lost species should not reappear, most naturalists will admit; but that they do not in themselves sufficiently explain why such reappearance may not *occasionally* occur, may be reasonably contended. Thus, in the case of the birds referred to, it would by no means be unreasonable to suppose, even if such instances are a decided rarity, that the parent rock-pigeon, through some special adaptation to surrounding conditions, might have long survived the generations of fantails that were primarily derived from it, and that, at some future period, after a process of selection, a second series of fantails might have been produced practically identical with the first. The irregular duration in length of time of species is well known to geologists, and its importance as a factor in evolution has been insisted upon by Mr. Darwin. It is probably true, as Darwin affirms, that "the parent-form will generally be supplanted and exterminated by its improved offspring;" but it does not necessarily follow that the offspring will invariably prove to be of an improved stock, and, where this is not the case, there would be nothing very surprising in the survival of the parental form. The tenacity with which certain specific characters adhere to some of the older genera, especially of mollusks, is so great that paleontologists are frequently at a loss to determine just what characters separate practically the newest from the oldest species of a given genus. This is the case with the Nautili, and with a number of the brachiopods. Now, either we have here a retention through almost indefinite periods of primary specific characters, or a reversion (or a series of reversions) to a type once formed, but which has at one or more periods been effaced. If the former, would it be illogical to suppose that some of the numerous varieties or species that may have been evolved during a long lapse of ages from the parental stock should have proved less hardy than it, and should have, therefore, suffered much earlier extinction or modification ? And, if so, what is there that we know that should absolutely prevent the same early extermi- nated forms from being re-evolved ? It will be naturally assumed that the new and the old species, which appear to be so closely connected, are in reality distinct, and this may be true in all cases; but, if the distinguishing characters separating the two are almost imperceivable, it will require not much stretching of scientific prin- ciples to conceive that at least some of the resulting varieties may be strictly indistinguishable.

Again, reverting to the case of the pigeons, it seems by no means clear, although such results may be of very rare occurrence, why a fantail, identical in every respect with the common form, may not be produced from a species of pigeon other than the rock- pigeon. It is true, as Darwin states, that newly-formed varieties would be almost sure to inherit from their progenitors certain dis- tinguishing traits or characters; but, as the formation of a species will depend upon the overbalancing by newly-acquired characters of those that may have been left by inheritance, there seems to be no reason why, through a process of selection and adaptation, such a convergence might not occasionally take place as would unite the ends of very distinct lines of descent. Paleontologists have long held to the opinion that the line of descent which leads up to the horse (genus Equus) in America is different from the similar line of Europe, and if it has been contended that the existing spe- cies, Equus caballus, was not in itself a product of the American line, but a modern European importation, its recent discovery in the Post-Pliocene deposits of this country proves the erroneousness of such an assumption. If, therefore, we are permitted to assume that two distinct lines of descent, the one passing through Eohip- pus, Orohippus, Mesohippus, Miohippus, Protohippus, and Plio- hippus, and the other through Palæotherium, Anchitherium, and Hipparion, lead up to an equivalent genus, the genus horse or Equus, would it be unreasonable to suppose that, with a continu- ance of the process of evolution, even the same species may be evolved ? * And, if this is possible, seeing that under apparently identical physical conditions the resulting genus became extinct on the American side of the Atlantic, while it still flourished on the European, might it not have so happened that the genus should have actually become extinct on the one continent before it even came into existence on the other ? Assuming this condition, we should then have a true case of extinction and reappearance. Similarly, in the case of the genus Tapirus, the tapir, it is by no means certain that the eastern and western forms are not the products of two distinct ancestral lines, independent in their evolution of such influences as might have been brought about by migration and interassociation.[55a]

It may, however, still be urged that the apparent development of parallel lines on the two continents is not in reality such, and that their independent convergence is merely the result of an intermixture of Old and New World types, during a period of land connection. That a union between the northern parts of the two hemispheres may have existed during the Eocene, and again in the Miocene and Pliocene periods, can very well be true, but that such a union need not necessarily imply the absolute inter-derivation of the two continental faunas can equally well be true. The complete absence from one continent of some of the more abundant fossil types occurring in the other, as the genera Palæotherium, Anoplotherium, and Lophiodon from America, and Oreodon from Europe, proves that there must have been (granting land connection) some formidable check to migration ; and this assumption is further borne out by the circumstance of the irregular appearance in time, on the two continents, of such animals as might readily be supposed to have been able to grapple with a northern barrier.

---

* Orohippus has been identified by many paleontologists with Hyracotherium, and Miohippus with Anchitherium.—Waagen, from his studies of the Indian Jurassic Ammonitidæ, has arrived at the conclusion that the genus Aspidoceras (Ammonites auct. pars) has descended from at least two distinct generic roots, and further maintains that the same duplex or multiple origin can be traced in other genera as well ("Palæontologia Indica," ser. ix., 4, 1875, p. 241). He also affirms that certain species of Phylloceras, as, for example, P. ptychoicum and P. Benacense, common to both Europe and India, are the product, in the two countries, of distinct ancestral lines, convergent modification, as "dependent on laws which were innate" in the species, having brought about an identical result (*loc. cit.*, p. 243).

Among the Invertebrata we have also, as has already been intimated, well-marked instances of apparent reappearance. Mr. Davidson, in his review of the British fossil Brachiopoda,[56] affirms that the resemblance between the recent Rhynchonella nigricans, which has nowhere as yet been found in any Tertiary deposit, "and some Cretaceous and Jurassic forms is so great that we are at a loss to define their differences." And, further, that certain varieties of the Mediterranean Terebratulina caput-serpentis, whose range extends downwards only to the Pliocene, so closely resemble the Cretaceous Terebratula striata as to render the two barely, if at all, distinguishable. It may be that at some future date intermediate links will be discovered to unite the forms of the two periods; but who can, in the meantime, affirm that these are not true cases of reappearance ? If it is true, as M. Barrande asserts, that the Triassic Nautilidæ show less affinity to existing species than do the primitive forms — in other words, that the recent species are more closely related to the original forms than the forms of a half-way intermediate period—are we not justified in assuming that there is here a reversion, or tendency towards a reversion, to specific characters once lost ? Among the Foraminifera we have several notable instances of apparent specific longevity, as, for example, the recurrence in the modern seas of some of the cretaceous species of Globigerina, Cristellaria, and Glandaria; and, not unlikely, seeing how very slow must of necessity be the variation in this class of animals, by reason of their ready adaptability to their surroundings, the same forms will, on future investigation, be found to date considerably further back in the geological scale.

It must not be construed, from the preceding argument, that we have attempted to prove the frequency of specific reappearances. These, if they actually have existed, which is not unlikely, were, doubtless, of exceptional occurrence, and in no way affect the problem of progressive development. It might be objected that, if the views here set forth are correct in their application to species, they must also apply in a corresponding degree to some of the higher animal groups—genera, for example. This is certainly true, and, indeed, it may well be asked, What insuperable obstacle is there, that we know of, that should absolutely prevent the occasional reappearance of a lost genus ? Assuming, with many paleontologists, that Goniatites is a genus evolved from Nautilus, either directly or indirectly, what is there, in view of the persistence of the nautiloid type, to preclude the possibility of the same genus being re-evolved ? There may be a sufficiency of reasons with which we are not acquainted; but if, with our present limited knowledge, we are unable to indicate what these may be, it is scarcely fair to insist, *a priori*, upon their existence. It is true, we have no special reasons for assuming why the Nautilus should become modified into a Goniatites; but this of itself is no proof that it may not. Numerous instances of genera widely separated by formations in which no representative of their tribe is to be met with are known to every paleontologist, but perhaps no more remarkable example is presented than that of the genus Nummulites. It has been well said by Mr. Brady [57] that "there are few time-marks in the geological record that have been regarded as better established, or more definite, than the first appearance of the Nummulite, at or near the commencement of the Tertiary epoch," and so little, in fact, is known of any of the antecedent forms, that many, if not most, paleontologists of the present day still hold to the correctness of the view here stated. The researches of Gümbel and Brady have, however, placed beyond all doubt the existence of at least one Jurassic species (Nummulites Jurassica, Franconia) and one Carboniferous (N. pristina, Belgium), and not unlikely a somewhat doubtful form, N. variolaria, var. prima, described from the Cretaceous rocks of Palestine, will, on further investigation, prove to be a true member of the genus to which it has been referred. That we have here an instance of generic reappearance it is impossible to affirm, but it is certainly almost inconceivable, whichever way it be considered, that a group of animals, so extensively developed as are the Nummulites in the Tertiary deposits, should have left practically no traces of their existence behind them in the deposits next preceding the Tertiary, the Cretaceous, when their ancestry dates so far back as the Carboniferous epoch. It is scarcely possible that at no period of time between the Carboniferous and Tertiary epochs should the conditions for their development have been favourable; and equally improbable does it appear that, if such development actually did take place, we should so thoroughly lose sight of their remains. Granting the absolutely unfavourable conditions, however, can it be readily imagined that a few miserable forms, evolved at an entirely unpropitious moment, should have battled through the struggle for existence to develop after an interval of millions of years? This seems very improbable, but yet it may be so; but why may it not also have been that, in the Carboniferous form, we had premature evolution, with subsequent extinction, and that at a much more recent period a re evolution of the same form, under more favourable circumstances, took place? The case is certainly very extraordinary, and probably has no parallel in the history of paleontology.

On Appearance and Extinction.—It would naturally be supposed on the hypothesis of evolution that the introduction of all species must be a very gradual one, for it can scarcely be conceived that the laws governing the formation of new species through descent and modification could be anything but very slow in their action. So true is this, that Darwin has himself admitted, that "if numerous species, belonging to the same genera or families, have really started into life at once, the fact would be fatal to the theory of descent, with slow modification through natural selection" ("Origin of Species"). Yet if we glance over the geological record we cannot fail to note a very considerable number of seemingly flagrant contradictions, groups of allied species appearing in almost every formation with apparently the greatest possible abruptness. We have but to instance as examples the genera of articulate brachiopods, and the tabulate corals of the Silurian period, the ganoid fishes of the Devonian period, the Tertiary placental mammals, and the foraminiferal genus Nummulites, already referred to. Probably there exists no more striking illustration of the abrupt or sudden development of a family than is furnished by the Nautilidæ. a group of animals for whose elucidation we are principally indebted to the labours of M. Barrande. Of this family of cephalopods, which comprises probably upwards of two thousand distinct species, no less than four hundred and sixty-three species, referable to some fifteen or more genera, are already represented in the Lower Silurian formation, although from the preceding Cambrian deposits at best only two well-authenticated forms (Cyrtoceras præcox and Orthoceras sericeum) are known. Of this number about two hundred and sixty belong to the genus Orthoceras itself, ninety to Cyrtoceras, and forty-six to Endoceras, the last, a genus restricted absolutely to the Lower Silurian deposits. Facts such as these have been eagerly seized hold of by the advocates of the doctrine of independent and successive creations as proving the fallacy of any slow modification theory of transformism, and were it not that they are in themselves fallacious, would alone be sufficient to overthrow any such theory. For not alone must of necessity the development of a group of forms, all of which were descended from some one progenitor, have been an extremely slow process, but the "progenitors must have lived long ages before their modified descendants." The experience of every paleontologist proves to him, however, how misleading are those apparent abrupt appearances, and how very frequently groups of forms, supposed to be restricted to a definite horizon or formation, suddenly appear in a region perhaps not hitherto worked over, or, even where the work of the geologist has been accomplished with a sufficient amount of care, in a horizon of considerably older date. Almost every large group of animals furnishes such instances of antedating. No fact was at one time considered to be more firmly established than that the Mammalia belonged exclusively to the Tertiary and Post-Tertiary epochs, and yet we now know of their existence, even in considerable variety, in the deposits of both Jurassic and Triassic age; the serpents, which until quite recently were thought to have their earliest ancestors in the deposits of the Tertiary age, have been traced back to the Cretaceous (France); the Insecta, whose supposed earliest appearance in the Carboniferous rocks was considered to mark an epoch in the faunal development of that period, have, in a series of impressions left in the Devonian shale of New Brunswick, proved their existence at a much earlier date, and only within the past year, 1884–'85, the announcement is made of the discovery of scorpion remains in the Silurian rocks of both Sweden and Scotland. It will still be in the memory of many geologists and paleontologists with what startling effect the announcement of the discovery of the first air-breathing vertebrates in the deposits of the coal was made, at the very time when the absence of such forms was ascribed to the impossibility of their breathing an atmosphere supercharged with carbonic acid ! In the case of special genera we have equally well-marked instances of antedating.

Waagen's discovery of an ammonite in the Carboniferous rocks of the Salt-Range of India was for a long time discredited, so firmly had the notion that the ammonites were restricted to the Mesozoic era been engrafted on the minds of paleontologists; but now, several individuals, belonging to two or more species, have been obtained from the same deposits, and one closely related form has been quite recently described from the nearly equivalent deposits of Texas.[58]   M. Barrande has laid great stress upon the sudden appearance, side by side, and in the full plenitude of their characters, of the more distinctive genera of cephalopods (Orthoceras, Cyrtoceras, Nautilus, Trochoceras, Bathmoceras) in the first aspect of his second Silurian fauna (Lower Silurian of geologists generally), and their complete absence from the Primordial Zone (Cambrian); but we have seen that at least two of the genera, Orthoceras and Cyrtoceras, have since been traced back to the earlier formation. Almost every chapter in geological history indicates some such case of antedating, and proves to the paleontologist how very cautious he should be in his limitation to time of particular groups of organisms.   It cannot be expected that in any portion of the earth's surface will there ever be found a complete sequence of the geological formations, nor can we hope satisfactorily to bridge over in all cases the gaps that occur in one locality with the deposits found in another.   The unequal period of time during which land areas have been laid dry or been kept submerged beneath water, in conjunction with the devastating effects of denudation, render such a complete restoration of the series impossible, and as long as this is so, the work of the paleontologist must inevitably be riddled with "breaks" in the geological record.   It is surprising, in view of these facts, which are too obvious to be overlooked, with what tenacity some paleontologists insist upon absolute limitation of species, or groups of species, and how slow they are to accept any new facts bearing upon distribution in time that might in any way disturb the harmony of their preconceived notions.   There is perhaps no more patent fact in the history of the physical development of our planet than the imperfection of the geological record, the full realisation of which could not fail to dispel many of the singular notions that still prevail relative to the support which paleontology brings to the doctrine of evolution.

It has been objected that, in assuming the universality of breaks, we are drawing largely upon our fancy, and that conditions which do not exist in fact are arranged to suit the views of the evolutionist. This may be true in a limited number of cases; but there is every reason to believe that the constancy of breaks is even far greater than the most enthusiastic advocate of the doctrine of imperfection would be ready to admit. For, even in such areas where the rock-masses through a general uniformity of character would seem to indicate continuous sedimentation, have we always definite proof that the sedimentation was really continuous? Far from it. If, for example, certain parts of the Atlantic border of the United States were depressed beneath the sea, and a new Post-Tertiary deposit imposed upon them, we might be not a particle the wiser, as far as stratigraphical and lithological evidence went to show, for the enormous period of time that intervened between the formation of the newest and next newest (Miocene) series of deposits. The strata would lie practically conformably on one another, and it would require but little degradation to plane down these inequalities, which would otherwise indicate an eroded land surface. The members of the Cretaceo-Eocene series of the State of New Jersey, and elsewhere along the same coast, are so intimately related to one another, by conformability of position and lithological structure, that it might readily have been assumed that we had here an instance of continuous sedimentation; and, indeed, for a long time no division-line was supposed to exist. But the unmistakable evidence of paleontology proves that here, as well as at most parts of the earth's surface which have been made accessible to the geologist, a break marks the junction of the Cretaceous and Tertiary formations.

It must be admitted that there are certain anomalies connected with the occurrence of breaks which have not thus far received an adequate explanation. Their broad distribution—it might, indeed, almost be said universality—in equivalent periods of time, has long been noted as a surprising fact, and one that still remains in the nature of a puzzle to the geologist. Nowhere on the surface of the earth has there as yet been found a distinct connection between the Paleozoic and Mesozoic series of deposits, and only at a very few points (India, New Zealand, California) what may be considered to be an unequivocal link between the Mesozoic and Cainozoic series (Cretaceous and Tertiary). It is true that the field surveyed by the geologist is of comparatively limited extent, when compared with that which still remains to be explored—the greater part of the continents of Asia, Africa, South America, and Australia—and it is but reasonable to expect that, in some of the regions here indicated, the connections will be found that elsewhere are wanting. But, even granting the justice of this plea, the facts, such as they are, are of themselves sufficiently remarkable, as indicating how very far-reaching in their action must have been the forces that were directly concerned in the causation of breaks. It is a little difficult to conceive of secular elevations and depressions of the land-surface extending simultaneously over nearly the entire circumference of the earth, even in the restricted area of the Temperate Zone; but such must undoubtedly have been the case to account for the phenomena that are presented to us. Otherwise complete passage-beds would be of much more frequent occurrence than we know them to be. This does not preclude the possibility of the existence of local areas showing a differential or contrary movement; such, however, do not seem to have in any way interfered with the grand scheme that was involved. But it is very unlikely that elevation or subsidence either was, or could be, universal at any given period; on the contrary, it appears far more rational to suppose that every very extensive elevation was accompanied by a corresponding depression somewhere else, and *vice versa*, and thus some sort of balance maintained. If this is so, then we would naturally look in some distant quarter for the counterpart of the effects which either elevation or subsidence* may have produced in any one region of the globe. Seeing how very general throughout the vast expanse of the Northern Hemisphere are certain breaks in the geological series, are we not justified in looking to the region farther south for the evidences proving uninterrupted sedimentation and continuous organic evolution?

If we attach full weight to the imperfection of the geological record, it is not difficult to account for the apparent abrupt appearance of certain animal groups or faunas; indeed, the problem would have been far more difficult to solve had the case been otherwise. But there is one special instance of such appearance which is not so readily accounted for, and which, under any hypothesis, is almost inexplicable. We refer to the sudden appearance of the numerous forms of life which characterise the oldest fossiliferous formation with which we are at present acquainted, the Cambrian,

* Reference is here made to the more extensive movements of the crust, producing the profounder breaks.

when no unequivocal traces of preexisting life are anywhere to be met with in the formation next preceding. So absolutely universal is this condition that it almost staggers belief. It cannot rationally be conceived that the varied Cambrian fauna could have come into existence *de se*, without there being a line of progenitors to account for its existence; but, if such progenitors did exist, which was doubtless the case, what has become of their remains? Can it be that all over the world, as far as we know, every fragment of such a pre-Cambrian fauna should have been so completely wiped out as to leave not a determinable vestige behind? It must be confessed this seems very incredible, seeing with what absolute perfection many of the oldest, and in many respects the most delicate, structures have been preserved through all the vicissitudes of geological time. The hexactinellid sponges of the Cambrian and Silurian periods, the Silurian Foraminifera, and scarcely less so the graptolites, bear ample testimony to a most astonishing power of resistance. To account for such a wholesale obliteration, we must invoke the aid of a kind or degree of metamorphic action very different from that which has since been made known to us, for it can scarcely be supposed that the ordinary action extending back through only one more period of geological time could have produced such profound results. And it is not only from a comparatively brief period of time that we must explain the utter absence of organic traces, but from a period which, in the opinion of many geologists, may have been of equal duration with the entire interval that has elapsed since the deposition of the Cambrian sediments. But, even granting this unknowable form of regional metamorphism, it still remains a mystery how its effects could have been so universal as to wipe out every vestige of an indisputable pre-Cambrian fauna. It is very possible that the limestone of the Laurentian rocks owes its existence to organic agencies, and therefore represents in part this earlier fauna; but even admitting this to be so, it helps the matter very little, since the limestone is overlaid by younger crystalline rocks, which are no less destitute of organic traces than the deposits underlying it. For the same reason the existence or nonexistence, as an animal, of the much-debated Eozoon, does not affect the point at issue; on the contrary, the total absence of determinable organic traces, either above or below the Eozoon line, would, in itself, apart from all other evidence, constitute strong grounds for relegating that *quasi*-organism to the class of mineral deposits.*

Darwin has sought to explain the anomaly on the supposition that possibly the most ancient fossiliferous deposits lie buried deep beneath the floor of our existing oceans, and that they may have lain there ever since the Cambrian (Silurian) period. They would then have been kept out of sight, and would, at the same time, have offered no opportunities for their remains to become intermingled with those of any subsequent formations. That there is no insuperable objection to this explanation every one must admit, and that it, at least, partially meets the case, is more than probable. But it is still far from being in the nature of a demonstration. The doctrine of the permanency of land areas and oceanic basins has much in its favour, and if true, would go far towards supporting Darwin's proposition; but, unfortunately, the absolute proofs of such a condition are still wanting, and may forever remain wanting. The land surfaces from which the Paleozoic rocks derived their sediments, either in part or in whole, may or may not have occupied the position of the present seas. If the former be the case, the problem remains in its original form; if the latter, it must be assumed that a broad hiatus exists between the Laurentian and Cambrian series, and that the gap is filled by vast submarine deposits, upon which massive accumulations of continental and organic *débris* have been superimposed. Fossils of a pre-Cambrian type may be abundant in these deposits. Manifestly, however, the assumption of large land areas depressed beneath the sea carries with it the implication of an alternation of oceanic and continental surfaces,

---

* The question of the animal nature of Eozoon has been practically settled in the negative through the researches of King, Rowney, Julien, and Möbius; the elaborate memoir on this subject by the last-mentioned scientist will probably be considered conclusive by most impartial zoologists. The present author has himself examined masses of Eozoon rock in which the network of green mineral, supposed to fill the chamber-cavities of the giant foraminifer, coalesce and merge into a broad band of serpentine. Now, either we have here a true Eozoon structure or not. If yes, then on what zoological basis, it may be asked, can the gradual convergence of the infiltrating mineral and its final coalescence with a broad band of serpentine be explained? If the contrary, what necessity is there for invoking the aid of organic forces in the explanation of a structure, when one fully as intricate, and piactically undistinguishable from it, can be shown to be of purely mineral formation?

and if this could have happened once, why may it not have occurred again ? Or might it be assumed that this, a primary oscillation, first marked out the existing boundaries of land and sea ? The problem, in whichever light it may be viewed, is beset with innumerable difficulties, and, it must be confessed, lies beyond the probability of a near solution. The evidence appears strong, however, for concluding that the Archæan rocks, so recognised—*i. e.*, those of the Laurentian and Huronian series—are by no means the immediate predecessors of the Cambrian series. These may still be found at some places underlying the last, or they may forever remain hidden from view beneath the aqueous deep.

Extinction.—It has very generally been remarked that the extinction of species, or groups of species, appears to have been a much more gradual process than their introduction. This is, doubtless, in great part true, and agrees well with the theory of natural selection. The formation of a new species usually implies favourable conditions for the development of that species, and it is, therefore, not surprising that when once formed the species should spread very rapidly. When, however, through certain causes— the alteration of the physical properties of the surroundings or inferiority in the general struggle for existence — the conditions for existence are no longer as favourable as they were before, we should naturally expect to meet with a decline in the development of that species, and its possible ultimate extinction. But unless the change in the conditions of life were very abrupt, we should nowhere look for immediate or sudden extermination. Every one is familiar with what prodigious rapidity certain weeds, as the wild carrot, for example, have spread in regions into which they had but recently been introduced, but how very much slower has been the extermination of the species of native plants which they may have supplanted. The English sparrow has developed with surprising rapidity in the Eastern United States, and, although since the period of its introduction scarcely twenty years have elapsed, it has so far multiplied and become master of the newly acquired situation as to have practically appropriated for itself a large portion of the domain formerly occupied by the native birds of the same family, and to the exclusion of those birds. These, if they eventually prove weaker in the race, may in course of time completely disappear, but, before that period will be reached, will doubtless have effected a foothold in some neighbouring region, and struggle on as best they can under what might be less favourable conditions for existence. Ultimately the race will be thinned out and extinction of the species follow. A parallel case of sudden appearance and much less sudden disappearance is afforded by the two common species of house-rat, the black and the brown, both of which, through introduction, have become more or less cosmopolitan in their range. Everywhere where the latter has succeeded in obtaining a foothold the former is gradually, but steadily, fading away, relinquishing piece by piece the territory of which it was at one time in full possession. But it still lingers on, and, no doubt, will still continue to so linger for some time in the future, a relic of a once formidable race. Yet its own march of conquest was probably no less rapid than that of its more successful competitor, the brown or Norway rat, which appears to have been unknown west of the Volga River prior to about the middle of the last century, and which has since spread so extensively as to render it one of the commonest pests of both continental and insular Europe. The species was first observed on the Pacific coast of the United States subsequent to 1850, but it is scarcely less common at the present time in California than anywhere else. The relation of natural selection to extinction and persistence is clearly stated by Darwin thus ("Origin of Species") : "The competition will generally be most severe, as formerly explained and illustrated by examples, between the forms which are most like to each other in all respects. Hence the improved and modified descendants of a species will generally cause the extermination of the parent species; and if many new forms have been developed from any one species, the nearest allies of that species, i. e., the species of the same genus, will be the most liable to extermination. Thus, as I believe, a number of new species descended from one species, that is, a new genus, comes to supplant an old genus belonging to the same family. But it must often have happened that a new species belonging to some one group will have seized on the place occupied by a species belonging to a distinct group, and thus cause its extermination; and, if many allied forms be developed from the successful intruder, many will have to yield their places, and it will generally be allied forms, which will suffer from some inherited inferiority in common. But, whether it be species belonging to the same or to a distinct class which yield their places to other species which have been improved and modified, a few of the sufferers may often long be preserved, from being fitted to some peculiar line of life, or from inhabiting some distant or isolated station, where they have escaped severe competition. For instance, some species of Trigonia, a great genus of shells in the secondary formations, survive in the Australian seas; and a few members of the great and almost extinct group of ganoid fishes still inhabit our fresh waters. Therefore the utter extinction of a group is generally, as we have seen, a slower process than its production."

There are instances, however, in which the extinction of certain animal groups is generally considered to have been very sudden. The trilobites and ammonites among invertebrates, and the dinosaurian reptiles among vertebrates, may be taken in illustration of such cases. But even here a careful examination of the premises shows that the suddenness of extinction is probably much more apparent than real, and that, as the facts now stand, they by no means sustain the inferences that have been drawn from them. The trilobites, for example, are frequently stated to stop suddenly at the close of the Paleozoic era, whereas, as a matter of fact, no trace of trilobites has ever been found in deposits unequivocally newer than the Carboniferous. A whole period (Permian)—true, a comparatively insignificant one—therefore, still intervenes between the extinction of the order and the close of the Paleozoic era. But, again, even with the Carboniferous period the extinction is far from being sudden. Of the numerous genera which so eminently characterise the Silurian fauna, only two genera, Phillipsia and Proetus, both of them restricted to a comparatively insignificant number of species, survive the Devonian period.* These, together with two other genera now for the first time introduced, Griffithides and Brachymetopus, constitute the entire known Carboniferous trilobitic fauna; and of this limited number only one genus, Phillipsia, and that apparently in America alone, passes up as high in the Carboniferous series as the Coal-Measures. We thus see how very gradual, rather than abrupt, has been the final —if final—extermination of this order of animals.

Nor do we have that sudden downfall, either in or from the

* Since the above was written, Professor Claypole has announced the discovery of Dalmania in the Waverly Group (Lower Carboniferous) of Ohio.

Silurian period, which many authors are in the habit of insisting upon. Thus, if we take the very elaborate tables of Mr. Etheridge [39] as our guide, we find that the rise and fall of the British trilobitic fauna has been, on the whole, gradual, the greatest break occurring between the Llandeilo and Caradoc on the one side, and the Caradoc and Llandovery on the other, or along the horizon which, by many geologists, is considered to mark a "Middle" Silurian division. The following scheme will exhibit the numerical, generic, and specific values presented by the different horizons of the Paleozoic series, from the base of the Silurian to the Carboniferous, inclusive:

FORMATIONS.	Genera of Trilobites.	Species.
Arenig........................	6	9
Llandeilo	18–20	45
Caradoc–Bala................	27	123
{ Lower Llandovery............	13	$35 \begin{cases} 25 \\ 24 \end{cases}$
{ Upper Llandovery............		
Wenlock....................		23–25 (?)
Ludlow.....................	10	20
Devonian....................	6	11
Carboniferous................	3	13

In the Bohemian basin, where the transition between the Cambrian and Silurian faunas is very abrupt, the decline is, if anything, still more gradual than in Great Britain. Barrande's Étage C (Cambrian) holds, according to its illustrious monographer, 27 species of trilobites; Étage D (base of Silurian), 118 species; E, 83; F, 88; and G and H together, 66.*

In the case of the ammonites the disappearance is somewhat more rapid than with the trilobites; but even here it is not nearly so abrupt as it is very generally conceived to be. Thus, if we take

* J. Barrande, "Trilobites," Prague, 1871. Of the 66 species contained in faunas G and H, G alone possesses 64, and H only 2; it might, therefore, be assumed that the extinction was here very rapid. But, as M. Barrande himself informs us, the deposits of Étage H are of insignificant development when compared with either E, F, or G, and have, in fact, practically disappeared through erosion from the greater portion of the territory which they formerly covered. "Système Silurien," i., 1851, p. 81.

the typical Cretaceous area of the peninsula of India as representing a series of nearly continuous depositions from the Neocomian to the Danian, inclusive, we find that there has been a very sudden diminution in the number of species before the top of the series is reached. In the Lower Cretaceous division, the Utatur group, corresponding to the Cenomanian of continental geologists, the number of species of coiled Ammonites is, according to Stoliczka, 67; in the middle division, Trichinopoly group (Turonian), 20; and in the upper division, Arialur group (Sennonian), 21. In the uppermost Arialur beds of Ninnyur, which probably correspond to the "Maestrichtian," or Danian, not a single species is found. The seeming anomaly that the upper and middle divisions of the Cretaceous (Arialur and Trichinopoly groups) should contain an equivalent number of species is not exactly in harmony with the law of gradual numeric diminution; but its explanation is, doubtless, found in the fact that the development of the Arialur deposits is double that of the underlying deposits of the Trichinopoly group.[60] In the typical Cretaceous areas of England there is an equally well-marked reduction in the number of species before the end of the series is reached, and the same, although to a less extent, is also the case in France. In California, where the breaks in the Cretaceous series appear to have been of comparatively insignificant value, and where a nearly continuous sedimentation tides over the gap which elsewhere exists between the Cretaceous and the Tertiary, the disappearance of the ammonitic fauna is a very gradual one. In the lowest member of the system, the Shasta group, the coiled forms number ten species; in the Chico group, six; and in the Martinez group, the top of the system, two. But, to draw out the line still further, we have here the indisputable passage of one species into the deposits of the Tejon group, the base of the Tertiary series (Eocene). This unique case of an ammonite surviving the Mesozoic fauna, which will be, doubtless, repeated at many parts of the earth's surface not yet explored by the geologist, finds its parallel in the similar survival in Australia of a solitary belemnite, if the organism so described really proves to be such. Less gradual in their disappearance, apparently, than the true ammonités, are the uncoiled forms of the same family, whose remarkable development, just before the close of the Mesozoic era, must be considered one of the most striking facts in paleontology. But

10

here, as elsewhere, it should be remembered that the break almost everywhere existing between the Tertiary and Cretaceous series of deposits is a profound one, and covers not impossibly a lapse of time fully equal to that which is measured by either of the two periods here mentioned, or even greater. It is, therefore, in no way very surprising that a family still in its prime at the beginning, or even near the end, of the Cretaceous period should have become practically extinct before the beginning of the Tertiary, in the interval between which there may have been, as Darwin characteristically observes, "much slow extermination."

The doctrine advanced by many of the earlier geologists and paleontologists, and still held by a few, that the duration of species, or groups of species, is uniformly defined the world over by a sharp and inflexible line, is, in the light of our present knowledge of facts, untenable, and will not stand the barest test of logical examination. It cannot for a moment be conceived that the conjunction of the physical and vital forces should have so acted as to simultaneously convert favourable into unfavourable conditions of existence for the entire surface of the earth; as far as the widely-distributed marine forms of life are concerned, nothing short of a complete upheaval and laying dry of the sea-bottom could have brought about such a condition, and, even granting such a consentaneous upheaval to have actually taken place, which probably no geologist will for an instant admit, the overflow of the oceanic waters would have afforded a safe harbour to many forms that might have been displaced from their native habitat. Extermination or extinction of the larger animal groups, especially if wide-spread, will have been almost invariably preceded by displacement; unfavourable conditions for existence, whichever way they may have been brought about, will tend to promote migration, and, consequently, when a certain group of animals becomes extinct at a given period of time, at any one locality or region, we may confidently look for its survival somewhere else, possibly in some very distant region. Instances of such survival, in favoured localities, after the nearly complete extermination of the race, we see in the Californian ammonite and Australian belemnite already referred to, in the genera Trigonia and Pholadomya among the acephalous mollusks, in the genus Pentacrinus, and in Ceratodus, among fishes. The dinosaurian reptiles, whose range in Europe only exceptionally extends to the top of the Cretaceous series, the Maestrichtian,[61] are carried in the Western Territories of the United States to a horizon probably considerably higher, the Laramie, whose position is now generally recognised as Post-Cretaceous. Indeed, if the reported association in these beds of one of the Hadrosauridæ with the remains of the mammalian Meniscoëssus be taken in its full value, we may not unreasonably assume this fact to be further evidence, in addition to that which has already been adduced, in favour of uniting at least a portion of the Laramie with the Eocene. Surely the circumstance that dinosaurian remains are found in these deposits is not of itself, as against all other evidence, sufficient to establish the age of the formation. What is there that we know of that should prevent the animals of this group from continuing up into the Eocene, any more than the Carboniferous and Permian labyrinthodonts into the Trias, and the Cretaceous Crocodilia into the Tertiary ? It is not to be supposed that a common catastrophe awaited all the members of this numerously-represented order of animals, when by their organisation they seem to have been fitted to such varying conditions of environment.*

* Were these reptilian remains the only fossils of a distinctive character found in the Laramie formation, then, naturally, we should conclude that the formation in question was of Cretaceous age; at least, the only evidence we had would be in favour of such a conclusion. But the case, as it here stands, is quite different. The fossil plants of the Lignite, as is well known, are almost altogether of Tertiary types, and many of the species, even, have been identified, by Starkie Gardner, Lesquereux, and others, with characteristic Lower Tertiary forms occurring in various parts of Europe (Island of Sheppey, &c.). The shell-fauna, as a whole, can scarcely be said to approximate very much more to the one side than the other, although a few species have been recognised by both Conrad and Meek as being more nearly Tertiary than Cretaceous. These, however, do not indicate much. The recent discovery of Meniscoëssus adds a much more powerful link to the evidence which favours the Tertiary side of the question; indeed, by itself it argues about as much for the Tertiary age of the formation as the reptilian remains do for the Cretaceous, for, were there no such conflicting testimony as in reality exists, its evidence would be accepted as conclusive. Further evidence in this direction is afforded by the reptilian genus Champsosaurus, which, as a member of a fauna, the Puerco, originally supposed to represent the Laramie, has been identified as an Eocene genus in the north of France. While, perhaps, it may be admitted that paleontology has not thus far given us absolute data by which to determine the question at issue, yet, on the whole, its facts appear to lean more towards the Tertiary side. From stratigraphy we learn but little, as, unfortunately, marine deposits of Post-Cretaceous age are wanting in the interior of the continent, thereby rendering the

Because they became extinct at a given period of time in one region is no reason why they should have faded out during the same period in every other; nothing could be more illogical than such an assumption, and the facts, in numerous parallel cases, clearly demonstrate its utter fallaciousness. Then why lay such stress upon the occurrence of these animals as an indication absolute of geological time ? Every day's lesson teaches the geologist how unstable are the limits that have been assigned by him to the duration of life in species, or groups of species; it ought to be, therefore, a matter of no surprise, but the reverse, to find certain so-called " distinctive " or " characteristic " forms becoming less and less distinctive with the progress of investigation. The discovery of dinosaurian remains in Tertiary deposits, or of trilobites in the Permian, should give far less cause for surprise than a positive announcement that they did nowhere so occur.

What the proximate cause of the extinction of species or groups of species may have been, it is in most cases impossible to determine. The process is such a gradual one, and its manifestation so casual, that we fail to see what it is just exactly that acts. If, as Darwin puts it, " we ask ourselves why this or that species is rare, we answer that something is unfavourable to its conditions of life; but what that something is, we can hardly ever tell. On the supposition of the fossil horse still existing as a rare species, we might have felt certain, from the analogy of all other mammals, even of the slow-oreeding elephant, and from the history of the naturalisation of the domestic horse in South America, that under more favourable conditions it would in a very few years have stocked the whole continent. But we could not have told what the unfavourable conditions were which checked its increase, whether some one or several contingencies, and at what period of the horse's life, and in what degree, they severally acted. If the

necessary correlation of strata impossible. At Laredo, on the Rio Grande, Texas, the Claiborne beds (Parisian) are stated by Professor Cope to rest "immediately on the Laramie" ("Proc. Am. Phil. Soc.," 1884, p. 615). If this is really the case. then it is more than likely that the latter is at least in part the equivalent of the basal Tertiary, otherwise it would be difficult to account for the sudden disappearance here of the vast thickness of sub-Claibornian deposits (Buhrstone ; Eo-Lignitic), measuring hundreds of feet, which elsewhere along the Atlantic and Gulf borders forms the base of the Tertiary series.

conditions had gone on, however slowly, becoming less and less favourable, we assuredly should not have perceived the fact, yet the fossil horse would certainly have become rarer and rarer, and finally extinct; its place being seized on by some more successful competitor." The case of the horse here stated strikingly illustrates the mystery in which the subject of extinction is still involved, and how very limited is our knowledge in this direction. Why the animal should have become extinct on the American continent, and at so early a period after its evolution, when under apparently identical, or, at any rate, very similar, physical conditions, it continued to develop and thrive in the Eastern Hemisphere, is a problem towards the solution of which we can offer but vague conjecture. Nor is the difficulty in the matter a whit lessened, but rather the contrary, by the circumstance of the ready adaptability to its apparently unfavourable environs which the more recently introduced, or modern, horse has shown. Instances of this kind, anomalies to our existing knowledge, are by no means rare. The extinction of the mammoth in the far north of the Eastern Hemisphere, and the survival of its first cousin, the elephant, in the south, are equally inexplicable. Both, as far as we are permitted to judge, appear to have been in harmony with their surroundings; vegetable-feeders, they inhabited regions of sufficiently luxurious vegetation, the one, provided with a shaggy coat of hair to protect it from the rigours of the frozen north, and the other, more nearly naked, suited to a home where little or no protection from climatic extremes was necessary. Both, again, were inhabitants of regions where a struggle against the attacks of savage Carnivora was a part of their existence, and if any advantage favoured the one side above the other in such internecine warfare, it was on the side of the northern species. It may just be, in the case of the mammoth, that the extreme cold of the Glacial epoch, combined with a continuous submergence of the land surface beneath an ice-cap, so far reduced the plant-growth of the north as no longer to provide adequately for the sustenance of these monsters, and that, as a consequence, they gradually diminished in numbers, and eventually completely vanished. Such a supposition, however, must needs remain in the nature of a mere hypothesis, until more facts than we now possess shall have been gathered, indicating for it a high probability. Similarly, the theory which accounts for the disappearance of the mammoth on the supposition that it was swept off by unknowably great glacial floods following in the wake of the northern ice-sheet, has as yet little to support it. Why just the mammoth should have been thus swept away, when other animals, like the reindeer, contemporaries of the mammoth, and like it animals of the frozen north, survived, is not very comprehensible.* The extinction of the musk-ox in Europe and its survival in America offers a no less remarkable puzzle to the biologist.

That very frequently what may appear to be insignificant causes are sufficient to bring about extermination is shown in the case of many of the largest, and seemingly most resisting, animals. The arrested numerical development of the Indian elephant has been attributed by Dr. Falconer, a competent authority, to the unceasing harassings of insect pests, a view which was also shared by Bruce with respect to the elephant of Abyssinia. And we are assured, on the authority of Darwin, that "insects and blood-sucking bats determine the existence of the larger naturalised quadrupeds in several parts of South America." The ravages of the tze-tze among the South African ruminants has long been commented on by travellers, and the "plague" of the mosquito is only too familiar to require special consideration. Humboldt has graphically delineated the numerous circumstances, including inundations, parched vegetation, ravages of wild beasts, and the like, which in many of the grassy regions of South America threaten the destruction of both cattle and horses, and these, or similar ones, exist over most parts of the earth's surface. In short, a perpetual check is placed upon the free increase of all classes of organisms, the overcoming of which displays the measure of success in the universal struggle for existence. We know of no law which by itself determines the duration of life in any group of organisms, or which explains why

* The mammoth, although its remains are most abundantly found in the far north, cannot rightly be classed as an exclusively northern animal, as is proved by the discovery of its bones in regions as far south as Santander, in Spain, and Rome, in Italy. Indeed, it is not exactly impossible, as is claimed by Boyd-Dawkins ("Early Man in Britain," p. 108), that the modern Indian elephant is only a varietal form of this species which, through long habitation in the tropical or semi-tropical forests, has lost some of those minor characteristics, such as the coating of hair or wool, which serve in a general way to distinguish the northern animal. If this be true, then the mammoth would still be a member of our existing fauna.

certain groups are much longer or shorter lived than others of a very closely related nature.

Persistence of Type-Structure. — The persistence of certain type-structures is very remarkable. Not only have they in a measure resisted all the modifying influences which Nature has brought to bear upon them during a period of hundreds of thousands or millions of years, but in such a manner as to render a specific separation of their newest and oldest representatives a matter of considerable difficulty. The Lingula of to-day differs but little from the Lingula of the oldest fossiliferous formation with which we are acquainted, the Cambrian, although the interval of time separating the two has been variously put by geologists and physicists at from one hundred to three hundred millions of years ! The pearly Nautilus is but little removed from some of its most ancient representatives of the Silurian and Devonian seas, and, indeed, it may be considered doubtful whether some of the existing Foraminifera are at all different from forms that occur deep down in the Paleozoic series of deposits. The number of generic types that have survived the Paleozoic era to the present day are sufficiently numerous, but these belong as a rule to the lower classes of organisms. Among the Brachiopoda we have at least five such genera —Lingula, Discina, Crania, Rhynchonella, and Terebratula—and two of these, Lingula and Discina, date back to the Cambrian period. The Acephala and Gasteropoda furnish an equally large number—Pecten, Mytilus, Nucula, Leda, Pinna, Lima, among the former, and Pleurotomaria, Capulus, Turbo, Natica, Dentalium, and Chiton, among the latter—and it is by no means improbable that many of the older genera, now recognised as distinct by reason of our imperfect knowledge concerning their true relationships, have in reality representatives living in the modern seas. Of the Paleozoic Cephalopoda we have but a single surviving genus, Nautilus. Several of the recent genera of entomostracous Crustacea (Estheria, Cypridina, Apus) range back in time to the Devonian or Carboniferous periods, and a more limited number (Bairdia, Cythere) even to the Silurian; but of the higher decapodous types we meet with no (even doubtful) modern generic representatives until the Carboniferous limestone is reached (Astacus Philippi, a supposed species of crayfish, from the mountain limestone of Ireland). The king-crab (Limulus), not unlikely a descendant from the more an-

cient line of Trilobites, appears for the first time in deposits of Jurassic age.   Contrary, perhaps, to what might have been expected,
following out the law of the persistence of lower over higher organic types, not one of the numerous genera of Paleozoic corals
has survived up to the present period; and what is still more surprising, even the broad structural type which they embody has,
as far as we now know, almost completely disappeared.   What particular conditions tended towards their extermination, and to their
being supplanted already in the early part of the Mesozoic era by
entirely new structural forms, we know not, and probably never
shall discover.

In sharp contrast to the more persistent types of animal life are
certain groups whose appearance and disappearance are alike sudden, and whose whole existence is measured by a very brief period
of geological time.   Such, for example, are the Rudistæ, a family
of acephalous mollusks which attains an extraordinary development
towards the close of the Mesozoic era, but all of whose members
appear to be restricted to the Middle and Upper Cretaceous periods.
Indeed, by far the greater number of representatives of this family,
constituting the genus (with several sub-genera) Hippurites, are
limited exclusively to the deposits of the Chalk and Chalk-marl.
An equally remarkable example of limitation in range is furnished
by the Graptolitidæ, one of the most widely distributed families of
invertebrates with which we are acquainted, not a single undoubted
representative of which is known either before or after the Silurian
epoch.

It is a singular fact that all of the more ancient terrestrial
air-breathing Mollusca that have thus far been discovered belong
not only to modern groups, but mainly, also, to modern genera.
Leaving out of consideration the more than doubtful Palæor-
bis, which by some is considered to be a land-snail, there have
been described, all in all, some seven or eight Paleozoic species, beginning with the Middle or Upper Devonian, all of them
from the deposits of the North American continent.   Three of these
have been referred to the genus Pupa, one to Zonites, and three
others to genera that have been created for them, respectively:
Dawsonella, Anthracopupa, and Strophites.   The first of these is
not unlikely a true Helix, while the last, based upon a single imperfect specimen, is too ill-defined to permit of its being classed

with certainty among the terrestrial pulmonates.[62]   It must be pre-
mised that these remains constitute only an infinitesimal fraction of
the entire pulmonate fauna of the period, and therefore it is impossible to say in what ratio the recent generic types stand to types
that are not recent, but which may have lived and flourished and
left no traces of their existence behind them.   The case as it
stands, however, is sufficiently interesting, and permits us to assume that the process of modification among the land Mollusca
was an exceedingly slow one, probably very much slower than
among the corresponding marine forms of life.   This seems also
to have been the rule with the fresh-water Mollusca, whose development in time runs about parallel with that of the terrestrial
Pulmonata.   It is now practically certain that the range of the genus Anodonta extends at least as far back as the Devonian period,[63]
and not improbably the forms described as Naiadites, from the
Coal-Measures of Nova Scotia, are true Unios.   No unequivocal
fresh-water Mollusca are as yet known from the Permian formations, and even in the Trias the number of such forms, doubtfully
referred to the genera Unio and Myacites, is very limited.   Only
with the succeeding formation do we have the first considerable
development; but from that time onward to the present day the
number of species, in most cases referable to existing genera, rapidly
increases.   The earliest fresh-water gasteropods date from the Jurassic period, and are comprised almost altogether in the modern genera
Neritina, Planorbis, Vivipara, Valvata, Hydrobia, and Melania.   It
should here be observed that, while from the preceding data it may
appear that, with few exceptions, all the earlier (as well as later)
fluviatile mollusks, whether lamellibranchs or gasteropods, belonged to genera which still flourish in our fresh waters, this needs
not necessarily have been the case as a matter of fact; for many
forms that, by reason of their association with marine organic types,
have in themselves been classed as marine, may have been of a distinctly fresh-water habit.   It is inconceivable that the only evidences
of life in the ancient waters of the land should be centered in the
few organic remains that we recognise to be of an indisputably
fresh-water character.   Rivers, then as now, discharged into the
sea, and deposited large quantities of sediment along the continental borders.   It could scarcely have happened otherwise than
that more or less perfect parts of shells, swept down by the currents,

should have accumulated in these river deposits in considerable numbers; and, when once there, there is no reason for supposing that their destruction would have been very much, if at all, more rapid than in the case of marine shells. But, if none of the forms that occur associated with distinctively marine types are in reality of a fresh-water nature, what has become of these remains? Surely it cannot be that they have suffered complete, or even nearly complete, destruction. It seems far more reasonable to assume that many of the forms described as marine, from indisputably oceanic deposits, are in reality not such; and this may be the case not only with such forms as, in their generic characters, are but barely distinguishable from known fresh-water types, but even with those which have a strictly marine facies. For there can be little or no question that the primitive fresh-water fauna was a derivative from the marine; hence, the earliest fresh-water types must have been of a structure but little different from that of their oceanic progenitors, and barely, if at all, distinguishable by external characters from them. From the first, however, they will have been subjected to the modifying influences which result from a change in the physical conditions of the environs, and which have wrought in the course of ages (and rendered more or less permanent) those structural features which, at the present day, serve to distinguish the marine from the fluviatile type of organism.

If direct proof of the ready adaptability of marine or brackish-water organisms to fresh-water conditions, or the reverse, were needed, no more decisive testimony in this direction could be had than is furnished by the ancient lake region of the Western United States, which marks the position of the Laramie or Lignitic formation. The lacustrine deposits of this formation, which attain a maximum development of some four thousand to five thousand feet, have evidently been laid down in lake-basins formed through the land-locking and slow desiccation of a continental arm of the sea, which projected completely across the United States during the Cretaceous period. The exclusively marine character of the organisms which flourished at this time clearly indicate what was the condition of the waters which they inhabited. Through a gradual elevation of the land, which appears to have set in about the close of the Cretaceous period, and the consequent formation of barriers, the waters of this vast inland sea, by reason of their severance from the oceanic basins, and the indraught of fresh water from its drainage area, progressively lost their salinity, becoming more and more fresh with the advance of time. Many or most of the molluscan types that flourished during the period of greatest salinity slowly disappeared from the region — through extermination or migration—the result of an innate incompetence to adjust themselves to the changing conditions of their surroundings. Others, more fortunate, by slow degrees accommodated themselves to the newly-imposed conditions, and found a congenial home in harmonic association with such forms as the inflowing fresh water may have thrown in with them, and which were in their way undergoing a reversed modification. That this must have been the actual history of at least a considerable part of the Laramie region there can be no question, seeing what a remarkable commingling of fresh water and marine, or brackish water, molluscan types is there exhibited. Thus, we have frequent associations in the same stratum of the genera Corbicula, Corbula, Unio, Neritina, Vivipara, and Goniobasis, or Corbicula, Corbula, Ostrea, and Anomia, and, as Dr. White [64] informs us, "the commingling of brackish - water [Ostrea, Anomia, Corbula] and fresh-water forms occurs in some portions of the Laramie deposits under such conditions as to compel the belief that some of them at least lived and thrived together." In further association with these forms are representatives of the genera Leda, Pectunculus, and Odontobasis, which, otherwise, are known only from marine deposits.

It is true that, in the case of the Laramie fauna, we have no evidence proving that any of its distinctive fluviatile types of organisms have been descended by modification from marine forms, inasmuch as all of them may have been primarily introduced into the sea by the inflowing streams. That this was the case with many of the genera is indisputably shown by their great antiquity, compared with that of the fauna of the Laramie lake-basin. But it is by no means unreasonable to assume, seeing how very heterogeneous in its character was the fauna of the period, that, had the freshening and desiccation of the primitive inland lakes been less rapid, a complete transformation of marine into fresh-water types with the retention of permanent characters, might have been brought about. Such transformation appears to have taken place in the case of a portion of the molluscan fauna of Lake Baikal,

and in that of two or three genera of mollusks recently described by
Smith and Bourguignat from Lake Tanganyika—Syrnolopsis, Lim-
notrochus, and Rumella—which have an undeniably marine facies.[65]
But this adaptability of animals to what might be called opposite conditions of existence is not confined exclusively to the class
of Mollusca; indeed, were this the case we should have but a small
fragment of our fresh-water fauna accounted for.   The metamorphosis of Branchipus into Artemia, and of Artemia into Branchipus,
is a well-known application of the law to the class of Crustacea.
The existence of seals in Lake Baikal, whose salinity is practically
nil, is a remarkable instance of much the same kind among the highest animals, the Mammalia, while as a noteworthy offset to this,
coming from very nearly the lowest of all organisms, is the occurrence of marine, or non-statoblastic, types of sponges (Lubomirskia)
in the same lake, as well as in the Upper Congo (Potamolepis).[66]

Variation in Persistent Types.—It is frequently asserted, even
by those who are considered competent of forming an opinion,
that the fact of our having in the existing faunas a number of very
ancient types is a proof positive against the validity of any slow
modification theory of descent.   For, it is contended, we are here
brought face to face with certain structural types which have resisted all sensible modification during a period of millions of years,
and have, consequently, baffled all evolutionary tendencies towards
reorganisation.   Thus, we have, as has already been stated, the
Lingula of to-day practically identical with the oldest Lingula
known; the modern Nautilus, but little different from the Nautilus
of nearly the most ancient Palæozoic deposit, and the modern Pupa,
practically identical with the Pupa of the Carboniferous period.
A little reflection will show how illogical is the position which
is here assumed.   That certain species should have come down
to us with but slight modification from the earliest periods known
is about as much proof of non-modification in the group to which
they belong as would be furnished in the case of a house, where
the retention of the primitive type, the hut, might be taken in
evidence of non-modification in the whole class.   If a want of
adaptation to the surroundings be the primary cause of variation,
then, manifestly, the rate of variation among the members of a
given group of animals cannot be a uniform one, for while some
of the members will be forced to extreme measures by reason of

their inability to cope with the less favourable conditions of existence to which, through one cause or another, they may be subjected, others, under more favourable circumstances, and requiring no adaptation to new conditions, will continue unchanged as before. Hence, the same stock may continue in a direct line, and yet, at the same time, throw off a number of side branches, whose ulterior development may, or may not, keep pace with the main stem. That the simple transformation of a group *requires* either the immediate or the ultimate obliteration of that group, has nothing to support it. Few naturalists at the present time question the descent of at least some of the races of the domestic dog from one or more species of wolf or jackal, yet these are living side by side, and apparently without interfering with one another's wants. That a decadence in the one group or the other may ultimately set in—in fact, has already set in—cannot be denied, but it may be safely doubted whether, if it had not been for man's intervention, extermination of either the wolf or jackal would have preceded that of the dog. A struggle with competitors in a certain quarter of the globe, or the necessity of conforming to new conditions of existence, may have developed specific characters in some of the wolves which they had not hitherto possessed, and which would not be necessary for more favoured forms occurring elsewhere, or, possibly, living even in the same region. In the same way that modification of the nautiloid type which, it is assumed, has resulted in the formation of the genus Goniatites, does not appear to have materially affected the parent stem, for we find the genus Nautilus itself developing in almost equal abundance in the periods succeeding the introduction of the goniatite as in those preceding it. In the deposits of the Carboniferous formation, which represent the period of maximum specific development in this genus, we have, according to Zittel, eighty-four species represented; in the Trias about seventy, in the Jura fifty, and in the Cretaceous deposits still sixty to seventy. The actual downfall begins only with the Tertiary, where through the entire series there are but fifteen known species. What should have brought about this sudden decadence it is, of course, impossible to determine. Cases of undoubted transformism and extreme persistence like the one here instanced are undeniably rare, but this seeming rarity is to be attributed in great part to our ignorance respecting the genetic relationship which binds together the more ancient forms of life. The Brachiopoda, however, offer partial examples of this kind, more particularly the family Terebratulidæ, and, where but a comparatively limited persistence is demanded, such instances are by no means rare.

Geographical Distribution.—In comparing the past with the present distribution of life upon the globe, one cannot fail to note a well-marked difference. Broad distribution appears to have been far more prevalent in the early period of the earth's history than now, and argues strongly for a predominance of more uniform conditions. Thus, if we take the class Brachiopoda by way of illustration, we find that of some one hundred and thirty-five recognised species and varieties living in our modern seas, there is scarcely a single species which can be said to be strictly cosmopolitan in its range, although not a few are very widely distributed, and, if we except boreal and hyperboreal forms, but a very limited number whose range embraces opposite sides of the same ocean. On the other hand, if we accept the data furnished by Richthofen [67] concerning the Chinese fossil Brachiopoda, we find that out of a total of thirteen Silurian and twenty-four Devonian species, no less than ten of the former and sixteen of the latter recur in the equivalent deposits of Western Europe; and, further, that the Devonian species furnish eleven—or nearly fifty per cent. of the entire number—which are cosmopolitan, or nearly so. Again, of the twenty-five Carboniferous species North America holds fully fifteen (or sixty per cent.), and a very nearly equal number are cosmopolitan. The evidence furnished by the fossils of the Arctic regions is equally conclusive in this direction. Not only are the vast majority of species of Paleozoic (Silurian and Carboniferous) fossils of the far north identical with forms occurring in the deposits of both temperate North America and Europe, but many of them are distinctive types of formations which have been recognised in almost all parts of the earth's surface where such formations have been themselves identified. Thus, among the fossils obtained by the officers of the late Polar expedition under command of Sir George Nares, Mr. Etheridge [68] has identified Atrypa reticularis (Silurian) and Productus semi-reticulatus and P. costatus (Carboniferous), the first from Cape Hilgard, in nearly the eightieth parallel of north latitude, and the last two from Fielden Isthmus, latitude

82° 43′. These species, moreover, appear to have been just as abundant in their respective positions north of the Arctic circle as they were south of it. The coral fauna of the same region comprises, among other forms common to both Europe and North America, such wide-spread species as Favosites Gothlandica and Halysites catenulata, both from latitude 79° 45′ (Cape Frazer); Lithostrotion junceum has been found as far north as latitude 82° 43′ (Fielden Isthmus).

Turning our attention to a somewhat antipodal portion of the earth's surface, Australia, we find that by far the greater number of fossils that have been catalogued from its Paleozoic deposits are species that were originally described from regions lying well within the limits of the north Temperate Zone. In fact, the relationship existing between this southern fauna and the faunas of Europe and North America is so great as to practically amount to identity. This correspondence is perhaps as well exhibited by the graptolites as by any other group of animals, for we find that of the twenty-four species recorded by Mr. Robert Etheridge, Jr., in his "Catalogue of Australian Fossils" (1878), no less than eighteen are species belonging to the United States and Canada. When we seek to explain the broad distribution of life in the early periods of the earth's history compared to what it is at the present time, it will naturally be concluded that greater facilities for dispersion, and the prevalence of more equable conditions of climate, especially the latter, were the prime factors involved in this distribution. If, as may be contended, other conditions were also largely instrumental in bringing about the general result, we are entirely ignorant of their nature. That a different disposition of the land and water areas than now obtains may have facilitated distribution in a manner now no longer possible must be conceded, and there are abundant proofs that considerable alterations of one kind or another, whether in favour of, or against, dispersion, have at various times taken place. But it is beyond question, as is shown by the distribution of our existing marine fauna, that while the relations existing between land and water have much to do with distribution, yet the determining factor in such distribution is after all the matter of temperature. The comparatively limited north and south extent of any fauna, even along a continuous coast line, contrasted with its broad east and west range, sufficiently proves this to be the case; and if further or more convincing proof were needed, we have the fact, made pregnant by the recent deep-sea dredgings, that many of the forms contributing to the surface faunas of the north, and hitherto recognised as being essentially northern, are in reality inhabitants of the southern zone as well, only that they here constitute a part of the deep-sea (cold water) fauna instead of the more superficial one. Naturalists have been in the habit of recog- nising four or more distinct zoological provinces along either border of the Atlantic, each one characterised by a more or less well-marked assemblage of animal species; and about an equal number of such provinces have been assigned to the Pacific littoral. In all these provinces the admixture of forms belonging either north or south, it must be admitted, is very great, so much so as to render the drawing of a line of division a matter of the utmost difficulty; nevertheless, taken in their entirety, the faunas are sufficiently dis- tinct, and serve to mark the climatic influences which limit dis- tribution. Of some five hundred and sixty-nine species of Mollusca recognised by Fischer, in 1878, as occurring on the Atlantic shores of France, no less than four hundred and twenty-seven, or seventy- five per cent., also belong to the British coast, and about an equal number to the Mediterranean.[69] On the other hand, of three hundred and fifty-three species obtained by M'Andrew from the southern coast of the Iberian Peninsula, only fifty-one per cent. were common to Britain, and a much smaller number, twenty-eight per cent., to Norway.[70] The molluscan fauna of the Canary Isl- ands numbers about three hundred species, of which sixty-three per cent. belong to the coasts of Spain and the Mediterranean, thirty- two per cent. to Britain, and seventeen per cent. to Norway. The relation existing between northern and southern, or cold and warm water, faunas is perhaps still better marked on the Western Atlantic border. Thus, of a total of about three hundred species of mollusks belonging to the "Transatlantic Province"—i. e., the eastern coast of the United States included between Florida and Cape Cod—only sixty recur north of the peninsula of Cape Cod. Again, of seventy- nine species of shells collected by D'Orbigny from the coast of North- ern Patagonia only twenty-seven were common to Uruguay and Brazil. The rich molluscan fauna of the Japanese Archipelago, com- prising four hundred and twenty-nine species, has only one hundred and eighty-five representatives in the Chinese and Philippine waters.

In sharp contrast to the limited faunas of a north-and-south extension is the fauna of the Indo-Pacific region, whose domain covers about forty-five degrees of latitude, mainly comprised within the Tropical Zone, and extends over fully three-quarters of the circumference of the globe. The influence of an equable climate is here plainly manifest. The molluscan species occurring in this tract have evidently a very broad distribution, for we find that, out of an estimated total of five or six thousand species, the Philippine Islands alone possess twenty-five hundred species, and New Caledonia an equal number. Upwards of one hundred of the East African species have been identified by Cuming in the faunas of the Philippines and the coral seas of the Pacific, or over an expanse of seventy to one hundred degrees of longitude, and Fischer enumerates twenty-one species whose range takes up practically every part of the province. On the other hand, only one hundred and sixty-five species are known to connect this fauna with the Japanese, although the two are separated from each other, in a north and south line, by an interval of only ten to fifteen degrees of latitude.

Facts such as have been here presented clearly demonstrate how all-powerful in its influence upon distribution is temperature, and warrant us in assuming that it was this agent, likewise, which primarily controlled distribution in the past as well as in the present. Were evidence of a nature other than that which is derived from purely zoogeographical considerations needed to prove the existence of more equable climatic conditions in the earlier periods of the earth's history, we have the testimony to this effect of the ancient reef-building corals, whose remains are so abundantly implanted in the deposits of the temperate and frozen north, and, in no less striking degree, cf the flora of the coal.

It is, however, a significant fact, that many parts of the oceanic surface which may be said to enjoy practically identical climates hold at the present day very dissimilar faunas (viewed from the stand-point of species). Thus, of the four hundred or more species of mollusks inhabiting the Japanese waters, it appears that not more than twenty are found on the west coast of North America (Oregon, California, Mexico). Nearly all the east-coast species of the United States, south of Cape Cod, differ from the species of the corresponding regions on the opposite side of the Atlantic, probably not more than fifteen or twenty being common to Europe. Evidently the oceanic abysses, with their deep layers of cold water, constitute an almost insuperable barrier to the free migration of the animals belonging to this class. It is a little remarkable that no larger proportion (about fifty per cent.) of species should be common to the European and American " Boreal " sub-regions. The influence of land-barriers in shaping distribution is still more marked than that of the sea. This is best seen in the case of the Mediterranean and Red Sea molluscan faunas, where, of eight hundred and eighteen species dredged by M'Andrew in the Gulf of Suez, only three were found to be identical with forms occurring in the Mediterranean ! For a long time it was supposed by naturalists that not a single molluscan species occurring on the west coast of the Isthmus of Panama reappeared on the Atlantic side, and, if it is now known that this supposition was not absolutely in accordance with the facts, it must be admitted that the number of recognised transgressional forms, thirty-five out of five hundred to six hundred,[71] is very insignificant. With the facts here stated before us, it cannot be doubted that the broad dispersion of animal life in past periods of the earth's history was not only conditioned by favourable climatic circumstances, but, in a marked degree, by the absence of barriers to a free migration. That the amount of land-surface permanently exposed during the Paleozoic era was insignificant, in comparison with that exposed at the present day, is strongly indicated by the vast extent covered by the more ancient marine deposits, from which it would appear that the seas were at that period practically continuous throughout their broadest expanse. But this general supposition is based upon the hypothesis of the permanence of continents and oceanic basins, which, as has already been remarked, has much in its favour, but is still far from being in the nature of a demonstration. Granting the probability of continental upheavals from the oceanic abysses, then, naturally, must all conjectures regarding the relationship between land- and water-surfaces, and the existence of interposing barriers, be valueless. It may be hastily concluded that the circumstance of very broad distribution is, in itself, strong evidence tending to show that such continental elevations did not take place, and that there was a true permanency in the position occupied by the sea. This need not necessarily have been the case, however, for, with an equable temperature over the greater part of the earth's surface, it does not appear why species, or groups of species, should not have readily found their way to the most distant localities by simply following the line of coast. The marine Antarctic fauna of the present time is more nearly related to the Arctic than to any other, and, if the species of the two are almost altogether different, yet a large proportion of the generic types, wanting in the intermediate region, are the same. Such are, for example, the genera Trophon, Buccinum, Margarita, Astarte, Admete, &c. A limited number of the species, too, are identical with species occurring in the waters of the north Temperate Zone. There can be little doubt that the distribution of these forms to the antipodal regions of the earth's surface was effected by way of the deeper zones of cold water, and there seems to be no necessity for invoking the aid of a "spontaneous" Antarctic fauna to account for the separation that exists between it and the Arctic. If this explanation is the correct one, it is then a little surprising that we do not meet, in the two regions, with a larger number of identical forms; but it can readily be conceived that the necessary accommodation to new conditions of existence, imposed by pressure, absence of light, and a different food-supply, may have brought about, in the course of such a lengthy migration, variation in specific characters, without sensibly interfering with the structural type of the group.

If it now be assumed that, during the Paleozoic era, the broad distribution of species was effected in pretty much the same manner —i. e., by following the trough of the sea rather than the continental border—we are at once confronted by the anomaly that the number of identical species occurring at the most widely separated localities, instead of being very limited, is just the reverse. It is true that, with a high temperature extending to the bottom of the sea, specific modification resulting from a transference of abode from the surface to the greater depths may have been less marked than appears to be the case with a low temperature; but we have no reason for supposing that any great variation in this respect did exist, and almost certainly not enough to account for the differences that present themselves. It would thus seem that the broad distribution of former periods was effected principally along predetermined coast-lines. One circumstance, however, which has thus far received but little attention from physiographers, must not be lost sight of in this connection, and that is, a difference in depth of the oceanic abyss. On the theory of the permanency of oceanic and continental areas, and the strong probability that the oceanic basins really represent areas of subsidence, it may be confidently assumed that the floor of the ocean has been pretty steadily subsiding, from first to last; and, further, that the continental off-flow, produced by the rise and development of the land-masses, will have just as steadily tended to increase the depth of water in these basins. At what rate this subsidence and gradual deepening may have taken place it is impossible, in the present state of science, to determine, and therefore we possess no means of ascertaining what might have been the difference in depth for any two widely separated periods of geological time. But it is by no means improbable that the depth of sea during the greater part of the Paleozoic era was very much less than it is at the present time; indeed, we are almost irresistibly led to this conclusion by our knowledge that the water formerly occupied a much greater lateral extension, measured on our present sphere, than it now does, and this, in addition, at a period when the equatorial circumference of the globe was probably considerably in excess of its present twenty-five thousand miles. With only a moderately deep sea, the difficulty in accounting for a phenomenally broad specific distribution is greatly lessened.*

When we seek to determine at what period the existing conditions of distribution, or approximately such, were first introduced, we find that it was not until the beginning of the Tertiary, although the gradual modifications leading up to this change are clearly traceable in the successive periods following the Paleozoic era. Of some thirty-six species of fossils described by Coquand and Bayle from the Jurassic deposits of Chili, South America, no less than twenty, or more than one-half, are forms which are also found in the equivalent deposits of Europe, and a fair proportion of these have been identified at various points on the continent of Africa, in the peninsula of India, the Himalayas, and elsewhere. The Jurassic Cephalopoda of Kutch, India, comprise, according to Waagen,[72] one

---

* It is true that, through terrestrial absorption, the oceanic mass is undergoing diminution in bulk; but it may be questioned whether the shallowing produced thereby, since the beginning of the Paleozoic era, has very materially affected the general depth of water.

hundred and fifty-six species, of which number at least forty-seven
(or thirty per cent.), and probably considerably more, belong equally to the fauna of West Central Europe. Altogether, the Jurassic
faunas of the world, even those most widely separated, are most
intimately related to one another in both generic and specific characters. The same may be said, although to a less extent, of the faunas
of the Cretaceous period, many of whose most distinctive types,
especially of cephalopods and lamellibranchs, have a world-wide
distribution. The cephalopodous fauna of the Cretaceous deposits
of India, we are informed by Stoliczka,[73] holds, among its one hundred and forty-eight species, at least thirty-eight that are common
in Europe (twenty-five and one-half per cent.), a high proportion,
but somewhat less than what we have seen obtains in the case of
the fauna of the period preceding, the Jurassic. Roemer[74] has determined fourteen out of one hundred and twenty-eight species
(eleven per cent.) of fossils coming from Texas to be identical
with European forms, while Morton has recognised no less than
seventeen such from among one hundred Invertebrata belonging to
the State of New Jersey. This proportion has since been raised
by Credner.[75] The researches of D'Orbigny in South America,
of Coquand and Rolland in Africa, of Stoliczka and Waagen in
India, and of M'Coy, Etheridge, and others in Australia, show
very clearly how far specific identity is carried over the earth's surface.

With the beginning of the Tertiary epoch (Eocene period) a new
era in zoogeography sets in. The broad dispersion of species has
become much more of a rarity than heretofore—indeed, an exception—so that the widely separated regions of the earth's surface,
while they may yet be mutually related to one another in their
general faunal characters, are no longer bound together by that multiplicity of identical specific types which is distinctive of the earlier
periods. The conditions which prevail at the present day have
already become accentuated. This fact is clearly brought to light
by a study of the deposits occurring on opposite sides of the Atlantic. Thus, it is doubtful whether, out of some four hundred to
five hundred species of mollusks belonging to the Eocene formation
of the Atlantic and Gulf borders of the United States, more than
twenty-five or thirty (six to eight per cent.) can be absolutely identified with forms occurring elsewhere. And yet we have seen that

the proportion among the Cretaceous species for the same region is fully the equivalent of seventeen per cent. In the Miocene the number of such equivalent forms appears to be still further lessened. We find that among the Australian Tertiary Mollusca by far the greater number of species are forms which have been for the first time described from that region, although the European contingent of the Cretaceous fauna is by no means an inconsiderable one.

Climatic Zones.—In a very ingenious paper on the Jurassic fauna,[76] Professor Neumayr, of Vienna, following the views that had already been expressed by Trautschold and Marcou, has attempted to show that climatic zones were already well differentiated in the Jurassic period of geological time, and that homoiozoic belts, corresponding to these, existed then pretty much as they do now. Thus, on the Eurasiatic continent he recognises three distinct Jurassic zones, a "Boreal," a "Central European," and a "Mediterranean" —the first comprising the region of Northern and Central Russia, from the Petchora to Moscow, and the deposits of Spitzbergen and Greenland; the second, the region lying north of the Alps, with France, England, Germany, and the Baltic provinces; and the third, the region of the Pyrenees, Alps, and Carpathians, with a probable continuation in the Crimeo-Caucasian belt. The faunas of these different regions are most intimately related to one another, so much so that at first sight they would appear to form a homogeneous whole. But Professor Neumayr has shown that, despite this apparent homogeneousness, there are certain well-marked differences which impress a distinct individuality on each of the three. Thus, as one of the most distinctive biological characters of the Mediterranean zone, we have the great development among the Cephalopoda of the ammonitic forms belonging to the groups (or genera) Lytoceras and Phylloceras, which are but feebly represented in the deposits of the Central European zone, and whose substitutes there are the genera Oppelia and Aspidoceras. Both these series of forms are wanting in the boreal zone, which also lacks the coral-reef structures of the last. Inasmuch as the differences here noted occur in regions which appear to have been in open-water communication with each other, it is contended that the absence from one fauna of the forms most distinctive of the other could only have been conditioned by direct climatic influences.

It is more than doubtful, however, whether the evidence that we possess is sufficient to prove this position. Were the conditions such as they are claimed, we should naturally look for some correspondence between the disposition of the ancient isothermal belts and the existing lines of latitude; but no such disposition appears to have obtained. Dr. Waagen, in his elaborate analysis of the Jurassic cephalopod fauna of Kutch, India, unreservedly admits its much greater affinity with the equivalent fauna of the Central European region than with the Mediterranean, although geographically it is placed in much closer juxtaposition with the former than with the latter. The genera Oppelia and Aspidoceras are here both abundantly developed, and scarcely less so the genus Phylloceras; Lytoceras, on the other hand, is restricted to two species. The fact that Kutch is itself situated only twenty-three degrees from the Equator, and fully twenty degrees to the south of the northern limit of the Mediterranean zone in Europe, is scarcely consistent with any theory upholding the zonal distribution of oceanic temperature. Neumayr maintains that the elevated temperature of the Mediterranean zone was induced by an equatorial current possibly flowing from the southeast, or, at any rate, in open communication with a sea in that quarter; but if this were so, surely the same temperature (and not improbably a considerably higher one) would have affected the Indian basin as well, whereas, on the theory set forth, we have here evidence of just the opposite character—*i. e.*, of a lower temperature. Moreover, the evidence obtained from the fossils of Australia and the east coast of Africa seems to point to the conclusion, in the opinion of Dr. Waagen,[77] that very probably "one large Indian Ocean existed during the Jurassic period, of which only the very outskirts have been preserved up to this day in India, the east coast of Africa and the west coast of Australia existing in nearly the same area as the Indian Ocean exists at the present time." It can hardly be conceived that colder currents flowing from the north should have so far antagonised the influence of the more southerly heated waters as to have brought about this singular discordance between the European and Asiatic faunas. And to make the problem still more intricate, it would appear that a part of the Himalaya Jura, the "Spiti shale," "from the prevalence of the genus Cosmoceras and the large number of Aucellæ," really belongs to the boreal zone.

The assumption that because certain marine genera, of whose habits we know nothing, are found in one region and not in another there must of necessity be well-marked differences in the thermometric conditions of the waters which they inhabit, is at best a precarious one, inasmuch as it is founded almost exclusively on negative evidence. This is clearly proved by the status of the Indian Jurassic fauna, where, as we have already seen, one of the distinctive "Mediterranean" genera of ammonites, Phylloceras, is abundantly intermingled with those supposed to represent a more northern facies, while the other, Lytoceras, is almost altogether absent. If it be assumed that the presence of Aucellæ in the Himalaya fauna sufficiently demonstrates the cold-water facies of that fauna simply because the genus is most abundantly developed in the northern regions, Central and Northern Russia, Siberia, Spitzbergen, and Greenland, why may it not just as well be assumed that the northern fauna was of a warm-water facies, from the fact that the same genus is abundantly represented in a region lying on the immediate confines of the tropics ? The great north and south range of the genus Aucella, which is also found along the thirty-seventh parallel of latitude in California (Cretaceous), far from proving the existence of homoiozoic belts, indicates rather the contrary.

But very little dependence can be placed upon the genera of Mollusca as indicating the thermal conditions of the waters which they inhabited. There is at the present time on the east coast of the United States, south of the forty-first parallel of latitude, only one species for each of the three boreal, or Arctic (so recognised), genera Astarte, Leda, and Nucula. Yet, during the Eocene and Miocene periods, these same genera were abundantly developed in the waters situated seven and ten degrees of latitude farther to the south, and where, consequently, we might conceive a stratum of cold water to have existed. But, unless certain astronomical or physical conditions prevailed at the time with which we are not acquainted, it is practically certain that the temperature of the water was at least as high, if not higher, than it now is. Nor can it be urged that in these Tertiary deposits we are dealing with a deep-sea (cold water) fauna, or that a low temperature might have prevailed simply as the result of the non-existence of a Gulf Stream—a continuous sea separating the North and South American continents—since we have just as conclusive testimony favouring the supposition of an elevated temperature in the presence of such tropical or sub-tropical genera as Voluta, Cypræa, and Oliva, whose remains occur associated with the other genera already referred to, and whose actual range extended at least as far north as at the present day, and in the case of Voluta still farther.

The conclusions reached by European geologists as to the exist-ence of two distinct belts or provinces in the Cretaceous area—a Mediterranean and a Baltic—have also been applied by Roemer [78] to the equivalent deposits occurring on the east and south American borders. The New Jersey greensands (Sennonian), for example, are petrographically and oryctographically correlated with the Upper Cretaceous deposits of Northern Europe, and more particularly with those of Northwest Germany, whereas the Texas basin is placed in a similar relation with the " Mediterranean " zone of Southern Eu-rope. The much more southerly position of the American beds relatively to the European is taken as conclusive evidence, not only that the present climatic variation on the opposite sides of the At-lantic, for the same parallels of latitude, had already then existed, but also the existence of a well-defined northeasterly trending current of warmer water, or Gulf Stream. That the New Jersey Cretaceous fauna departs most widely from the Texan cannot be denied; but the evidence is far from conclusive that this variation is the result of direct climatic influences. At the present day the fauna of the Gulf may be considered as a homogeneous whole, being represented by much the same forms along its northern, western, and southern contours. On the other hand, it differs almost wholly from the fauna of the east Atlantic coast, although there can be but little doubt that, were it not for the peninsula of Florida, a consid-erable number of the forms now occurring exclusively on the east coast would have found their way into the Gulf basin as well. During the Cretaceous period the intermixture of east and south coast species was very large, but, contrary to what might have been expected, and directly opposed to Professor Roemer's conclusions, the Gulf fauna, or that corresponding to an enlargement of the present Gulf area, appears to have been in itself of a very clearly recognisable multiple character. From the list of Cretaceous fossils of the United States, prepared by Mr. Meek, in 1864, [79] it would seem that, of some one hundred and sixty species belonging to the State of New Jersey, no less than fifty, or nearly one-third, are also com-

11

mon to the Cretaceous area comprised within the States of Alabama and Mississippi (forty-eight species to Alabama, twenty-seven to Mississippi), lying some four hundred miles farther to the south, and thus separated by an interval considerably exceeding that which separates the Mediterranean and Baltic zones of the continent of Europe. On the other hand, of about two hundred species catalogued as belonging to the State of Texas only some ten or twelve are indicated as forming a part of the Mississippi and Alabama faunas, although the last occupied a position but little removed, either geographically or climatically, from the Texan. These facts are hardly in consonance with the conclusions which have been reached respecting the past distribution of climatic zones, and, if broad differences do obtain between certain contiguous north and south faunas, as is undoubtedly the case, their explanation must be sought in causes other than those directly connected with climatology. The nature of the sea-bottom, depth of water, &c., may have had much to do in the matter.

Despite what may appear as overwhelming evidence, proving exceptionally broad distribution in the past periods of the earth's history, even as late as the Jurassic and Cretaceous periods, Professor Neumayr [80] and others have attempted to show that this condition did not actually exist, at least as far as the Mesozoic era is concerned, and that in reality specific cosmopolitanism is as well-pronounced at the present time as it was formerly. Thus, it is contended that the modern pelagic Mollusca, more particularly the Pteropoda, are of exceedingly wide distribution, and that even among the Cephalopoda a considerable number of more or less cosmopolitan forms— among them Argonauta Argo and A. hians—are to be met with. Furthermore, it is urged that, if well-marked faunal variations are manifested along the oceanic border, such variations do not obtain in the oceanic abyss, where a uniform type dominates the faunal facies, and with whose fauna many of the ancient life-series are to be compared. The facts relating to the distribution of the Mollusca, modern and ancient, that have already been given, are sufficient to prove the general erroneousness of the proposition here stated, and, if any more proof in this direction were needed, none more significant could be had than that drawn from the distribution of a member of one of the very groups of animals cited by Neumayr, the Cephalopoda. The nautilus, which to-day appears to be restricted to the tropical seas, was, even as late as the Eocene period, found abundantly as far north as the fortieth and fiftieth parallels of latitude. That some of the ancient faunas may represent faunas of the deep sea cannot be denied; but it appears far more probable, in the light of recent investigation, that they are in the main of a littoral character. At any rate, that they are not comparable to the cold-water fauna of the deep, which alone maintains somewhat of a uniform character, is indisputably proved by the abundance of forms indicating a high temperature. Furthermore, even the general identity claimed by Sir Wyville Thomson for the abyssal fauna of the world has quite recently been contested by the late Gwyn Jeffreys.[81]

Synchronism of Geological Formations.—It is well known that the order of deposit of the various formations, from the oldest to the newest, is constant the world over, and that nowhere, except where there may have been a reversal of the strata themselves, is there evidence of a reversed position. Corresponding strata, as indicated by the contained fossils, have, therefore, been considered to belong to the same age, even though occurring in widely-separated regions. This view, for a long time maintained undisturbed by the earlier geologists and paleontologists, has been dissented from by Edward Forbes, Huxley, and other advocates of the doctrine of faunal dispersion from localised areas or centres of distribution (opponents of independent creation), on the obvious ground that faunas, starting from a given point of origination, could only spread by migration, and that such migration must consume time, proportioned to the distance travelled and the physical and physiographical facilities afforded for travelling. Hence, it was argued, that widely-separated formations, showing an equivalent faunal facies, as, for example, the Silurian of America and the Silurian of Europe or Eastern Asia, or the Cretaceous of Europe and of South America, could not be of identical age, and, probably, not even approximately so. In support of this position, it has been urged that during the present age of the world the faunas of the several continents are widely distinct, and could, under geological conditions, be considered as indicating different zoological (geological) eras. In conformity with this view, Professor Huxley has proposed[82] the term "homotoxis," indicating similarity of arrangement, in place of synchrony, to describe the relation of distant areas of the same formation.

Pushing his conclusion to what appeared to be its furthest legitimate point, Professor Huxley deduced therefrom two important considerations: 1. That formations exhibiting the same faunal facies may belong to two or more very distinct periods of the geological scale as now recognised; and, conversely, formations whose faunal elements are quite distinct may be absolutely contemporaneous; *e. g.*, "For anything that geology or paleontology is able to show to the contrary, a Devonian fauna and flora in the British Islands may have been contemporaneous with Silurian life in North America, and with a Carboniferous fauna and flora in Africa." 2. That, granting this disparity of age between closely-related faunas, all evidence as to the uniformity of physical conditions over the surface of the earth during the same geological period (*i. e.*, the periods of the geological scale), as would appear to be indicated by the similarity of the fossil remains belonging to that period, falls to the ground. "Geographical provinces and zones may have been as distinctly marked in the Paleozoic epoch as at present, and those seemingly sudden appearances of new genera and species, which we ascribe to new creation, may be simple results of migration."

These views are still held by a very large body of geologists. But it can be readily shown by a logical deduction that at least one of the conclusions arrived at (1) is, almost certainly, erroneous; and that the second, based upon this one, derives no confirmation from the supposed facts. If, as is contended, several distinct faunas—*i. e.*, faunas characteristic of distinct geological epochs— may have existed contemporaneously, then evidences of inversion in the order of deposit ought to be common, or, at any rate, they ought to be indicated somewhere, since it can scarcely be conceived that animals everywhere would have observed the same order or direction in their migrations. Given the possible equivalence in age, as is argued, of the Silurian fauna of North America with the Devonian of the British Isles and the Carboniferous of Africa, or any similar arrangement, why has it never happened that when migration, necessitated by alterations in the physical conditions of the environs, commenced, a fauna with an earlier facies has been imposed upon a later one, as the Devonian of Britain upon the Carboniferous of Africa, or the American Silurian upon the British Devonian? Or, for that matter, the American Silurian might have just as well been made to succeed the African Carboniferous.

Reference to the annexed diagram, where D represents a Devonian area, say, in Europe, S a Silurian one in America, and C a Carboniferous one in Africa—all contemporaneous—will render this point more intelligible.   Now, on the proposition above stated, reasoning from our present knowledge of the antiquity of faunas, and accepting the doctrine of migration, as maintained by Professor Huxley and others, to account for the possible contemporaneity of

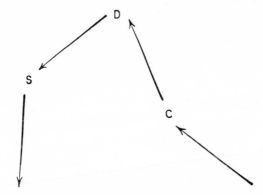

distinct faunas, it may be assumed that S (or America) will receive its Devonian fauna from D ; D (Europe) its Carboniferous from C ; and C (Africa) a later fauna from some locality not here indicated. In other words, migration, as indicated by the arrows, would set in from D to S ; one from C to D ; one from S to some possibly South American Cambrian locality, and one, bringing a Permian or some later-day fauna, from an unknown locality towards C.   Were this order of migration to continue here, or at other portions of the earth's surface, in this or in a similarly consecutive manner, the results obtained would be in perfect consonance with the facts presented by geology.   But is there any reason whatever for the continuance of this order of migration ?   Surely no facts that have as yet been brought to light argue in favour of a continued migration in one direction.   Why, then, it might justly be asked, could not just as well a migration take place from S to D, and impose with it a Silurian fauna upon a Devonian ?   What would there be to hinder a migration from S to C, placing the American Silurian fauna upon the Carboniferous of Africa ?   Why has it just so happened that a fauna characteristic of a given period has *invariably*

succeeded one which, when the two are in superposition, all over
the world (as far as we are aware) indicates precedence in creation
or origination, and *never* one that can be shown to be of a later
birth ?   Surely these peculiar circumstances cannot be accounted
for on the doctrine of a fortuitous migration.   And it certainly cannot be supposed that, through a process of transmutation or development, depending upon the evolutionary forces, a fauna with an
early-life facies will, in each case, at the point of its arrest have
assumed the character of the later-day fauna which belongs to that
position.   Therefore, it appears inconceivable that a very great
period of time should have intervened between the deposition of the
fauna of one great geological epoch at one locality and that of the
same or similar fauna at another locality distantly removed from the
first.   In other words, the migrations, for such must undoubtedly
have been the means of the distant propagation of identical or very
closely related life-forms (unless we admit the seemingly untenable
hypothesis that equivalent life-forms may have been very largely
developed from independent and very dissimilar lines of ancestry),
must have been much more rapidly performed than has generally
been admitted by naturalists.   The facts of geology and paleontology are decidedly antagonistic to any such broad contemporaneity
or non-contemporaneity as has been assumed by Professor Huxley;
and their careful consideration will probably cause geologists to
demur to the statement that " all competent authorities will probably assent to the proposition that physical geology does not enable
us in any way to reply to this question : Were the British Cretaceous
rocks deposited at the same time as those of India, or are they a
million of years younger or a million of years older ? "

But what applies to the broader divisions of the geological scale
also applies to the minor divisions.   Thus, the subordinate groups
of a formation are almost as definitely marked off in the same order,
the world over, as are the formations themselves.   After breaks in
formations the appearance of characteristic fossils is largely the
same, whereas, on the theory of synchronism of distinct faunas,
such a succession of forms would certainly not be constant.   Taking
the facts in their entirety, the conclusion appears irresistible that
formations characterised by the same or very nearly related faunas,
in widely separated regions, belonged, in very moderate limits, to
approximately the same actual age, and were practically synchronous

or contemporaneous. The singular uniformity in the lithological character of many of the equivalent formations also favours this conclusion. It is true that a very limited number of cases are known, or at least have been cited, where an old-type fauna is found intercalated with a fauna of newer date, reversing, as it were, the general order of succession or deposition. Such are the "colonies" which M. Barrande has indicated as existing in the Silurian basin of Bohemia, and which have been so forcibly dwelt upon by that investigator as standing in opposition to any slow-modification theory of descent and progression. Of a similar character are the colonies to which attention has been called by Marcou; and the occurrence, which has been observed in some parts of Scotland, of an Upper Silurian fish-fauna in rocks of unquestionably Devonian age, may be placed in the same category. It appears practically certain, however, that in none of these cases is there a true reversion. In the Siluro-Devonian faunal association of Scotland we have, doubtless, only one of those common occurrences where the range of certain species of animals has to be extended beyond what was supposed to be its farthest limit; for, as Mr. Geikie informs us, there can be no question that the Silurian element was present in the seas immediately adjoining the Devonian basin during the period of the first deposition of the Devonian rocks, and that it was subsequently admitted into the last by the rupture of a separating barrier. The case of the Bohemian colonies is, perhaps, not as readily explained; but, if they exist at all, it is very likely that they represent a fraction of a fauna which, through specially favourable conditions, has been permitted to linger beyond the period marking the extinction of the fauna as a whole, and which, through a short migration, has been transported to its present quarters. It must be confessed, however, that our knowledge respecting these colonies is very imperfect, and it is a significant fact that their existence, in the sense which has been preferred by their interpreter, M. Barrande, is completely denied by Lapworth, Marr, and others.

In a recent address, delivered before the British Association for the Advancement of Science (1884), Mr. Blanford has brought out numerous facts tending to prove, in the language of the author, "homotoxial perversity," or the want of synchronic correspondence existing between certain identical, or closely related, assemblages of fossil remains. It is shown, for example, that the famous mammalian beds of Pikermi, Greece, with Hipparion, Dinotherium giganteum, &c., which are generally classed by geologists as Miocene, or Mio-Pliocene, actually overlie marine deposits of almost unquestionably Pliocene age.   In the Gondwana system of rocks of the peninsula of India the Damuda group, whose flora is stated to be more nearly allied to the Jurassic flora of Europe than to any other, is overlaid by deposits holding an apparently Rhætic flora, and these, again, are succeeded by beds whose faunal characters partake of the Triassic period.   The coal-bearing beds of Southern and Eastern Australia are claimed by paleobotanists to be typically Jurassic, while the interstratified marine beds, in the character of their animal remains, are just as unequivocally Carboniferous.   In the Laramie formation of the Western Territories of the United States we have a somewhat similar association of an Eocene Tertiary flora, many of whose species are identical with forms occurring in the Island of Sheppey, and elsewhere in Britain, with a vertebrate fauna of a distinctively Cretaceous type.

That there should be no direct correspondence existing between the chronological facies of the marine and terrestrial faunas and floras is not very surprising, seeing that there exists no reason why their special development should have covered equal periods of time.   Nor could it be rationally expected, in view of the variable physical conditions prevailing over the land-surface, and the interposition of impassable water-barriers, that the development of a land fauna or flora should, in itself, be equal for all parts of the earth's surface, even though these parts enjoy approximately the same climatic influences.   This negative condition is beautifully exemplified in the utter dissimilarity of the modern mammalian faunas of South America, Africa, and Australia, which, had we known them in a fossil state, might have been taken to indicate three very distinct periods of geological time, and the same might have been said, in great measure, of their floras.   Yet, were the marine faunas belonging to these different regions examined, there could be but little doubt as to their representing chronological equivalents.   It is, therefore, not very remarkable, and not specially indicative of the existence of climatic zones, that the Australian coal-flora should be of the type which is elsewhere represented in the Jurassic deposits, and that it should be associated with a distinctively Carboniferous marine fauna.   And no more remarkable is the association in the Gondwana system of a Jurassic flora with a Triassic fauna, or of a Tertiary flora and Cretaceous fauna in the Laramie deposits of the United States. Less intelligible is the reported superposition in the Gondwana rocks of a Rhætic flora upon one of a Jurassic facies, and of a Triassic or Permian fauna upon one of Jurassic age. As far as the faunal evidence is concerned, both as relates to the "Triassic" or "Permian" and the "Jurassic" types, it is so slender as to permit no safe conclusion being based upon it. Thus, the only animals thus far recorded from the Damuda (Jurassic) beds are an Estheria and two labyrinthodonts, one of which, Brachyops, is claimed to be allied to a European genus from the Oolites. Equally unsatisfactory is the Permian relationship that has been attached to the vertebrate fauna of the overlying Panchet beds, which comprises three genera (four species) of labyrinthodonts, whose nearest allies are found in the European Trias, one species of dinosaur, and two dicynodonts, believed to be nearly allied to forms described from what are now recognised to be Permian rocks of the Ural Mountains. The very limited number of species found in these deposits, together with their uncertain relationship, is scarcely sufficient to identify a fauna, and it must, therefore, be concluded that we are still far from having evidence of a positive character proving an inversion in the faunal series, or as indicating any important difference between homotoxis and synchrony, where marine fossils are taken as the determining guide.* The advantages possessed by these last over others of a terrestrial nature for the purposes of geological classification are obvious from their broad distribution and equal chronological development.

* Mr. Lydekker, discussing the remarkable labyrinthodont Gondwanosaurus from the Bijori group, maintains that the balance of evidence is in favour of regarding the Panchet beds as of Triassic age; the age of the Damuda beds is left in doubt (Palæontologia Indica, 1885).

PART III.

GEOGRAPHICAL AND GEOLOGICAL DISTRIBUTION.

I.

The present and past distribution of individual animal groups.—Foraminifera. —Corals.—Brachiopoda.—Mollusca generally.—Crustacea.—Insecta.

FORAMINIFERA.

DESPITE the simplicity of their organisation, and their apparent ready adaptability to the most varied phases of existence, it appears that the distribution of these animals is influenced, although perhaps to a less degree, by much the same conditions which govern the distribution of higher forms. While practically the class is of world-wide distribution, and cosmopolitanism—at least among groups of supra-specific value (genera, &c.)—the rule rather than the exception, yet the influence of special conditions, as pertaining to different sections of the earth's surface, is markedly manifest. Thus, we find that in the warmer regions of the globe the foraminiferal fauna, as compared with the corresponding fauna of the regions lying to the north and south, is very much richer both as regards specific and individual development; and, further, that it comprises a disproportionately large number of forms characterised by unusual size and structural complexity (Orbiculina, Orbitolites, Cycloclypeus, Tinopora). Of some seventy or more genera of Foraminifera calcarea recognised by Bütschli,[83] thirty-eight, or about one-half, are wanting in the Arctic seas, twenty-five are wanting in the British and North seas, and fifteen in the Mediterranean; on the other hand, all the Arctic and north temperate genera are represented in the tropical and sub-tropical seas. The number of generic forms thus made peculiar to the tropics is about twelve. That all the northern genera should also be found in the central waters is not specially surprising, seeing that the same conditions of temperature which prevail over the one region in the surface or shallow waters are met with in the other in the deeper layers, or upon the oceanic floor. The numerical relation existing between species and genera does not appear to vary according to any known law of distribution. Many of the genera are most abundantly represented in specific types in the tropics, while, again, others attain their maximum development in the regions of high latitude. Bütschli recognises ninety-nine species of Foraminifera calcarea from the Arctic province, one hundred and eighty-five from the north temperate, and one hundred and ninety-eight from the Mediterranean, an estimate which makes the Arctic province slightly deficient as compared with either of the other provinces, and the north temperate slightly in excess of the Mediterranean. A more thorough investigation of the Mediterranean waters, however, will doubtless increase the number of forms occurring there very considerably, seeing that upwards of one hundred and sixty species have been found on the British coast alone. Despite the seeming diversity given to the character of the Arctic foraminiferal fauna by the ninety-nine or more * species occurring there, the fauna is strictly a uniform one, the vast mass, about ninety-five per cent., of its component material being made up of only a very limited number of species—Globigerina bulloides, Cassidulina lævigata, C. crassa, Polystomella striatopunctata, &c.—a feature in a measure also distinguishing the foraminiferal accumulation of the deeper parts of the sea, the Globigerina ooze, in which the individuals of Globigerina, Pulvinulina, Orbulina, Spheroidina, and Pullenia, especially of the former, preponderate to a very marked degree. As in the case of other marine animal groups, the foraminiferal fauna of the high north is most intimately related to its antipodal fauna of the south. Of a total of fifty-three genera and one hundred and eighty-nine species occurring in both the Southern (below the fiftieth parallel of latitude) and Arctic oceans, thirty-two genera and sixty species are held in common by both areas; [84]

* Brady, in his report on the " Challenger " expedition, recognises all in all one hundred and eleven species.

eleven of the southern genera are wanting in the north, and ten of the northern genera in the south.

The relation of the pelagic or surface fauna to that found in the bottom deposits of the sea, which has given rise to much diversity of opinion among naturalists, has had much light thrown upon it by the recent deep-sea dredging expeditions. It appears now practically certain that a vast majority, probably ninety-eight or ninety-nine per cent., of known foraminiferal forms, including all of the porcellanous and arenaceous groups, are permanent inhabitants of the oceanic floor, being endowed with no swimming or floating powers. The surface and mid-water forms are limited to some eight or nine genera and twenty species,[65] many or most of which (species of Globigerina, Orbulina, Spheroidina, Pullenia, Pulvinulina) are identical with forms living in the lowest oceanic layers and found in the Globigerina ooze, to the formation of which they doubtless contribute very extensively through their prodigious development.

In their importance as rock constituents, the Foraminifera stand second to no other group of organisms; in their geological development they are probably coeval with the entire period during which the sedimentary deposits were being formed, although it is not till beyond the middle of the Paleozoic series that their unequivocal remains are met with in any abundance. Barring the much-disputed Eozoon, whose inorganic character may now be accepted without much hesitation, the earliest indication that we possess of this class of animals is in the Lower Silurian rocks, where thus far but a single genus, Saccamina, has been positively determined. Two species of Lagena occur in the Upper Silurian of England. There still exists too much uncertainty relative to the forms known as Receptaculites, Ischadites, and their allies, to permit of their being absolutely classed with the Foraminifera, and likewise in the case of the various structures which have been referred by Ehrenberg to the Silurian of Russia. This paucity in the Silurian rocks, and not less so in the succeeding Devonian, and their complete absence from the Cambrian, is not a little surprising, but may perhaps be explained on the hypothesis that the earlier members of the class were in the main devoid of tests, or of such parts as could be readily preserved in a fossil state. The marked development of the genus Fusulina in the Carboniferous rocks is one of the most striking of the faunal features of the period, and, aside from the general importance attaching to the genus itself as a rock-builder, acquires special significance from the circumstance of the barrenness in foraminiferal remains of the formations immediately preceding.   From the close of the Paleozoic period to the present day there is a steadily increasing development of distinctive types, and an equally steady approximation to the structural type which dominates the modern seas.   Almost every period is marked by some extensive foraminiferal accumulation, and in nearly all cases the distinguishing character is given to these accumulations by genera which acquire successive importance.   Such are the Gyroporella limestones of the Triassic and Rhætic formations, the miliolite and nummulitic rocks of the Eocene and Oligocene, and the orbitoide rocks of the Oligocene and Miocene periods.   In the newer Tertiary deposits the representative genera are mainly identical with those inhabiting the modern seas, and even of the species a fair proportion are identical.   Of the most widely distributed and most numerously represented modern genera the one which has most thoroughly left its impress upon the rocks of past ages is the genus Globigerina.   Beginning in the Trias, it already attains to considerable importance in the Rhætic, where, in association with Textularia, Orbulina, and Quinqueloculina, it forms massive limestones (Dachsteinkalk, near Hallstadt, in Salzburg).   The chalk is made up in large part of the three foraminiferal genera Globigerina, Textularia, and Rotalia, and the first of these, as has already been seen, is the principal constituent also of the modern Globigerina or Atlantic ooze.

  Much diversity of opinion has existed among naturalists as to the mutual relationship of these two classes of deposits—the chalk and ooze.   The general similarity of lithological character, heightened by a close identity existing between the contained faunas, has led most scientists to believe that the present oceanic bottom is but a continuation of the bottom of the Cretaceous seas—in other words, that the Cretaceous epoch is continued up to the present time without any very material change marking the progress of organic life in the ocean's deepest parts.   The majority of the foraminiferal types occurring in the one formation are represented in the other, and the identity is carried even to a fair proportion of the species.   Further, Sir Wyville Thomson has shown that the biological relationship is not restricted to the lower or microscopic forms of life, but, on the contrary, manifests itself in almost as marked a degree in some of the higher groups. Thus the sponges of the Atlantic ooze appear to have been most closely related to the ventriculite sponges of the Cretaceous period, and there is also a predominance in both classes of deposits of the cidaroid type of sea-urchin.

As opposed to the biological relationship, on the other hand, it has been pointed out that the physical constitution of the two classes of deposits is very different. Thus, while Cretaceous chalk almost invariably contains from ninety-four to ninety-nine per cent. of carbonate of lime, and, consequently, at the utmost only about six per cent. of foreign substances, analyses of Globigerina ooze show it to contain only from forty-nine to eighty per cent., allowing a very considerable percentage for impurities. Such a difference in chemical composition is certainly very striking, the more especially as the composition of true chalk is very constant, and appears to point to a mode of formation different in the two cases. The correspondence in chemical composition existing between true chalk and the Oahu chalk (coral *débris*) of the Sandwich Islands has suggested to Mr. Wallace the notion that not improbably both deposits have much the same formation—in other words, that the chalk was deposited in a comparatively shallow sea, probably not exceeding one thousand fathoms in depth, in which numerous islands were scattered about in a manner somewhat similar to what is observed in the coralline zones of the Pacific Ocean. It has been further urged, in support of this view, that in the opinion of many conchologists, notably, Mr. Gwyn Jeffreys, all, or nearly all, the Mollusca of the Cretaceous sea represent comparatively shallow-water forms, there being a total absence of such shells as could with positiveness be considered as pointing to a deep-sea habitat. Numerous objections, however, interpose themselves to the views so ingeniously framed by Mr. Wallace. In the first place, it would be purely gratuitous to assume that a comparatively shallow coralline sea extended into a continental area whose expanse equalled the tract covered by the chalk deposits of Eurasia; secondly, had such a sea existed we should naturally expect to find, as has been urged by Mr. Starkie Gardner, the remnants of ancient coral reefs, such as are at the present day being formed in the

Atlantic, Pacific, and Indian oceans, but, singularly enough, no such reefs have thus far been detected in the deposits in question; indeed, even the scattered coral fragments are in themselves remarkably scanty, and far from sufficient to give a coralline aspect to the formation; thirdly, while many of the chalk Mollusca may appear to represent shallow-water forms, there are yet a number, including some of the most characteristic forms, concerning whose habit nothing positive can be stated, since not only have many of them no living representatives in the faunas of the present seas, but even the families to which they belonged have completely died out. While possibly, then, we may not have as yet arrived at an exact comprehension of the nature of chalk, it must be confessed that the biological facts already indicated point to a very close relationship with the Globigerina ooze.   Nor is it at all unlikely that the difference existing in chemical composition is one more apparent than real.   Thus, the deficiency of silica—and, consequently, the surplus of carbonate of lime—in chalk can be readily accounted for on the supposition that the free silica originally present in the Cretaceous seas may have sifted itself during the formation of the chalk into those irregular nodules which we now recognise as flints, and likewise into the irregular fissures in the chalk, to form the chalk-veins.   Again, it may be assumed that chalk, during the long lapse of ages that has intervened since the period of its formation, may have through various causes undergone considerable alteration in its chemical composition, and sufficient to account for the dissimilarity existing between it and the Globigerina ooze.

Perhaps the most remarkable feature connected with the history of the Foraminifera is the long period of time through which primary characters have been retained; persistence of type-structure is, indeed, immeasurably better marked in this group of organisms than in any other.   It is true that the views of naturalists are very much at variance as to the proper limitations to be assigned to the beings composing this most difficult group of the Invertebrata, and that some of the most eminent authorities are disposed to unite the greater number of both recent and fossil forms into a comparatively limited number of diverging or central types; but, even from the more conservative standpoint, enough is patent to indicate a most extraordinary specific longevity.   This is most clearly brought out by the data furnished in Mr. Brady's report on the "Challenger"

Foraminifera, whence it would appear that, of some seven hundred recognised species, the range of not less than three hundred and thirty-eight extends as far back as the Tertiary period; of one hundred and twelve to the Cretaceous; fifty-six to the Oolite; forty-one to the Lias; sixteen to the Trias; eight to the Permian; six to the Carboniferous (Haplophragmium agglutinans, Ammodiscus incertus, A. gordialis, Truncatulina lobatula, and the two Silurian forms); two to the Devonian (also Silurian); and two to the Upper Silurian (Lagena lævis and L. sulcata). A number of other doubtful forms occurring fossil may perhaps also be referred to the category of recent equivalents, as, for example, the Girvanella, described by Nicholson and Etheridge from the English Silurian deposits, which not improbably is the recent Hyperamina vagans.

CORALS.

The more important corals of the present day belong to the groups Zoantharia and Alcyonaria, the former of which, frequently designated the six-tentacled corals (Hexacoralla), embrace the naked sea-anemones (Actiniæ), and nearly all the familiar stone-corals, and the latter (Octocoralla, eight-tentacled corals), the sea-pens, sea-shrubs, red-coral, and organ-pipe. In the group of the comb-bearers (Ctenophora) are comprised a limited number of forms, Beroë, Venus's girdle, &c., which are entirely destitute of a corallum, and in many essential points of structure depart widely from the other members of the class. A fourth division, the rugose corals (Rugosa or Tetracoralla), which, in the early periods of the earth's history, constituted such a marked feature in the successive faunas (Silurian, Devonian, Carboniferous), are limited at the present day, as far as our existing knowledge goes, to possibly not more than two generic types, Guynia, from the Mediterranean and the Gulf of Mexico, and Haplophyllia, from the coast of Florida.

The sea-anemones are, collectively, cosmopolitan in their distribution, and inhabit the sea to very nearly the profoundest depths that have thus far been reached by the dredge, although, both as regards specific and numerical development, they more distinctively characterise the littoral and laminarian zones. The deep-sea species are, however, not exactly scarce, and it would appear, from the "Challenger" observations, that the numerical decrease corresponding to the increase in depth is not nearly as great as might have been expected. In ninety-seven hauls, made in depths ranging from ten to five hundred fathoms, positive results, with a capture of some twenty specimens, belonging to thirteen or more species, were obtained eleven times; in one hundred and sixty-five hauls, made in water of from five hundred to twenty-nine hundred fathoms, similar results, with a capture of sixty specimens, representing twenty-one species, were obtained fourteen times. In the deeper dredgings, however, the casting of the net was protracted over a greater period of time, and, consequently, covered more space than in the shallower ones, and due allowance should be made for this circumstance. The relative abundance of the deep-sea Actiniæ is shown by the fact that not infrequently two or more species, or individuals belonging to a single species, are found associated in the same locality. The greatest depth from which any species has thus far been obtained is twenty-nine hundred fathoms (Antheomorphe elegans).

Professor Hertwig, from the data collected by the "Challenger," believes it may be safely assumed that, the greater the depth of water the greater is the variation between the deep-sea forms and those of the coast-line. Thus, it is shown that, of the thirteen species and twelve genera obtained from a zone of ten to five hundred fathoms, only five species and two genera were found to be new, or as not essentially belonging to the coast; on the other hand, of the twenty-one species and seventeen genera obtained from a depth of from five hundred to twenty-nine hundred fathoms, eleven of the genera, and, with one exception, all of the species, were new.[86] The same authority finds that increased depth exerts a remarkable influence in modifying the organisation of many of the forms. Thus, the tentacles exhibit a distinct retrograde formation, or degeneration, being first transformed into tubes, and ultimately into simple openings in the oral disk. "In Paractis tubulifera (depth eighteen hundred and seventy-five fathoms) the tentacles have the same constitution as in the majority of Actiniæ, except in one point, that the terminal opening, which is usually small or entirely wanting, gapes widely. In Polysiphonia tuberosa (five hundred and sixty-five fathoms) the tentacles have become short, slightly movable, wide-mouthed tubes; in Sicyonis crassa (sixteen hundred fathoms) they are small, wart-like rings, and in Polystomidium patens (eighteen hundred and twenty-five fathoms) and Polyopis striata (twenty-one hundred and sixty fathoms) the walls have almost entirely disappeared, so that the terminal opening forms a fissure in the oral disk, the last remains of the tentacle being represented by a circular margin surrounding the fissure, and so we come finally to the genus Liponema (eighteen hundred and seventy-five fathoms), in which the points at which the tentacles were actually placed are merely indicated by openings in the oral disk." The very limited number of specimens obtained by the expedition prevents any definite conclusions being arrived at as to the horizontal distribution of the individual species. The majority of the forms, as Anemone, Actinia, Metridium, &c., inhabit a superficial zone, especially the rock-pools, situated at about low-water mark, where they attach themselves to some foreign body by means of their muscular pedal-disks; others, like Peachia, Edwardsia, Cerianthus, and Halcampa, lie more or less buried in the sand, while a very limited number are free-swimming or pelagic (Arachnactis). No fossil remains indisputably belonging to this group have as yet been discovered.

Much the larger number of coral-depositing zoophytes of modern seas belong to the group of the Madreporaria, or, as they are frequently termed, from their dermal corallum, the Zoantharia sclerodermata, as distinguished from those which, like the "black corals" (Antipathidæ), secrete an internal corallum or basal skeleton (Zoantharia sclerobasica). Two divisions of these corals may be conveniently recognised, the "solitary" (whether simple or compound) and the "massive," or reef-building corals, the former of which are essentially deep-sea forms, rarely coming within the littoral zone, while the latter are just as distinctively shallow-water forms, extending from low-water line to twenty or twenty-five fathoms.

The deep-sea corals are spread throughout nearly the whole of the oceanic expanse, from the confines of the frozen sea on the north to the Antarctic barrier on the south, and from very nearly the surface of the water to depths of at least twenty-nine hundred fathoms. The distribution of the individual species appears to be largely, if not almost altogether, independent of considerations connected with temperature. Sars obtained Fungiacyathus fragilis, a member of the family Turbinolidæ, off the Loffoden islands, from a depth of three hundred fathoms. Bathyactis symmetrica, apparently the most widely distributed of all known corals, has been dredged off the Bermuda islands in thirty fathoms of water; off the Virgin islands in three hundred and ninety fathoms; off Martha's Vineyard in two hundred and fifty; and off the coast of Japan in twenty-nine hundred. Deltocyathus Italicus has been obtained off the Bermuda islands from depths of two hundred and ten hundred and seventy-five fathoms, respectively; off the West Indies from a depth of seventy-five fathoms; off the coast of Cape Cod from one hundred and forty-two; and in the South Pacific (latitude 32° 36', longitude 137° 43' west) from a depth of twenty-three hundred and seventy-five. The total number of genera which have been found at depths exceeding fifty fathoms is, according to Mosely,[87] forty-eight. Of these, only five—Caryophyllia, Deltocyathus, Flabellum, Bathyactis, and Leptopenus—reach or pass beyond the fifteen hundred fathom line; the last has thus far been obtained only in water exceeding fifteen hundred fathoms, whereas the other genera range to within fifty or one hundred and fifty fathoms of the surface. By far the greater number of the genera range to within the fifty to one hundred fathom line, and a limited number (Caryophyllia, Paracyathus, Flabellum, Balanophyllia, Dendrophyllia, &c.) into still shallower water. Most of the species have a very broad horizontal distribution, and a fair proportion of them approach cosmopolitanism. The range of Bathyactis symmetrica, above noted, is practically world-wide, individuals of the species having been obtained from the Azores, the Massachusetts coast, the Bermudas, the east and west coasts of South America, Kerguelen Island, the coast of Australia, the Molucca seas, and the coast of Japan. Scarcely less extensive is the range of Deltocyathus Italicus. Caryophyllia communis has been taken off Nova Scotia, the Azores, the Bermudas, and the Cape of Good Hope. The only genera which appear to be restricted in range are Stephanophyllia and Sphenotrochus, which as yet have been obtained only from the shallow seas of the Malay Archipelago. The compound Madreporaria of the deep-sea are very limited, both specifically and individually, the most abundant form dredged by the " Challenger " being a species of Lophohelia (L. prolifera).

In their relations with fossil forms the deep-sea corals have revealed very little of special importance. Somewhat more than one-third of the genera are represented in the Tertiary formations, and a considerably smaller number (Caryophyllia, Trochocyathus, Thecocyathus, Parasmilia, Lophohelia, &c.) in the Mesozoic series. No Palæozoic genera are represented, and, if we except the Cyathon-axidæ, apparently not even any of the distinctive families. A few of the forms have been identified with Tertiary species, as Delto-cyathus Italicus, Caryophyllia communis (Sicily), and, among others, possibly one or more species of Flabellum and Stephanophyllia. Caryophyllia cylindracea, a well-known European Cretaceous spe-cies, has been dredged in eleven hundred fathoms water off the coast of Portugal, in associations with other forms of an equally ancient aspect.[88] One species of Stephanophyllia (S. complicata, from the Ki Islands), although having its nearest ally in the S. dis-coides from the London Clay (Eocene), is, remarkably enough, in certain peculiarities of structure, most intimately related to a form from the Jurassic formation of Germany (S. florealis). At the pres-ent day, as far as is known, the genus survives only in a remote spot of the East Indies, completely severed from its former Euro-pean habitat. Desmophyllum ingens, from the fjords of Western Patagonia, appears to be specifically identical with an undetermined species from the Quaternary deposits of Messina, Sicily.

It will be manifest, from what has preceded, that the modern distribution of the deep-sea corals affords little satisfactory evidence as to the special conditions—light, temperature, or depth—which affect the development of this class of animals. Indeed, it would appear at first sight as though none of these conditions were di-rectly involved in the distribution of the group as a whole, or of its individual members. The great bathymetrical range of many, or the majority, of the forms, from the surface waters of a mild tem-perature to the water of icy coldness, clearly proves the ready adapt-ability of these organisms to extremes of temperature, or, what might also be true, the want of appreciation, on their part, of ther-mometric conditions. It is to be noted as a remarkable circumstance in this connection, seeing through what an extensive range of tem-perature their vertical distribution extends, that so few of the forms penetrate within shallow water, or water of less than fifty fathoms. Professor Fuchs has attempted to explain this anomaly on the as-sumption that these, as well as other strictly deep-sea organisms, were animals of darkness, and that they rarely penetrated within the zone of light-penetration. But it may be questioned, as has already been done when treating of the life of the sea, whether the

interpretation here given of the abyssal fauna is a valid one. Indeed, it might be doubted altogether whether there is such a thing
as a true fauna of darkness; or, if existing, it may be considered
questionable whether the members composing it have not become
habituated to present conditions through force of accident or a
process of degeneration, rather than through selection as guided by
individual instinct or volition. It appears highly improbable on its
face that animals so feebly endowed with perceptive powers as
these appear to be, and which at the same time possess a most
extraordinary adaptability to extreme conditions of temperature and
pressure, should be so constituted in their relations to illumination
as not to be able to endure the quantum of light which passes
through even the shallowest stratum of water.

The present broad distribution of the deep-sea Madreporaria
appears, likewise, to have obtained in the earlier geological periods.
Thus, Professor Duncan has shown, from his researches on the fossil
coral fauna of the West India islands, that a number of the forms
occurring there, in both the Eocene and Miocene formations, are
such as had been already previously described from the equivalent
deposits of Europe, e. g., Paracyathus crassus, Trochocyathus cor-
nucopiæ, T. laterospinosus, Ceratotrochus duodecimcostatus, the last
three from Italy, and the second also from the Vienna basin. Flabellum appendiculatum, from the Oligocene beds of the island of
St. Bartholomew, is a species from Biarritz and Ronca; Trocho-
smilia subcurvata, from the same island, occurs in the Eocene beds
of Oberburg, in Styria, and T. arguta, at Castel Gomberto, in
Venetia, and other Oligocene localities.[89] Other Tertiary species
have since been identified as being trans-Atlantic, and, doubtless,
many forms will be found whose range is still very much greater.
The Italian Conotrochus typus, Balanophyllia cylindrica, and Deltocyathus Italicus, are found also in the Australian Tertiary strata.[90]
In view of the very extended range of so many of the species, the
distinctness of the Indian (Sindh) coral fauna, which holds scarcely
any species in common with any distant region, at least as far as
has yet been determined, is not a little remarkable. Considering
the apparent independence of the animals of this class, of the varying conditions of temperature and pressure which a habitation of
the deep-sea presents, and not unlikely also of other physical
conditions as well, it is difficult to account for the comparatively

short range in time of the different species and genera. It is indeed true, as far as the genera are concerned, that the very limited number of areas of deep-sea deposits may satisfactorily explain the absence, as fossils, of the greater number of the modern genera, but this circumstance will scarcely account for the distinctness of the species. None of the species of the English Tertiaries, from the Crag to the Eocene, are living forms, and the same appears to be the case with the species from the German and Italian Oligocene. In his enumeration of the species from the Miocene deposits of Austria, Professor Reuss admits but a single recent form, Caryophyllia clavus.[91] Not a single one of the species of the West India Miocene deep-sea Madreporaria is, according to Duncan, a member of the recent coral fauna, and yet just here a certain stability of physical conditions would have been supposed to ensure a steady perpetuation of the species. Similarly, if we except two or three species no longer found in the region, the extensively distributed Deltocyathus Italicus, and Flabellum Candeanum, from the China seas, and F. distinctum, from the Red and Japanese seas, all of the Australian Tertiary (Miocene or Pliocene) Madreporaria are extinct, and, what is very remarkable, only a very insignificant fraction of the living local genera is represented specifically in the Tertiary deposits of that region. Conocyathus sulcatus, a species from the Oligocene beds of the Maintz basin, is recognised by Duncan as a member of the recent Australian fauna.[92]

The restricted specific longevity here indicated is certainly in marked contrast to what obtains among the Mollusca, where, as is well known, the recent forms constitute a no inconsiderable percentage of the Miocene fauna, and are not even wholly absent from the fauna of the Eocene series.* And yet the Mollusca would seem to be much more dependent for their existence upon special physical conditions of their surroundings than are the corals. The surprising persistence of the Foraminifera is, by way of contrast, not a little remarkable.

The limited range in time of the species and genera of deep-sea

---

* The late Mr. Gwyn Jeffreys has recently attempted to show that none of the Eocene molluscan species are identical with living forms. Whether this be so or not, there can be no question as to the existence of a limited number of representative types, which unmistakably bind together the past and present faunas, and in a manner which we do not find among the corals.

corals is likewise exemplified in the case of the reef-builders; but here the absolute limitations to existence that are set by the physical aspects of the environs render such a condition in no way surprising; indeed, it is just what might have been expected. Only a very limited number of the existing generic types date from the Cretaceous period (Porites, Diploria, Mæandrina, Goniastræa) and a still smaller number from the Jurassic (Favia, Heliastræa, Cladocora), although not unlikely all, or very nearly all, of them belong to some part or other of the Tertiary period. On the other hand, the number of generic types that range through the whole series of the Mesozoic deposits (Isastræa, Thamnastræa, Rhabdophyllia, Thecosmilia, Cladophyllia, Latimæandra, Stylina) is considerable, and some of these, as Thamnastræa, Thecosmilia, and Rhabdophyllia, also pass over into the Tertiaries.

In their geographical distribution the modern reef-builders are confined to a zone extending on either side of the Equator whose outer limits are bounded by the isocryme of 68° Fahr., or the line which marks a lowest average temperature of 68° for all months of the year. No species, apparently, can endure a lower temperature, while most of them require for their greatest development a temperature considerably more elevated. Professor Dana has divided the coral-reef seas into two primary sections, the torrid and the sub-torrid, the former of which, whose delimitation is fixed by the isocrymes of 74°, is included in principal part between the twentieth or twenty-third parallels of north and south latitude, although reaching in the Red Sea and the Gulf of Mexico considerably beyond the normal limit. The greatest profusion and wealth of coral growth is exemplified in this region, and particularly in the waters of the Central Pacific (Fiji islands, &c.). The astræas, mæandrinas, porites, and Pocilloporæ attain here their fullest perfection, and with them are associated large beds or masses of madrepores, Pavoniæ, Fungiæ, and tubipores, and hosts of other forms of the most diverse outline and brilliancy of colouring. A deficiency in the variety of species and genera becomes apparent as we proceed eastward. Much the same types as occur in the Pacific are represented in the coral islands of the Red Sea and the Indian Ocean, and in the East Indies; and in the limited fauna which is developed along the western coast of America, between Guayaquil and the peninsula of Lower California, the Pacific element is almost exclusively represented (Porites, Pavonia, Pocillopora, Dendro-phyllia, Fungia, &c.). The West Indian coral fauna, on the other hand, is deficient, either wholly or in great part, in many of the more distinctive forms of the Pacific Ocean (Astræa, Pavonia, Pocillopora, Fungia), a deficiency in part made good by a special development of forms (Diploria, Agaricia, Siderina, Oculina, Cladocora, Astrangia) which are either wholly wanting, or have only a relatively feeble representation, in the islands of the Pacific. The madrepores of this region attain to prodigious size. Clumps of Madrepora palmata, a foliaceous species, have been found to measure two yards in width, while the branches of the tree-like M. cervicornis not rarely reach a height of from ten to fifteen feet. Professor Verrill has pointed out the somewhat remarkable fact that none of the West Indian species of coral are specifically identical with the species of the Panama coast, although most, if not all, of the Florida reef-builders (species of Porites, Madrepora, Mæandrina, Manicina, Siderina, Agaricia, Orbicella) are also found on the coast of Aspinwall. Doubtless this difference is in great part attributable to the comparative brevity of the natural life of the species, as it is well known that direct communication between the Atlantic and Pacific oceans, in the region of Panama, was maintained during the middle or later part of the Tertiary period (Miocene or Pliocene), and, if this was so, there can be little doubt that at that time many of the forms on opposite sides of the present isthmus were identical specifically. The present differentiation, arising from isolation, would then date back at least as far as the permanent (or nearly that) elevation of the separating land-mass, and not improbably to a period considerably antecedent to that. For although, as has already been seen, the number of recent species that extend back to the Miocene period is very limited in most parts of the earth's surface, yet just in the West India region Professor Duncan has shown that very nearly ten per cent. of the Miocene coral fauna is made up of existing species. This being true, we should naturally expect to find, if the isolation of the Panamaic and Gulf faunas took place in the Miocene period, a considerable intermixture of identical species, which is not the case. It appears probable, therefore, that for some time previous to the final emergence of the isthmus, whether through the down-wash of sediment or otherwise, the region was in a measure rendered inimical to coral growth, and that but little, if any, transference of species from one side to the other was effected.* The principal reefs of the sub-torrid zone, other than the South Pacific, are those of the Bermudas and the Sandwich Islands, both of which are characterised by a comparative paucity of specific forms. Most of the species occurring among the former (Isophyllia, Diploria, Oculina, Siderastrea, Porites) are West Indian types; in the Sandwich Islands the predominating forms are Porites and Pocillopora, there being a marked deficiency in the representatives of the Astræa and Fungia tribes, and a complete absence of Madrepora. The point most distant from the Equator about which reef-structures have been noted appears to be Quelpaert's Island, situated south of Corea on the thirty-fourth parallel of north latitude.

In comparing the past with the present distribution of coral reefs, we are at once confronted with the not very surprising, although all-important, fact that the areas of such distribution in no way correspond with those distinctive of the modern seas. The extension of reefs northward to points far beyond any now occupied by such structures is practically proof positive of the existence of thermal conditions very different from those which obtain at the present day, and of a much more equable climate, with a more elevated temperature, than is now found in the higher latitudes of the earth's surface. Palæozoic reef-building corals have been found in Eurasia (Scandinavia, Russia) far above the sixtieth parallel of latitude, and a number of genera even in Spitzbergen, Nova Zembla, and Barentz Islands; Lithostrotion was obtained by the officers of the British North-Pole Expedition, under command of Sir George Nares, at a point beyond the eighty-first parallel of latitude. The reef-building corals of the Silurian and Devonian periods—Favosites, Heliolites, Halysites, Syringopora, Cyathophyllum, Acervularia, &c.—have left traces of their profuse development in seas as far north as Canada and Scandinavia, but it would perhaps be straining a point to infer from their occurrence there that climatic conditions in any way identical with those existing in the modern coral zones prevailed during those periods in high northern latitudes. Our knowledge respecting the habits and affinities of these ancient organisms is still much too limited to permit of a positive

---

* All the species described by Duncan from the Oligocene deposits of the island of St. Bartholomew are extinct.

12

determination of the question. Yet the recession southward of reef-structures, in the making of which the modern type corals were largely involved, during the succeeding periods of the Mesozoic and Cainozoic eras, seems to indicate that this must have been the case; for as far as the evidence from these later structures goes there can be no reasonable doubt as to a progressive lowering of the oceanic temperature correlatively with an advance in time. And if this was the case from the Triassic period onward, there is every reason to suppose that it has been so from a much earlier period.

Reef-structures appear to have been very extensively developed in South-Central Europe during the Triassic period, and not unlikely much of the giant dolomites of the Tyrol, whose abrupt and pinnacled masses so wonderfully diversify the face of the country, is the product of the unceasing labours of the minute polyp. In the succeeding Jurassic period a more or less continuous coral sea occupied a considerable portion of Western and Central Europe, as is evidenced by the vestiges of reefs which still remain in England, France, Germany, and Switzerland. During the deposition of the Oolites the reef-structures appear to have attained their maximum development, the shallow coral sea, with its atolls and barrier reefs, extending as far east as the Carpathian Mountains, and covering much of the region now occupied by the Alps. The British area was still favourable to the growth of the coral polyp. With the beginning of the Cretaceous period there would seem to have been a gradual deepening of the oceanic bottom to the north, and the introduction of conditions inimical to the proper development of coral life, for the number and extent of the reefs occurring there are comparatively very limited; a Southern European belt, on the other hand, is very coralliferous. No doubt the gradual lowering of the oceanic temperature had much to do with this recession, more, in fact, than the simple lowering of the oceanic bottom, since the latter, if gradual, while it would almost certainly check a lateral extension of coral structures, would scarcely tend towards their complete or wholesale obliteration. The coral formations would still persist with the conversion of connected land-masses into islands, and after the complete submersion of these last, just as we now find them over the deeper oceanic abysses. That the temperature at this period was not entirely too low, however, in all parts of the region under consideration, is proved by the Upper

Cretaceous reefs of Maestricht, Faxoe, and the neighbourhood of St. Petersburg, and by the limited vestiges of such which still mark the Eocene deposits of the typical English basin.   The reefs of the Eocene period find their greatest extension in Southern Europe—from the northern flanks of the Alps and Pyrenees southward, and eastward through the Crimea, Egypt, Syria, Arabia, and East India.   The more extensive European reefs of the Oligocene period are those of Oberburg, in Styria, and Northern Italy (Crosara, Castel Gomberto, &c.); Miocene reef-patches still exist in Spain, Southeastern France, Northern Italy (Superga), the Vienna basin, and Hungary, the larger structures, however, occurring in the region farther to the south (Malta, Asia Minor, the West Indies, Java). At the close of this period the reefs appear to have still further receded, and in the Pliocene they completely vacated the present continental area.   Dr. Duncan has remarked the existence in the Table Cape Tertiaries of Tasmania (Miocene ?) of reef-building corals (Heliastræa, Thamnastræa) at a point removed some fifteen degrees of latitude south of the coral isotherm of that region.[93]

The limited range in depth—not exceeding twenty fathoms—of reef-building corals is certainly extraordinary, and something that still remains in the nature of a puzzle to the naturalist.   That it is not, either wholly or in great part, dependent upon conditions of temperature is conclusively proved by the total absence of such organisms in depths beyond one hundred and twenty feet where a temperature considerably above that required for coral growth still prevails (Pacific Ocean, Red Sea).   Possibly a movement, or want of movement, in the oceanic waters has something to do with this abrupt limitation.

The Alcyonaria or Octocoralla, except in so far as some of the forms until recently classed with the Tabulata may be considered to belong to this group—as Halysites and Syringopora, supposed to be allied to the organ-pipe (Tubipora), Heliolites to Heliopora— acquire but little geological importance.   The pennatulids appear to have one or more representatives (Pavonaria) extending back to the Cretaceous period, and a limited number (Graphularia ?) also in the Eocene.   The sea-fans (Gorgonidæ), whose brilliantly coloured masses constitute such a striking feature of the coral patches of both the Atlantic and Pacific oceans, are not positively known before the Miocene period (Primnoa, Gorgonella).   The genus Isis occurs in the Cretaceous deposits, and Corallium, to which the red coral of commerce belongs, already in the Jurassic. The distribution of the latter genus appears to be at the present time confined principally to the Mediterranean Sea, where it ranges in depth from shallow water (twenty-five to fifty feet) to water over one thousand feet, and to the Atlantic off the northwest of Africa. A species of the genus (Corallium stylasteroides) has been obtained off the coast of Mauritius, and another from Japan (C. [Pleuro-corallium] secundum). The officers of the "Challenger" expedition obtained water-worn fragments at Banda and the Ki Islands, indicating the existence of the genus in the Malay Archipelago.

BRACHIOPODA.

The most detailed information that we possess respecting the geographical and bathymetrical distribution of the recent Brachio-poda is furnished by Davidson in his report appended to the narrative of the "Challenger" expedition (1880). Of the one hundred and thirty-five species and varieties (referable to some twenty genera and sub-genera) here recognised, whose distribution covers all parts of the oceanic surface, from Spitzbergen (Terebratella Spitzbergensis, Terebratulina caput-serpentis) and Franklin Pierce Bay (latitude 79° 25'; Rhynchonella pisittacea) on the north to Kerguelen Island and the Straits of Magellan on the south, there are few, if any, that can in any way be considered truly cosmopolitan, although species of broad distribution are not exactly uncommon. Terebratulina caput-serpentis is perhaps the most widely distributed of all the known forms, its habitat comprising the Arctic seas of both the Eastern and Western Hemisphere (Spitzbergen, Davis Strait), the Atlantic coast of Europe as far south as Spain, Jamaica, Corea, and Australia. Terebratula Wyvillii is found off the coasts of South Australia, the Falkland Islands, and Chili. As a rule it may be said that the north and south extension of a species is greater than the east and west extension, a condition doubtless due in principal part to the general disposition of the land areas in this direction. Arctic or circumpolar species have naturally a broad lateral dispersion. Species restricted in their habitat to shallow water are as a rule much more sharply circumscribed in their range than those inhabiting the greater depths, as might reasonably have been supposed. Localisation to limited coast lines, as Japan,

Corea, Florida, is of frequent occurrence, and not unlikely certain species may be even restricted to special bays or inlets of the sea. But few other than boreal or hyperboreal species are found inhabiting opposite sides of the same oceanic basin; notable exceptions are Platydia anomioides, whose range embraces the Mediterranean and West European coasts and the Florida reefs, and Thecidium Mediterraneum, from the northwest coast of Africa and Jamaica, both with a vertical range falling within six hundred fathoms. The most striking instances of areal discontinuity are furnished by Terebratella Frielii (near Halifax, one thousand three hundred and forty fathoms, and the Philippine Islands, one hundred and two fathoms) and Discinisca stella (Singapore, Philippines, Japanese and Chinese seas, and Bermuda), the former a deep-sea species, and the latter restricted to shallow water (seventeen to forty-nine fathoms). From such data as have been given, it would appear that the oceanic abysses form an insuperable barrier to the passage of the Brachiopoda; how the transference was effected in the case of the few exceptional species that have been indicated remains a matter of conjecture.

In their bathymetrical distribution the Brachiopoda affect the most diverse conditions of existence. While some forms (all or nearly all the species of Lingula, for example) appear to be incapable of living in greater depths than a few fathoms below water-line, others, again, seem just as incapable of leaving the greater depths. Discinisca Atlantica, a widely spread deep-sea species, has thus far been found only in water exceeding six hundred and fifty fathoms, and Terebratula Wyvillii, an equally wide-spread and abundant species, in water exceeding one thousand fathoms. Other species, on the other hand (Terebratula vitrea, 5–1456 fathoms; Terebratulina caput-serpentis, 0–1180 fathoms), seem capable of accommodating themselves to the greatest variety of depths. The species exhibiting the widest range of accommodation are such as have in part a boreal or arctic habitat; in other words, forms which find the same water-temperature at the surface and at varying depths beneath the surface. *Per contra*, the species having the most limited vertical range are those confined to the warm waters, tropical and sub-tropical. From this it follows that distribution in depth is effected primarily by conditions of temperature, and not by considerations of light, food-supply, &c., as has been urged by Fuchs and others. A marked exception to the above law is furnished by the type form of Terebratula vitrea, of the Mediterranean and adjacent seas, whose range varies between five and fourteen hundred and fifty-six fathoms. The greatest depth whence any brachiopod has been obtained is twenty-nine hundred fathoms (Terebratula Wyvillii).

Of about one hundred and seven species with whose range we are acquainted, fifty-seven, or more than one-half, are restricted to depths of under one hundred fathoms, and of these a large proportion properly belong to the shore-line, or to a zone of from five to fifteen fathoms water. There are twenty-one species (or varieties) whose range extends to, or above, five hundred fathoms; ten with a range of one thousand and upward; and three with two thousand. These facts indicate, as Mr. Davidson has pointed out, that the "greater bulk of known species live at comparatively small or moderate depths," and that "Brachiopoda are specifically rare at depths varying from five hundred to twenty-nine hundred fathoms." As to numerical distribution we find, as the result of the "Challenger's" explorations, that in ninety-nine dredgings, taken in water of from one to five hundred fathoms, Brachiopoda were brought to the surface about twenty-three times; in thirty dredgings, of five hundred to one thousand fathoms, four times; in ninety-four dredgings, of one thousand to two thousand fathoms, nine times; and in one hundred and seventy-six dredgings, of two to three thousand fathoms, only six times. It is seen here, therefore, that the numeric diminution keeps pace with the specific, and that practically the Brachiopoda cease to abound in depths exceeding five hundred to six hundred fathoms. The proper appreciation of these facts becomes of prime importance when discussing the nature of geological deposits containing brachiopod remains.

In respect of numerical development and broad distribution, both in time and space, the Brachiopoda constitute, for the geologist, the most important landmark in the determination of his horizons. Beginning with the very earliest fossiliferous formation, the Cambrian, and there already in nearly the bottom bed (St. David's), they continue throughout all time, and if, during the more recent geological periods, they have suffered rapid diminution, both in actual numbers and the variety of forms, they are still sufficiently abundant and varied to indicate that a long period must elapse before their final, or even approximate, extinction will have been reached. Indeed, if specific variation be considered a just criterion in the determination of the question of development or extinction, it may be reasonably doubted whether we are not now actually more remote from the apparent closing period of the existence of this group of animals than we were at the beginning of the Tertiary epoch, possibly a million or more years ago. Some sixty species and varieties, referable to ten genera, of brachiopods have been described by Davidson (1870) from the Tertiary deposits of Italy, and these constitute by far the largest number furnished by any one country of all the Tertiary forms that have thus far been described. England has but seven species, and France and Germany scarcely more, while the United States have not even as many. The number of distinct forms found in the present Mediterranean waters is about fifteen, or only one-fourth the number found in the Italian Tertiaries; but it must be recollected that the periods of time which we are here comparing are of very unequal duration, the "recent" period being only a mere figment of that indicated by the Tertiary formations. If to the recent and Quaternary species we add those found in the Upper Pliocene we will have a total of twenty-one species, which will then considerably outnumber the species from the entire Eocene series (thirteen or seventeen), and only fall eight or nine short of the total number from the combined Lower, Middle, and Upper Miocene. As to generic development, we find the Italian Tertiary species to belong to ten genera or sub-genera, to wit: Terebratula, Terebratulina, Waldheimia, Terebratella, Megerlia, Platydia, Argiope, Thecidium, Rhynchonella, and Crania, all of which, except Terebratella and Rhynchonella, are still found in the Mediterranean waters.* The number of Tertiary brachiopod genera and sub-genera thus far recognised is fourteen,** as against twenty or more of the present day.

The earliest known Brachiopoda, or those of the Cambrian period, belong almost exclusively to the group of the Brachiopoda inarticulata (Pleuropygia), forms in which the shell is horny-calcareous and devoid of a dental articulation—Lingula, Lingulella, Lingulepis, Obolella, Kutorgina, Acrothele, Acrotreta, Discina. The remains of Orthis (and Orthisina ?) alone of the Brachiopoda articu-

* Thecidium Mediterraneum appears to belong only to the African side of the Mediterranean.

lata occur associated with these lower forms. Up to 1868 there were catalogued some one hundred and twenty-six species (Bigsby) from the Cambrian deposits of the world, most of which belonged to Canada, Scandinavia, and the British Isles (twenty species in 1880, Etheridge). The prodigious development of the Brachiopoda in the Silurian period, apart from the mere consideration of numbers, is important, as representing the climax of development in this remarkable group of organisms. Henceforward they show a pretty steady decline, although the rate of decline for the various periods is different for different regions of the earth's surface. The fact that there are some fifteen hundred or more species in the Silurian deposits, of which considerably over five hundred are already represented in the Lower Silurian, and less than two hundred in the Cambrian, is extraordinary, whichever way it be considered, for, whether in the Cambrian we were somewhere near the beginning of life, or very distantly remote from it, as is much more likely to have been the case, the difficulty of explanation is in no wise affected. In either case the suddenness of the Silurian apparition is the same, and this is the more remarkable, seeing that there is scarcely any advance in the number of species of the Upper as compared with the Lower Cambrian. In the Bohemian Silurian basin Barrande enumerates six hundred and forty species, whereas in the Cambrian there are but two !

It is usually assumed that the Silurian species of brachiopods are vastly in excess of the Devonian, but the latest revised tables seem to indicate that this is not the case, and, indeed, it is not improbable that the numerical balance will be found to weigh on the other side. The number of genera is, however, greatly reduced, from about seventy to fifty, and this reduction is further carried into the Carboniferous period, where we have but forty genera, representing some eight hundred to nine hundred species. That there should have been, so soon after the climax had been reached, such a rapid decline is not a little surprising, but yet the suddenness of this decline is in no way comparable with the suddenness of the apparition already noticed in the Silurian. More surprising is the almost total absence of forms from the Permian deposits, where the deficiency can only partially be explained by the circumstance of the limited extent which these deposits occupy, and the special conditions under which their formation was effected.

The gradational passage between the brachiopod faunas of the Palæozoic and Mesozoic eras is probably more complete than in the case of any other group of the Invertebrata. Indeed, were it not for considerations drawn from other and higher groups of animals, the Triassic deposits would from this testimony alone be more properly relegated to the former than to the latter of the two eras, for with very insignificant limitations—Thecidium, Waldheimia, being exceptions—all the specific forms of this period belong, if not to such genera as are absolutely peculiar to the Trias—Koninckia, Cœnothyris—to types which are not only represented in the Paleozoic formations, but in most instances are eminently distinctive of them. It must be noted, however, that by far the greater number of the Palæozoic types had already ceased to exist, and such as still linger on, except the few more persistent types, like Lingula, Discina, Crania, and Rhynchonella, rapidly near extermination. The Cretaceo-Jurassic brachiopods constitute, strictly speaking, a single series, the members of which belong mainly to the genus Rhynchonella, and to a number of closely inter-related genera of the family Terebratulidæ.

Perhaps the most striking fact taken in connection with the geological distribution of the Brachiopoda is the remarkable variation shown in the adaptation of different groups to their surroundings. While certain generic types—those of the family Obolidæ, for example—appear to have been incapable of surviving for more than a comparatively brief period of time, dying out with the suddenness of their introduction, others, again, like the Lingula and Discina, have persisted throughout all time, and with such slight modification of structure as to render it difficult in some instances to determine specific differences between the most ancient and the most modern forms. That the least complex or most primitive forms of brachiopods should exhibit the greatest persistence might have been naturally expected, but it is nevertheless rather singular that the number of persistent types following these earliest precursors of a class should be so very limited in number. This persistence appears the more marked, too, when we reflect how very narrowly circumscribed in their vertical range are the majority of the species. Thus, out of the vast number of described forms, comprising possibly not less than five thousand distinct species, only a bare handful pass from one formation to another, and, indeed, a very large proportion of the species appear to be restricted to special zones of a given formation.*

In their geographical relations the ancient Brachiopoda differ essentially from those of the modern seas by reason of their broad horizontal distribution. Cosmopolitanism, or something approaching it, if not exactly the rule, was at least distinctive of a very large proportion of the species, a circumstance undoubtedly due to more equable conditions of environment, and the absence of interposed barriers to migration. This broad dispersion is perhaps nowhere better illustrated than in the case of the extinct fauna of China. Thus, out of a total of thirteen Silurian and twenty-four Devonian species described in Richthofen's work, no less than ten of the former and sixteen of the latter are also found in Western Europe; and, further, of the Devonian species, about eleven, or nearly fifty per cent., are cosmopolitan. Again, of about twenty-five Carboniferous species, some fifteen (or sixty per cent.) are common to North America, and about an equal number are cosmopolitan.

MOLLUSCA GENERALLY.

The more salient features connected with the distribution of the animals of this class have already been considered in our treatment of Geographical Distribution and Brachiopoda, and do not require restatement. It has been seen that the principal factors involved in this distribution are temperature and the presence or absence of continuous coast-lines along which migration might be effected; to these two categories may also be added the circumstance of light (or darkness), but to what extent this influence is exerted has not as yet been determined.

The attempted subdivision of the oceanic area into a number of distinct regions (provinces), each one characterised by a more or less peculiar assemblage of molluscan forms, while undoubtedly indicating a certain amount of faunal individuality, is still far from

---

* A more critical and impartial revision of the species will not unlikely materially increase the number of such connecting forms, and forms of even widely separated formations may be found to be identical. Thus, Davidson ("British Fossil Brachiopoda," "Palæont. Soc. Rep.," 1884, p. 396–398) admits that some of the forms of the Mediterranean Terebratulina caput-serpentis may only be varieties of the Cretaceous T. striata; and, likewise, that the recent Rhynchonella nigricans and some Cretaceous and Jurassic forms are so closely related to each other "that we are at a loss to define their differences."

satisfactory. The notion of absolute limitation, which was entertained when the provinces were first instituted, sees its own disproof in the records of almost every new exploration, and the annihilation of the very essentials which were considered requisite for the framing of zoogeographical boundaries. The more philosophical interpretation of the nature of species, and the more general recognition of the fact that certain forms considered to be peculiar to a definite region or district may also occur elsewhere, or where they are assumed not to belong, *even if they show no variation in their characters*, have done much to render the generally accepted provinces illusory. It is true that many of the regions now recognised by conchologists are strictly defined as such, but it is equally true, using the generally accepted criterion in the formation of provinces—the inclusion of a certain proportion of peculiar species—that new provinces, with entirely different boundaries, and with as much claim to faunal peculiarity, might be instituted in place of others that are also fully recognised. Until greater harmony is reached by malacologists in their realisation of species, and more regard paid to the facts of nature rather than to preconceived notions, any attempt at delimitation of zoogeographical boundaries must prove at best only half satisfactory.

The Mollusca proper—Lamellibranchiata, Gasteropoda, Pteropoda, and Cephalopoda—of which there are some 30,000, or more, recent species, have a world-wide distribution, being found in almost all parts of the earth's surface that have thus far been visited by man. From beyond the eighty-second parallel of north latitude to the Equator, and from the surface of the sea to a depth of sixteen or seventeen thousand feet, and to an equal height above it on the land, they are everywhere more or less abundant. Despite the extraordinary range which certain species are reputed to enjoy, there is yet scarcely a single one that is in any way entitled to the claim of cosmopolitanism, nor, indeed, any, if we except possibly some pelagic species (pteropods, cephalopods), and the members of the Arctic and Antarctic faunas, which occupy a complete circumferential zone of the globe.*

---

* A further exception may possibly have to be made in favour of some of the abyssal species, concerning whose distribution we know practically nothing. Mytilus edulis and Saxicava arctica are perhaps the most strictly cosmopolitan species known.

This condition is somewhat surprising in view of the free-swimming character of the molluscan embryo, and the fact that a number of forms of tropical habit have in some way managed to gain access to the opposite shores of the oceanic basin. Many of the forms belonging to the Caribbean province recur on the west coast of Africa, and the Dolium galea of the Mediterranean is also a member of the Brazilian and Antillean faunas. In what particular manner the transport was effected in the case of these few favoured species it is impossible even to conjecture, but it appears to point to a remarkable vital tenacity on the part of the embryo. Still more surprising, and entirely inexplicable, is the distribution of the fifteen or more species (including a dozen species of the genus Triton) whose common habitat is the Indian Ocean and the West Indian sea, when no connecting representative is found in the intermediate area. This is perhaps the most remarkable instance of specific areal discontinuity known, and contrasts sharply with what is observed on the west coast of America, where, of the very rich fauna of the Panamaic province, with no intervening barrier, an only equally small number of species is held in common with the Indo-Pacific region. The species of the far south, on the other hand, are largely similar, despite the absence of existing land-connection. From the occurrence of identical forms in New Zealand, the Magellan district, and the isolated tracts represented by the Kerguelen, Marion, Crozet, and Prince Edward islands, Fischer argues [b] that we have represented here the disrupted parts of a former Antarctic continent, along which specific diffusion was primarily effected. But there seems to be no reason why the present distribution might not at least in part be explained on the assumption of diffusion along the oceanic bottom—where the difference between surface and bottom temperature is no longer extreme —seeing that we have here also a number of distinctive Arctic or northern types represented (Chiton Belknapi, Lasæa rubra, Terebratulina septentrionalis, Terebratula vitrea, var. minor). The most southerly mollusk thus far met with is a pteropod, Limacina ? cucullata, which was obtained by the Wilkes Exploring Expedition on the sixty-sixth parallel of south latitude.

Instances of specific limitation are exceedingly numerous, and particularly characterise insular faunas. To such an extent is this the case that it may be said that almost every oceanic island or island group, when surrounded by deep water, has its own distinctive molluscan fauna. This we see in such islands as Cuba, Jamaica, Hayti, Madagascar, New Caledonia, the Philippines, &c., where the faunal peculiarity manifests itself not only among species, but extends largely to genera. Even the very much smaller islands of Malta, Cos, Naxos, Corfu, Zante, Lesbos, Rhodes, &c., in the Mediterranean, have a full share of species which are not found elsewhere. The Atlantic island groups of the Azores, Canaries, and Cape Verde are still more marked in their individuality. Both species and genera are here largely restricted to the several groups, and of the forms which ally them with the European fauna a large number are such as have been introduced by man. The common Holarctic fresh-water genus Unio is entirely wanting, and, indeed, in the Azores there is not a single fluviatile form represented. Of the eighteen species of Bulimus which constitute the entire non-marine molluscan fauna of the Galapagos Islands ten are restricted to single islands of the group. The genus Achatinella, which is restricted to the Sandwich Islands, has there some two hundred and eighty-eight species or varieties, or very nearly three-fourths the number of the entire molluscan fauna of the region.

This faunal individuality, the result of a long-continued isolation of the various island groups, permitting of a gradual but steady evolution of new and independent forms, does not extend to Great Britain, whose separation from the continent has been effected in a comparatively recent period, and after the constitution of the contiguous faunas. There appear to be but two molluscan species (Limnæa involuta and Assiminea Grayana) that are distinctive of this island group. Similarly, the identity existing between the Peninsular and North African faunas points, apart from all other evidence, to the recent formation of the Strait of Gibraltar.

The natural barriers which interpose themselves to the free migration of the terrestrial and fluviatile Mollusca, such as intercepting land and water areas, elevated mountain-chains, or deserts, are apparently much more numerous than those affecting marine forms, and account for the comparatively limited range of most species. Yet it is remarkable how far certain forms or types of forms have spread. Thus, the two most important genera of land and fresh-water mollusks, Helix and Unio, and among fresh-water pulmonates, the genera Limnæa, Physa, Ancylus, and Planorbis, have an almost world-wide extension, many of the species even appearing in widely separated parts of two or more continents. Still, we know of no truly cosmopolitan species of these groups, although through artificial transport many of the forms have been made to spread far beyond the limits of their natural domain. Helix similaris, indigenous to Eastern Asia, is now largely found in Malaysia, Polynesia, Australia, the Seychelles, South Africa, the Antilles, and Brazil, and as nearly an extensive range is claimed for Ennea bicolor. The common garden-snail of Europe (Helix aspersa) has been naturalised in Algeria, the Azores, Brazil, and California, and human agency has, doubtless, been largely involved in the dissemination of the small moss-inhabiting Helix pulchella, which now inhabits the greater part of the continent of Europe, the Caucasus, Madeira, the region of the Cape of Good Hope, and nearly all Northern North America.

Bathymetrical and Hypsometrical Distribution.—There are five zones of distribution usually recognised in the oceanic waters, as follows: 1, the Littoral Zone, or that existing between tide-marks —the habitat of the periwinkle, limpets, sand-clam, cockle, mussel, and barnacle; 2, the Laminarian Zone, extending from low-water mark to about fifteen fathoms, and characterised by a dense growth in many places of sea-weed and tangle—the haunt of the vegetable-feeding Testacea and of various nudibranchs, and the home of the oyster; 3, the Coralline Zone (from fifteen to forty or fifty fathoms), the zone of the encrusting Algæ (nullipores), and of the large car-nivorous Gasteropoda (Buccinum, Fusus, Pleurotoma, Natica, &c.); 4, the Deep-sea Zone, or that of the Brachiopoda and deep-sea corals (from fifty to two hundred and fifty or three hundred fathoms); and, 5, the Abyssal Zone, from three hundred fathoms to the oceanic floor, throughout which the shells are generally of small size, trans-lucent (thin), and white or but feebly coloured. The visual organs are here exceptionally devoid of pigment, and blindness has been noted in a few species of normally seeing gasteropods. It might be doubted, however, whether the last two divisions ought not more properly to constitute a single division, considering the large pro-portion of genera and species which pass from the shallower to the deeper parts. Dall, from the data furnished by the dredgings made in the Gulf of Mexico by the steamer "Blake" (1877–'78), affirms that fully twenty per cent. of the molluscan species obtained from that region enjoy a vertical range extending from less than fifty to two hundred and fifty and two thousand fathoms,* which would then give a very high proportion for those connecting the fourth and fifth zones.

At depths of from three hundred to one thousand fathoms mollusks are still numerically very abundant, although the number of species very rapidly diminishes. From an extreme depth of 2,435 fathoms (14,610 feet) the "Porcupine" obtained but five species, and from 2,900 fathoms (17,400 feet) the "Challenger" dredged only two—Semele profundorum and Callocardia Pacifica. In the greatest depths the Lamellibranchiata (principally represented by the families Arcadæ, Nuculidæ, and Pectinidæ, the genera Pecchiolia, Neæra, &c.) appear to preponderate over the Gasteropoda, whose dominating forms are tectibranchs and the Scaphopoda, and, among the prosobranchs, the genera Fusus and Pleurotoma.†

The question whether the abyssal fauna is of a generally uniform type, marked by identical or representative species extending from pole to pole, as was first suggested by Lovén, and subsequently admitted by Sir Wyville Thomson, still lacks the necessary data required for its solution. The extremely broad or antipodal

---

* "Bull. Mus. Comp. Zool.," vi., 1880. In a supplemental note only fifty-one species (out of four hundred and sixty-two) are recorded whose range covers both the littoral and the abyssal zones, thereby reducing the ratio to eleven per cent.

† The researches of Edgar Smith and Boog Watson upon the Mollusca obtained by the "Challenger" reveal some very remarkable instances among this group of animals of ready adaptability to the most varying conditions of depth, and of discontinuous habitation. Silenia Sarsii was dredged about 1,100 miles southwest of Australia in water of 1,950 fathoms, and again off the mouth of the Rio de la Plata, in 2,650 fathoms ; Verticordia Deshayesiana, found off Pernambuco in water of 350 fathoms, was also dredged off Cape York in 155 fathoms ; Petricola lapicida, a well-known West Indian form, recurs off the North Australian coast (seven fathoms) ; and Nuculina ovalis, a fossil of the Suffolk Crag, reappears in the waters of the Cape of Good Hope in twenty fathoms. Venus mesodesma, a shore species, descends to 1,000 fathoms, while the range of Lima multicosta extends from two to 1,075 fathoms, and of Arca pteroessa from 390 to 2,050 fathoms (the West Indies and the North Pacific, respectively). The total number of lamellibranch species obtained in depths under 100 fathoms was 384 ; in depths between 100 and 500 fathoms, 148 ; between 500 and 1,000 fathoms, 24 (ten stations) ; and between 1,000 and 2,900 fathoms (thirty-three stations), 70. ("Challenger" Reports, "Zoology," xiii., 1885.)

distribution of many of the species has led to the belief that such is the case, but, on the other hand, certain facts that have recently been brought to light seem to point in the contrary direction. Thus, it has been shown by Mr. Dall that of the species composing the abyssal fauna of the Gulf of Mexico only ten per cent. are such as may be termed boreal, a very small proportion to what might have been expected, were it to be assumed that the peopling of the cold bottom wastes was effected by a descent from the polar regions. On the other hand, thirteen per cent. were found to be tropical, and seventy-five per cent. uncharacteristic, forms. It is concluded from these facts, and from the circumstance that "the tropical forms belong to the same groups as those characteristic of the local littoral mollusk fauna," that in all probability "the abyssal regions have local faunæ proper to their various portions, and that a universal exclusive abyssal fauna, so far as mollusks are concerned, does not exist." This conclusion, which was concurred in by the late Mr. Gwyn Jeffreys, receives further support, it is claimed, from the distinctness of the "Challenger" Mollusca as compared with those of the "Blake."

The hypsometrical distribution of the Mollusca is governed almost exclusively by conditions of climate and food-supply, the influence of the latter being manifest in the intimate relation which binds many of the species to the plants upon which they habitually feed. Thus, in the Higher Kabylia, Aucapitaine has framed three molluscan zones, each corresponding to a particular plant growth : 1. The zone of the ash, olive, and pomegranate (450 to 2,100 feet); 2. That of the oak and pine (2,100 to 3,600 feet); and, 3. That of the cedar and green turf (3,600 to 7,200 feet). The upper limit to which mollusks attain on the continent of Europe (Alps) is about eight thousand feet, somewhat below the line of perpetual snow ; along the region of the equatorial Andes and the Himalayas the line is placed at about twice this height, also approximating the snow-level. Five species of fresh-water shells, of the genera Planorbis, Paludestrina, and Cyclas, were found by Morelet to inhabit Lake Titicaca at an elevation of nearly 13,000 feet, while from the Himalayas Anadenus Schlagintweiti has been obtained at a height of 16,500 feet, and Limnæa Hookeri at 18,000 feet. Of the North American land shells the representatives of extreme hypsometric range appear to be Pupa alticola and

Vallonia pulchella.[96]   The various hypsometric zones that have been
established by conchologists differ at almost all parts of the earth's
surface, and are of but local import.   The shells of the more
elevated mountain-summits are many of them, or mostly, of types
which are found at lower levels in regions of reduced average
annual temperature, following the well-known law of climatic
dispersion which we recognise among plants.   Vertigo alpestris,
an inhabitant of Scandinavia, reappears in the Alps of Switzerland,
although completely wanting in the intermediate regions; and,
similarly, many Alpine summits hold identical or representative
species which are wanting in the connecting lowlands.

Geological Distribution.—The most salient fact that presents
itself in connection with the past distribution of the Mollusca is
the reversed order to what might have been expected of the successive development of its primary classes.   Thus, almost everywhere, the Cephalophora, or head-bearing mollusks, antedate by
one full period the Acephala, or headless forms, which indisputably represent a lower grade of organism; and among themselves
the first to attain a maximum development are the cuttle-fishes
(Cephalopoda), which are structurally the highest.   But two species
of this group—an Orthoceras and a Cyrtoceras—are positively known
from the Cambrian formation, while in 1868 Bigsby enumerated no
less than one hundred and thirteen species of Gasteropoda as belonging to the same period of time.   On the other hand, the total
number of gasteropod species credited by the same author to the
Silurian deposits is about eight hundred, whereas Barrande has
described upwards of eleven hundred species of Nautilidæ from
the Upper Silurian deposits of Bohemia alone.   As has already been
intimated, there is a marked deficiency of Cambrian lamellibranchs,
and even in the Silurian formation the number of species is comparatively limited.   Bigsby, in 1868, enumerated, all in all, some
six hundred and thirty-six species, or but little more than one-half
the number of Upper Silurian cephalopods of the Bohemian basin.
The complete differentiation which the different classes of the
Mollusca had already attained in the Silurian period argues for a
great antiquity beyond that period of the members of this group.
To what extent the time measured by the Cambrian period, and
the interval intervening between it and the Silurian, may have been
effective in bringing about the various changes, cannot be at pres-

ent determined; it would appear at first sight as though the length of its duration were sufficient, for how otherwise could we intelligently explain the total or nearly total absence of the members of at least two of the molluscan group from deposits in which the representatives of other groups are sufficiently abundant? One fact, however, must not be lost sight of in this connection, and that is, that in these earliest deposits the obliteration of organic remains has been most excessive, and that not improbably the absence of the required forms is to be attributed rather to the destruction of parts than to an actual non-existence in the region. For the present it is impossible to affirm whether the Cephalophora came into existence before the Lamellibranchiata or not, but the evidence scarcely appears sufficient for considering the latter as a race derived by degeneration from the former, as has been presumed by Professor Lankester.

The sudden decline of the Cephalopoda (Nautilidæ) after the close of the Silurian period is very remarkable, and scarcely less so their rehabilitation under the form of their successors, the Ammonitidæ, in the deposits of the Mesozoic era; Barrande enumerates but two hundred and forty-two species from the Devonian formation, more than one-half of which belong to the genus Orthoceras, and the remainder principally to the genera Cyrtoceras, Gyroceras, and Gomphoceras. About an equal number are indicated from the deposits of Carboniferous age, where also we find much the same genera represented, although with different specific relations. The genus Nautilus now for the first time acquires any importance, and it and Orthoceras alone of the limited surviving members of the family of the Permian period transgress the boundaries of the Paleozoic era. The latter genus disappears early in the Trias, while the former steadily increases in number, until in the Cretaceous deposits it attains its maximum development, with a representation of some sixty or more species.

The displacement of the Nautilidæ by the Ammonitidæ is, if nothing more, certainly an interesting circumstance, and leads one to inquire what special advantage the latter may have possessed over the former in the struggle for existence, by means of which they triumphed over their predecessors. For there can be little question that the Ammonitidæ, despite certain peculiarities in their structure, which are not as yet comprehensible to us, are the truly modified descendants of the nautiloids, a transition to which appears to have been effected by way of the genus Goniatites and those forms of the Carboniferous period (India and Texas) which, like Arcestes, have the sutural plication intermediate between what is seen in Goniatites (Silurian-Permian) and Ceratites (Triassic). Just where the embranchment from the nautiloid line took place it has been impossible to determine, but it is significant that the most nautiloid form of the Ammonitidæ, the Goniatites, appeared after the Nautilidæ had attained their maximum development, and some time after the genus Nautilus had itself appeared. Sutner estimates that there are in the neighbourhood of four thousand species belonging to the group of the ammonoids, nearly all of which have the foliated sutures characteristic of the true ammonites. With the exception of the genera Goniatites and Clymenia, and the primitive ammonitic forms of the Carboniferous rocks already referred to (Sageceras, Arcestes, Xenodiscus, Medlicottia, Cyclolobus *), and a single form which occurs in the lowest member of the Californian Tertiaries (the Tejon group), all the species are restricted to the Mesozoic deposits, which by their great numerical development they might be said to characterise. Probably no other group of invertebrates exhibits such a remarkable series of developments corresponding with successive periods of time as do the Ammonitidæ, and in none do the species appear to be so distinctively characteristic of certain horizons (zones of ammonites). The singularly diversified types of the Triassic period, which combine all the various sutural modification seen in the goniatitic stage (Sageceras, Lobites), the ceratitic (Tyrolites, Celtites), and the ammonitic, from the simplest to the most complex (Pinacoceras), are almost wholly wanting in the Lias, where an entirely new series of forms begins (Ægoceras, Harpoceras, Amaltheus). These in turn are succeeded by groups more or less distinctive of the different Jurassic zones (Oppelia, Stephanoceras, Lytoceras, Phylloceras), which in the main die out before the close of the period. Most of the Cretaceous forms belong to genera or sub-genera which have not hitherto been represented, and here for the first time do we find any great development of the remarkable groups of uncoiled ammonites —Scaphites, Hamites, Turrilites, Crioceras, Ancyloceras (also Jurassic), and Baculites—whose advent seems to be foreshadowed by the

* Cyclolobus has the typical ammonitic sutures.

Triassic genera Choristoceras, Cochloceras, Rhabdoceras, &c., which, in their departure from the normal ammonitic type, resemble some of these forms, but which differ in the simple character of the sutural foliation. Considering the Ammonoidea to be the modified descendants of the Nautiloidea, we have presented the somewhat anomalous fact that while at the beginning of their existence *involution* was the order of development, towards the end this development was marked by a contrary *evolution*, with an accompanying approximation in outline to that of the primitive type-forms. No facts with which we are at present acquainted permit us to state what was the underlying principle involved in these reversed changes. It can merely be said that, involution having once set in, any broad departure from the type newly attained must almost necessarily have been accompanied by a certain amount of evolution.

The persistence of type-structure among the Nautiloidea, and the relation which the geographical distribution of the Ammonoidea bear to supposed climatic zones, have already been discussed in previous sections. Still more involved in doubt than that of the Ammonoidea is the ancestry of the Belemnitidæ, which, as the earliest representatives of the two-gilled order of cuttle-fishes, first appear in the Trias, and practically disappear with the close of the Mesozoic era, one species only, and that somewhat doubtful—Belemnites senescens, from Australia—being reported to pass beyond the boundaries of the Cretaceous period.* That the group, however, represents the ancestral line whence the recent Sepiophora (Sepia) have been derived there can be but little question, seeing how close is the relationship between the determining parts—internal skeleton —of the fossil and living species. In the Eocene genus Belosepia the phragmocone is of a somewhat transitional character. The modern pen-bearing cuttle-fishes or calamaries (Chondrophora) appear to have their direct ancestors in the various forms of Teuthidæ, whose remains, in a more or less perfect state of preservation, occur in the Liassic and Oolitic deposits.

Of the two other classes of Cephalophora, the Gasteropoda and Pteropoda, only the former acquire any geological importance. Beginning with a comparatively limited number of forms in the

---

* The reference of this form to the Belemnitidæ is considered more than doubtful by Branco (" Zeitschrift d. deutsch. geol. Gesellschaft," 1885).

Cambrian period, they steadily increase in number, until at the present day the number of known species is far in excess of that recorded from any other period of geological time. According to the estimates made by Bronn between the years 1862 and 1866, which may possibly still serve as a basis for the computation of successive ratios, although no longer abreast of the times, the numerical distribution for the several geological eras is as follows:

	Paleozoic.	Mesozoic.	Cainozoic.	Recent.
	No. of species.	No. of species.	No. of species.	No. of species.
Scaphopoda............	22	48	55	50
Prosobranchiata........	737	1,764	4,622	7,500
Heteropoda (inclusive of the Bellerophontidæ)..	141	3	1	54
Opisthobranchiata.......	1	152	185	825
Pulmonata *...........			530	5,700

With very few exceptions (Dentalium, Pleurotomaria, Capulus, Natica, Narica, Emarginula) all the Paleozoic genera are extinct, or at least generally considered to be so, and it is still questionable whether they include even a single siphonate form. The reference of Fusus, Pyrula, and a few other members of the Siphonata to this period, probably rests on unsatisfactory determinations, although, indeed, no special reasons can be assigned for the non-extension back of such genera. The absence of prominent characters in many of the species renders their determination difficult or impossible, and it is by no means improbable that a fair proportion of the genera which have received distinct names are in reality identical with modern genera otherwise designated. The generally holostomate character of the Paleozoic Gasteropoda imparts to the fauna a peculiarity which eminently serves to distinguish it from the similar fauna of the later Mesozoic and Tertiary eras, especially the latter, where the Siphonata largely preponderate. The dominant Paleozoic genera are Pleurotomaria, Murchisonia, Euomphalus, and Loxonema, whose special development throughout the greater part

---

* The Pulmonata are now known to be represented by a limited number of species in the Devonian and Carboniferous deposits. The number of Paleozoic species indicated by Bigsby (1868-'78) is about treble the figure given by Bronn.

of the Paleozoic series tends to link the different members together.

The earliest great differentiation in the ancient gasteropod fauna is seen in the Jurassic deposits, where a host of new forms, especially of the Siphonata (of the families Cerithiidæ, Nerinæidæ, Aporrhaidæ, Strombidæ, Buccinidæ, Purpuridæ, Columbellidæ, &c.), are for the first time introduced. One or two of these families appear to have had their representatives already in the Trias, but they were there of insignificant import. It is not, however, until the Cretaceous period that many of the more distinctive of the modern families (Cypræidæ, Cassididæ, Ficulidæ, Tritonidæ, Muricidæ, Volutidæ, Olividæ, Cancellaridæ, Terebridæ, Pleurotomidæ, and Conidæ) appear, and of these a fair proportion of the genera date back only to the Eocene period. No recent species is recognised as extending back beyond the Tertiary series, and even in the Eocene the proportion of living to extinct forms is very slight, averaging not more than three to five per cent. ; indeed, it is questionable whether any of the early Tertiary species can be identified with recent forms. The same is also possibly true of the Oligocene, but in the Miocene the percentage ranges as high as thirty-five, and in the Pliocene to seventy-five or more. Practically, all the Post-Pliocene forms are still living. It is impossible to arrive at any absolute estimate of the number of species occurring in each formation. Bigsby [97] enumerates some seven hundred to eight hundred species as belonging to each of the Silurian, Devonian, and Carboniferous periods, or very nearly that which is given by Zittel [98] for the Jurassic forms; the Permian, which is deficient in nearly all forms of life, has but about thirty. The Eocene-Oligocene Paris basin contains, according to Deshayes, upwards of eighteen hundred species, or more than double the number of the entire Eocene shell fauna of the Eastern and Southern United States. The Miocene basin of Vienna holds upwards of four hundred species of Prosobranchiata.

There are but very few truly cosmopolitan species of fossil Gasteropoda, although broad distribution was much more marked in the early periods of the earth's history than now. Thus, while from the American Tertiaries only a very insignificant number of forms could be selected which might in any way be correlated with contemporary European species, and but a fairly representative series from the Cretaceous, no less than two hundred and fifty out of a round six hundred from the Silurian deposits are stated to occur in Northern and Western Europe. Some seventeen or more species out of the fifty-two recorded by Etheridge from the middle and upper Paleozoic divisions of Australia find their analogues in the equivalent deposits of Europe, and not unlikely this identity will be increased on further comparisons being made. The Tertiary species of the two regions, on the other hand, are almost without exception distinct, and of the recent forms it may be doubted whether there is a single species held in common.

The geological distribution of the Lamellibranchiata may be considered to run parallel with that of the Gasteropoda, and in a general way to partake of its peculiarities. Most regions are entirely deficient in Cambrian forms, and even in the Lower Silurian formations the number of species is rather limited. Barrande enumerates upwards of eleven hundred species from the Upper Silurian formation of Bohemia alone, nearly double the number (636) that was assigned by Bigsby in 1868 for the Silurian deposits of the world generally, and considerably over that (918) which was claimed by Bronn in 1862 for the entire Paleozoic series. The oldest forms —*i. e.*, Silurian and Devonian—belong almost exclusively to the Heteromyaria (Aviculidæ, Mytilidæ) and the Dimyaria (Nuculidæ, Arcadæ, Astartidæ, Cardiidæ), although many of the Devonian forms that have been referred to the Heteromyaria may really belong to the Monomyaria. The Sinupalliata among the Siphonida appear to be completely wanting, as they are likewise from the Carboniferous deposits, but it is by no means unlikely that some of the commoner genera, as Grammysia, Allorisma, Sanguinolites, Edmondia, &c., in which no sinual impression has been detected, are true members of families whose modern representatives are all furnished with retractile siphons. This supposition, which is based upon external resemblances and habits as deduced from the shell, is in full consonance with the theory of evolution, which would lead us to suppose that the direct ancestors of the Sinupalliata were closely resembling forms devoid of a sinual inflection. The reference of the forms above mentioned to the family Pholadomyidæ may, however, still be considered uncertain.

The Carboniferous Lamellibranchiata do not differ very broadly from those of the preceding (Devonian) period, except in so far as pertains to the first considerable development of the Monomyaria (Pectinidæ, Limidæ). Here, too, if we except the still problematical Præostrea, we meet with the earliest indubitable remains of the oyster (Ostrea). As in the case of the Gasteropoda and Brachiopoda, the number of Permian species is very limited. The lamellibranch fauna of the lower and middle divisions of the Mesozoic series, Trias and Jura, is characterised by a remarkable development of the families Ostreidæ, Pectinidæ, and Limidæ among the monomyarians, the Mytilidæ and Pinnidæ among the heteromyarians, the Arcadæ among the Asiphonida, and the Astartidæ, Lucinidæ, and Cardiidæ among the integropalliate Siphonida. The Pholadomyidæ are especially abundant in the Jurassic deposits. Of the two most distinctive Cretaceous groups, the Chamidæ and the Rudistæ, only the former have their Jurassic representative (Diceras). Beyond the great specialisation of those two families, the Cretaceous fauna does not differ essentially from the Jurassic, although the number of true sinupalliate forms (Veneridæ, Tellinidæ, Solenidæ, Glycimeridæ) is very much greater.

Through the different divisions of the Tertiary, beginning with the Eocene, we see the gradual development which by almost imperceptible stages leads up to the fauna of the present day. The Monomyaria, which in the Mesozoic period constitute nearly thirty per cent. of the entire lamellibranch fauna, enter upon their decline, and are succeeded, as well as the Heteromyaria, by the Dimyaria, of both the sinuate and non-sinuate types. The relation of extinct to recent forms in the different divisions of the Tertiary holds much the same as with the Gasteropoda.

CRUSTACEA GENERALLY.

Of the recent orders of Crustacea the only ones that acquire any geological significance are the Phyllopoda, Ostracoda, and Decapoda, although representatives of some of the other orders occur sparingly in formations extending as far back as the Devonian period (Præarcturus, among the isopods [?]), and possibly even to the Silurian (Necrogammarus, amphipod). It is a rather surprising fact that of the first two orders, if we except one or two special types—Leperditia, Beyrichia—all the most abundantly represented genera of the earliest periods, as well as of the periods succeeding, are such as still hold considerable prominence at the

present day—Estheria, Cythere, Bairdia, Cypridina—thus presenting one of the most remarkable instances of the persistence of
type-structure known in the whole range of the animal kingdom.
The genus Estheria dates from the Devonian period, and attains
its maximum development in the Trias. In its modern distribution it may be said to be almost cosmopolitan, although it would
seem to prefer the regions of warm climate, and not to penetrate
much beyond the fifty-fifth parallel of latitude. The range of some
of the species is extraordinary. Estheria Dahalacensis, which occurs as far north as Vienna, is found from Sicily to the island of
Dhalak, in the Red Sea, or over an area whose extent is measured by about thirty degrees of longitude, and thirty-two degrees
of latitude. The range of E. tetracera comprises fully forty degrees of longitude (Oran—Kharkov), but is more than equalled
by the Carboniferous species E. Leidyi, which has been reported
from both England and the State of Pennsylvania. It is a singular
circumstance that while all the recent species of Estheria are inhabitants of fresh water, or of water which is but barely brackish,
the fossil forms are frequently, or generally, found associated with
distinctively marine types of organisms, indicating apparently for
these species also a marine habit. While this may have been
true, the association with fresh-water forms, which also occurs,
tends to show that it was only partially the case.

Possibly belonging to the order of the Phyllopoda, but by
some authors placed among the Malacostraca, are the singular
shield - bearing crustaceans of the Silurian period, Ceratiocaris,
Dictyocaris, Discinocaris, Peltocaris, &c., whose affinities have been
placed with the modern genera Nebalia (marine) and Apus (fresh-
water). In Discinocaris the shield in some individuals measures as
much as six inches across. Hymenocaris, which is exclusively
Cambrian, represents the oldest type of this order of crustaceans.
The genus Apus itself, which is almost universally distributed, and
has been observed even in Norway at an elevation exceeding three
thousand feet, appears as early as the Carboniferous period. Branchipus, although devoid of a head-shield, is stated by Woodward
to be preserved as a fossil in the Eocene (Oligocene ?) beds of the
Isle of Wight.[99]

Of the more important genera of recent Ostracoda, Cythere and
Bairdia both date from the Silurian period, and Cypridina from
13

the Devonian; the fresh-water genus Cypris is unknown prior to the Carboniferous. The species of the first two genera, whose distribution is almost universal, are particularly numerous in the Cretaceous and Tertiary deposits, and present some remarkable instances of reputed longevity. Two recent species of Bairdia, B. subdeltoides and B. angusta, are claimed by Gerstaecker [100] to be found also in the Carboniferous deposits, but it is by no means improbable that the identification rests on improper determinations. This is rendered the more likely, seeing how very few, comparatively, are the recent forms that have been identified as occurring in the Tertiary deposits. Of some seventy species or varieties of Entomostraca described by Rupert Jones from the British Tertiary deposits from the Eocene to the Pliocene inclusive, not more than seventeen or eighteen are recognised by him as constituting a part of the recent fauna. A re-examination of the forms may possibly increase this number somewhat, but it is certainly very remarkable that of two hundred and twenty or more species of Ostracoda dredged by the officers of the "Challenger" expedition only three or four had previously been described by paleontologists. While many of the modern species have a broad distribution, the number of forms which are known to be in any way cosmopolitan is exceedingly limited (species of Halocypris and Cythere). On the other hand, there are some very marked instances of antipodal reappearance. Several well-known British and northern forms have been identified from Kerguelen Island and other remote regions of the earth's surface. Although apparently penetrating to the profoundest depths of the sea, the number of both species and individuals rapidly diminishes beyond a comparatively shallow superficial zone. Only fifty-two species were obtained by the "Challenger" from a depth exceeding five hundred fathoms, and but nineteen from below fifteen hundred. The much greater diversity of the shore fauna as compared with that of the open sea is shown by the fact that among the "Challenger" Ostracoda only twenty-eight genera were represented, whereas on the British coast alone there are at least thirty-one. Among the oldest known representatives of the order are the Primitia prima and Leperditia Cambrensis, from the St. David's (Lower Cambrian) rocks of Wales.

The remains of decapod crustaceans in the Paleozoic rocks are exceedingly scanty, as indeed they are also in the earlier part of the Mesozoic formations. In North America they have been traced back to the Devonian period (Palæopalæmon ;—Macrura), but in the European deposits they are not known before the Carboniferous age (Brachypyge ;—Brachyura ?). Whether the astaciform crustacean found in the mountain limestone of Ireland, and described as Astacus Philippi, was of a fresh-water habit, may still be considered as more than doubtful. The same doubt extends to the various Astacomorpha of the Middle Mesozoic period—Eryma, Pseudastacus—and not until we reach the chalk of Westphalia do we meet with any undoubted remains of the genus Astacus itself. But even here the deposit in which the remains are imbedded is of a marine facies, and seems to argue that up to this time the crayfishes were by nature inhabitants of salt water. Too much weight cannot, however, be attached to the negative evidence afforded by the absence of known fresh-water forms of this group, inasmuch as the chances for their preservation in deposits of this kind among the older rocks is very slim. At the same time, it is not a little singular that the extensive deposits of this nature of Wealden age should be entirely barren of their remains. In the Tertiary deposits indisputably fresh-water forms are met with, and it is not unlikely that the full differentiation of these types from those of a marine habit was effected somewhere about the close of the Mesozoic era. The lobster or homarine type has probably its oldest representative in the genus Hoploparia, which occurs in the Cretaceous and older Tertiary deposits. Of the modern genera of prawns, Penæus appears to extend back to the Lias, and Palæmon to the Tertiary period.

The most important order of Crustacea, considered from the geological standpoint, is that of the Trilobita, which, apart from the simple fact of numerical development, acquires special significance from the circumstance of its representing the earliest animal group which attained to any prominence in geological history, although in point of actual appearance it would seem to have been preceded by the Brachiopoda and Annelida. Thus, from the basal portion of the St. David's beds of South Wales, which represent the oldest or very nearly the oldest of the fossiliferous rocks that have been thus far discovered, no trilobites are known, and it is not until a full thousand feet in the same series of deposits is passed that their remains are first met with. The number of species of trilobites found in the inferior division of the Lower Cambrian of Britain (the Longmynd and Harlech groups) is about ten, representing six genera — Agnostus, Conocoryphe, Paradoxides, Microdiscus, Palæopyge, Plutonia—and of the entire Cambrian series one hundred, distributed as follows :

Upper Tremadoc...... 12 (including 2 from the Upper Lingula).
Lower Tremadoc...... 19.
Upper Lingula........ 30 ("   5   "   Lower Lingula).
Lower Lingula........ 22 ("   14   "   Menevian).
Menevian .... 32 ("   4   "   Longmynd).
Longmynd and Harlech 10.

From the Primordial zone of Bohemia (Cambrian) Barrande recognised in 1871 only twenty-seven species, or but little more than one-tenth the number (252) which he claimed for the Cambrian deposits of the world generally. This paucity, as compared with the richness of the British fauna, is the more surprising when we consider how largely in excess of the insular forms are the Silurian species from the same region. The Bohemian basin contained at the period stated about three hundred and twenty species, whereas the total number of species recognised at about the same time from the entire British Paleozoic series of deposits—i. e., from the Cambrian to Carboniferous inclusive—only slightly exceeded two hundred and twenty.

Of the seventeen hundred species, representing seventy-five genera, tabulated by Barrande for the world at large we find (using his data) two hundred and fifty-two relegated to the Cambrian formations, eight hundred and sixty-six to the Lower Silurian, four hundred and eighty-two to the Upper Silurian, one hundred and five to the Devonian, fifteen to the Carboniferous, and one to the Permian.* While these figures would indicate a pronounced culmination of the group in the Lower Silurian period,

* The number of species has been very materially increased since the publication of Barrande's paper, but his estimates will still serve as a proper basis for the computation of ratios. The American Carboniferous species seem to fall not far short of a dozen by themselves. Considerable uncertainty still attaches to the stratigraphy of the forms that have been referred to the Permian of Germany and the United States (Guadalupe Mountains, in Texas and New Mexico), and it is generally assumed that there are no Permian species at all.

it is a significant fact that in Bohemia, which stands next to Scandinavia in respect of number of species, the numerical ratio of Lower Silurian to Upper Silurian forms is as one hundred and eighteen to two hundred and five, reversing, apparently, the order of development.

The very limited number of generic forms that pass from one major formation to another is remarkable. Barrande enumerates but seven of the twenty-seven Cambrian genera which pass'over into the Silurian, and twelve of the fifty-five Silurian genera which reappear in the Devonian. The Carboniferous genera are but three or four in number (Phillipsia, Griffithides, Brachymetopus, Proetus).* Of the fifty-five Silurian genera, with three exceptions, all the forms are already represented in the lower division. The number of genera that extend through two or more formations is reduced to two or three (Phillipsia, Proetus). In their geographical relations it may be said that broad distribution is the rule rather than the exception. Thus, of the forty-two genera of the Bohemian basin thirty are held in common with Sweden, and twenty-four with England. More than one-half (seventeen out of thirty) of the North American genera are also trans-Atlantic forms, and the greater number of these are widely distributed over the European continent. It has generally been considered that the most widely distributed genera are those which also have the greatest vertical range; but the exceptions to this supposed rule are so numerous—Paradoxides, Agnostus, Trinucleus, Asaphus—that it may be doubted whether any value is to be attached to it. Nor can it be maintained that in all cases the genera having the longest range are those which have the greatest number of specific representatives, although this is more often the case than otherwise. Probably the greatest number of species represented in any one genus is exemplified in the case of the genus Dalmanites (Lower Silurian–Upper Devonian), of which there were up to 1871 some one hundred and twenty-nine known. The nearest approach to this (some one hundred and fifteen or more) is seen in the genus Asaphus, which is restricted to the Lower Silurian formation.

The number of species that transgress the boundaries of any major formation is exceedingly limited. Thus, it is very doubtful

---

* Professor Claypole has more recently described a species of Dalmanites (Dalmania) from the Waverly Group (Sub-Carboniferous) of the United States.

whether there is even a single species from the Cambrian which also forms part of the Silurian fauna. There are apparently but three forms—Calymene Blumenbachii, Dalmanites caudatus, Sphærexochus mirus—which in England connect the Lower and Upper Silurian faunas, and an equally small number which unite the Silurian with the Devonian. In Bohemia the number of Lower and Upper Silurian connecting-forms is somewhat larger; but even here°the proportion to the entire fauna—nine out of three hundred and twenty-three—is very small. The limitation in most cases is even more pronounced, fixing the species to definite horizons of the broader formations.

With respect to horizontal specific distribution, instances of broad dispersion are not exactly uncommon. A fair proportion of the Bohemian species, for example, are spread throughout Sweden, Italy, Russia, and England, and a number have also been indicated as occurring in the equivalent deposits of the North American continent. Calymene Blumenbachii is the commonest form occurring on both sides of the Atlantic, in both regions being alike a Silurian and Devonian species.

Coincidently with the decline of the Trilobita we note the appearance of crustacean forms (Eurypterida) which hold a somewhat intermediate position between these and the modern king-crab (Limulus), whose remains are first met with in rocks of Jurassic age. Of this singular order, which comprises the giants of the class, some half-dozen or more genera are recognised, whose combined range includes the Upper Silurian and Carboniferous deposits. In the Ludlow (U. Silurian) rocks of Britain there are no less than thirty-two species of this order, representing six genera—Eurypterus, Pterygotus, Himantopterus, Slimonia, Stylonurus, and Hemiaspis, the last a transitional type connecting the group with the Carboniferous limuloid forms constituting the family Bellinuridæ (Bellinurus, Prestwichia, Euprööps). The most ponderous individuals of the order (Pterygotus Anglicus, Slimonia Scoticus), which are at the same time the largest of all known Crustacea, recent or fossil, measuring from five to six feet in length, do not appear until the Devonian period, or not until the group had attained to considerable development. The simultaneous appearance among the trilobites of the largest and smallest forms — Paradoxides, Agnostus—would seem to point to a contrary order of development, but there can be no question that the first appearance of these animals long anteceded the Cambrian period. The total number of eurypteroid forms occurring in the American deposits is twenty-three (representing the genera Eurypterus, Dolichopterus, Pterygotus, and Stylonurus), six of which (Eurypteri) are Carboniferous, two Devonian, and the remainder Upper Silurian.[101] The oldest known limuloid form is the Neolimulus falcatus, from the Upper Silurian rocks of Lanarkshire, whose early appearance would seem to indicate that while the family to which it belongs (Bellinuridæ) may stand in its relations intermediately between the eurypterids and the king crabs, its actual origination may be traceable, at least in part, to direct modification from the trilobitic type.

INSECTS.

The number of recognised species of insects is generally conceded to be upwards of 100,000, and by some authors is placed as high as 150,000, but it is very questionable whether these represent more than one-tenth of the number actually inhabiting the earth's surface. Probably not less than one-half of the indicated forms belong to the order Coleoptera, or beetles, which is by far the most numerously represented of all the orders. The Lepidoptera, or butterflies, have thus far yielded some 15,000 species—or about one-thirteenth of the total number (200,000) estimated by Speyer for the world at large—and an equal number may, perhaps, with a certain amount of accuracy, be credited to the Hymenoptera (bees, wasps, and ants), the Hemiptera (bugs), and Diptera (flies). The Orthoptera, or straight-winged insects, which include the locusts, grasshoppers, &c., are considerably less numerous, while the species of netted-veined forms (Neuroptera) probably do not much exceed 2,000, or perhaps do not even reach this figure.

Our knowledge of the general insect fauna of the globe is still too limited to allow of any satisfactory conclusion being drawn as to the geographical distribution of the class as a whole ; enough is known, however, to permit it being stated that practically every portion of the earth's surface harbours a more or less extensive insect fauna, so that the distribution of this class of animals may be said to be universal. While most numerously developed in the warmer or tropical areas, insects are by no means rare in the region of high latitudes, and, indeed, in some of the most northerly points reached by man they appear to be still remarkably abundant. The officers of the British North Pole Expedition, under command of Sir George Nares, brought home a surprisingly rich fauna from the region (Grinnell Land) lying between the seventy-eighth and eighty-third parallels of latitude, comprising no less than forty-five species of true insects and sixteen arachnids, the former distributed as follows : Hymenoptera, five species (two humble-bees); Coleoptera, one; Lepidoptera, thirteen; Diptera, fifteen; Hemiptera, one; Mallophaga, seven; and Collembola, three.[102] Among the Lepidoptera are a number of forms belonging to genera common in the temperate zones, such as Colias, Argynnis, Lycæna, &c., which appear the more remarkable, seeing that the species of this order are more limited in Greenland (with an insect fauna numbering eighty species), and that no forms are met with either in Iceland or Spitzbergen, although upwards of three hundred species of insects are represented in the former. On the other hand, M. Bonpland observed butterflies on the slopes of Chimborazo at an elevation of 16,626 feet, or but 1,600 feet below the highest level (18,225 feet) reached by insects (Diptera), as observed by Humboldt, on the same mountain.

Insect life flourishes also to a certain extent in waters of high temperature (hot springs); and even on the free surface of the ocean, most distantly removed from land, the officers of the "Challenger" circumnavigating expedition everywhere obtained one or more species of Halobates, a member of the Hemiptera, which is stated to live entirely at sea, and to carry its eggs about with it attached to its body.

The distribution of insects is determined largely by climatic and general physiographical conditions, and also to a great extent by the nature of the food-supply, many forms, as has already been seen, being dependent for their sustenance (whether in the larval or mature state) upon the development of a particular vegetable product. Indeed, localisation or restriction appears to be more frequently brought about as the result of changes in the character of the vegetation than as a condition depending upon the interposition of physical barriers which, through their powers of flight, the animals of this class are in great measure able to overcome. The effect of climate is, however, well marked, and is seen in the general restriction of numerous forms to particular climatic zones.

An extended meridional extension is more frequently to be observed among the tropicopolitan forms than among those which more properly belong to the temperate regions ; and this is especially the case among African insects, where frequently the same species is found to inhabit both the northern and southern parts of the continent. This condition is also to be observed in the case of northern forms, when mountain-chains, trending in a meridional direction, permit of easy access to regions of very varying physiographical features, which in their more elevated parts present conditions more nearly uniform with those exhibited elsewhere on the lowlands. Thus, we find in the Chilian fauna numerous forms that more properly belong to the north temperate zone—the preponderating element among the Carabidæ, for example, and the genera Lycæna, Colias, and Argynnis among the butterflies. These not improbably found their way southwards in successive migrations along the Andean mountain-system, where suitable habitations, corresponding in general physiographical features to those of their northern home, could readily be found. A broad horizontal or latitudinal distribution, *per contra*, characterises the insect fauna of the north temperate zone. A large proportion of the European species, for example, are spread over the far interior of the Asiatic continent, and, indeed, many of them reappear in America. Nothing more strikingly illustrates this broad diffusion of species than the case of the Japanese lepidopterous fauna, which, out of some 1,110 species of Macro-Lepidoptera, contains, according to Pryer,[103] not less than 123 species that are common to Great Britain, or about 16 per cent. of the entire British fauna.

Some remarkable and not wholly comprehensible anomalies of distribution are exhibited by the faunas of almost every region, which render unusually intricate the general problems of distribution presented by the class as a whole. One of the most interesting and instructive of these is the special relationship which unites the New Zealand and Chilian and Patagonian coleopterous faunas, and the distinctness of the fauna of the first-named region from the Australian,[104] a condition which, as Professor Hutton has shown, also characterises some of the other animal groups, and which would seem to argue in favour of some former direct land connection (trans-Pacific) between New Zealand and a portion of the South American continent.

The remains of insects in the older Paleozoic rocks are very scanty, and in the main they occur in such an unsatisfactory state of preservation as to have led to the most divergent views respecting their true relationship. The most ancient of these forms, and the only one that is thus far known from the Silurian rocks, is the Palæoblattina Douvillei, recently described by Brongniart from the Middle Silurian sandstones of Calvados, France, and referred by that naturalist to the orthopteroid group of the cockroaches. No other member of the orthopterous division of insects is positively known prior to the Carboniferous period, and it is not unlikely that the fossil in question, which is represented by a single impression of a wing, may on further investigation prove to belong to the group of netted-veins (Neuroptera or Pseudo-Neuroptera), which alone among the different orders has representatives in the Devonian rocks. These last comprise fragments representing some five or six species, belonging to possibly as many genera—Palephemera, Gerephemera, Lithentomum, Homothetus, Xenoneura, and Discrytus. The first two are referred by Hagen to the modern type of the Libellulæ (Pseudo-Neuroptera), and the remaining three, not counting the very doubtful and fragmentary Discrytus, to the likewise modern sialine tribe of the Neuroptera. By Mr. Scudder, on the other hand, several of these earlier hexapods are considered to represent synthetic types, and have accordingly been referred to families specially created for their reception—Palephemeridæ, Homothetidæ, Xenoneuridæ; indeed, the same authority, following in the footsteps of Goldenberg and Brongniart, insists that these forms, as well as all others, with possibly one exception, from the Paleozoic deposits, in a given departure from modern type-structures, and in the possession of combination ordinal characters, constitute a special group or order apart by itself, the Palæodictyoptera, whose diverging specialisation (effected at about the beginning of the Mesozoic era) outlined the various higher groups or orders now recognised by entomologists.[105] In conformity with this view the Paleozoic insect was a synthetic hexapod, in which ordinal differentiation had not yet asserted itself. The Carboniferous cockroaches (Palæoblattariæ), which constitute probably one-half of the entire Paleozoic insect fauna, and of which some sixty or more species are known, and the contemporaneous walking-sticks—Protophasma, the giant Titanophasma from Commentry, France,

&c.—are accordingly not true Orthoptera, but orthopteroid Palæo-dictyoptera; the Carboniferous and Devonian netted-veins not true Neuroptera or Pseudo-Neuroptera, but neuropteroid Palæodictyop-tera, and, similarly, the Hemiptera (Eugereon, Fulgorina), hemipte-roid Palæodictyoptera. Apart from the contradictory conclusions which have been reached from the study of these forms by Hagen, Gerstaecker, Eaton, and others, it may be reasonably doubted whether the extreme specialisation seen among the Palæodictyoptera will carry out the inference that the larger groups of Paleozoic times were more closely related to one another "than any one of them is to that modern group to which it is most allied, and of which it was with little doubt the precursor or ancestral type." [106] Surely, it will not be contended that Palephemera and the highly special-ised Titanophasma are more nearly related to each other than they are to the modern families Libellulidæ and Phasmidæ, not to men-tion the orders to which these belong ; and if this be so, why should they be referred rather to the one loose comprehensive group than to the several groups which they immediately represent ?

The remains of beetles (Coleoptera), if we except the very doubtful Troxites (which is considered by some naturalists to represent the fruit of a plant), are unknown in the Paleozoic depos-its; but it is by no means unlikely that the members of this order had already existed, since borings in wood, very like those made by the coleopteroid larva, have been discovered in various localities.*

The Triassic insect fauna, which is represented by some four or five European species, and by about twenty in America (Colorado), is almost exclusively orthopteroid, and exhibits a distinct passage between the ancient and modern types of cockroaches—Palæoblat-tariæ and Blattariæ. The first indisputable beetle (Chrysomelites), followed by several highly differentiated types in the Rhætic—Hydrophilites, Buprestites, Curculionites—is found in the deposits of this age. No truly metabolous insects, or those undergoing complete metamorphosis, other than Coleoptera, are known before the Lias, where, however, we have several species of Diptera (Chironomidæ, Tipulidæ), and at least one representative of the Hymenoptera, an ant (Palæomyrmex prodomus), from Schambelen,

---

* Different coleopteroid species have at various times been described from Carboniferous strata, but these are now known to be mainly referable to the Arachnida or spiders.

Aargau. The same Swiss deposits have yielded upwards of one hundred and ten species of Coleoptera, in addition to various forms of Orthoptera, Neuroptera, and Hemiptera (Cicadina). A true song-cricket (Cicada Murchisoni) has been described by Brodie from the English Lias. The most diverse types of existing Neuroptera and Orthoptera, which appear to have been fully differentiated at this period, acquire a further development in the succeeding Oolites, where, in addition to representatives of all the other orders that had thus far appeared, we meet with the earliest Lepidoptera (Sphingidæ).

The singularly deficient fauna of the Cretaceous period is succeeded in the Tertiary, more particularly in the Oligocene and Miocene divisions, by a most prolific development of specific forms. The Miocene deposits of Switzerland, and the immediately adjoining tracts, have yielded to Dr. Heer no less than eight hundred and seventy-six species—eight hundred and forty-four of which are recorded from Oeningen (Baden) alone—distributed as follows: Coleoptera, five hundred and forty-three; Orthoptera, twenty; Neuroptera, twenty-nine; Hymenoptera, eighty-one; Lepidoptera, three; Diptera, sixty-four; and Rhynchota, one hundred and thirty-six. Scarcely less famous as insect localities than Oeningen are Aix in France and Radoboj in Croatia; but apparently far surpassing either of these in respect to both individual and specific development are the tufa-beds of the Florissant region in Colorado, referred to the Oligocene period. The Hymenoptera from these deposits comprise several species of bees (Apidæ and Andrenidæ), about thirty species of wasp-like forms (Vespidæ, Sphegidæ, &c.), fifty species of ants (represented by about four thousand specimens, mainly Formicidæ), and some eighty species of Ichneumonidæ, besides numerous other forms. The Diptera individually make up nearly one-third of all the specimens found in the region (Culicidæ, Tipulidæ, Bibionidæ), and of the heteropterous division of the Hemiptera there appear to be no less than one hundred species. The Coleoptera are likewise exceedingly abundant, and comprise among other forms some thirty species each of the families Carabidæ, Staphylinidæ, and Scarabeidæ, and forty of the Elateridæ. The total number of species represented in this order is about three hundred, of which about one hundred and twenty belong to the rhynchophorus division.

Owing to the very limited nature of insect faunas in general, and the circumstance that only a few localities have thus far yielded insect remains in any abundance, it is impossible to draw any positive conclusions respecting either the geographical distribution or the genealogy of the members of this class of animals. It appears practically certain, however, that the metabolous types were descended from the ametabolous, and that the larvæ of the earlier forms were all, or mostly all, aquatic in habit. The geological and horizontal range of the majority of the species appears to have been very limited, but it must be recollected that in most instances the number of individuals found representing any one given species is altogether too small to permit of any logical inference being drawn from their occurrence. Thus, of the numerous species of Paleozoic cockroaches, by far the greater number are represented by single specimens, or by specimens coming from a single locality.

ARACHNIDA AND MYRIAPODA.

The earliest arachnoid remains (scorpions) occur in the Upper Silurian deposits of the island of Gothland, Sweden, and Lanarkshire, Scotland, and in the Helderberg rocks of Waterville, New York, from each of which regions a single specimen has been obtained. The Swedish species, Palæophoneus nuncius, with which the Scotch form appears to have been nearly related, ranks among the largest of its class, measuring about three and a half inches in length ; in its general characters it closely approximates the modern scorpions, although, as has been pointed out by Professor Lindström, the structure of the large and pointed thoracic limbs more nearly approaches what is seen in the embryonic forms of other Tracheata and in Campodea. The presence of stigmata and the whole organisation of the body clearly demonstrate the animal to have been an air-breather and an inhabitant of dry land. The American form, Proscorpius Osborni, is less clearly recognisable than the European, and some doubt has been thrown upon its arachnoid nature, which is maintained by both Whitfield and Thorell. Excepting these two or three earliest precursors of the Scorpionidæ, whose presence would naturally seem to indicate a rich insect fauna for the period, no traces of arachnoid remains are known antedating the Carboniferous deposits, where, however, several well-marked genera, singularly close in their relationship to modern forms, and representing spiders (Protolycosa, Architarbus, Anthracomartus), scorpions (Cyclopthalmus, Eoscorpius), and pseudo-scorpions (Microlabis), are met with. In the Mesozoic deposits but few arachnoid remains, in most cases in an imperfect state of preservation, have thus far been discovered, and it is not until about the middle of the Tertiary series, Oligocene, that they acquire by their numbers any importance. The amber deposits of Europe and the Florissant beds of Colorado have yielded the greatest abundance of such remains.

It is not a little remarkable with what degree of persistence the fundamental characters of scorpions have been preserved, for it appears, as has been claimed by Mr. Peach, that the Carboniferous forms, as represented by Eoscorpius, were as highly organised and specialised towards the beginning of this period as their descendants of the present day. The Paleozoic araneids, on the other hand, appear to have been all, or in the main, possessed of distinctly segmented abdomens, thereby forming a transitionary group between the arthrogastric and non-arthrogastric arachnoids.

Both the chilognathous and chilopodous types of Myriapoda may be said to be represented in the Carboniferous rocks, although from certain peculiarities of structure possessed by these early forms— the genera Euphoberia, Xylobius, Acantherpestes, Trichiulus, among the former, and Palæocampa among the latter — which are unknown in their modern representatives, special ordinal groups, the Archipolypoda and Protosyngnatha respectively, have been created for them. It may be doubted, however, whether such knowledge as we possess of the animals in question will permit of the retention of these groups; indeed, it appears by no means certain that all the forms referred here as Myriapoda are actually such at all.

The earliest myriapod remains, referred by Peach [107] to the Chilognatha, occur in the Old Red Sandstone (Devonian) of Forfarshire, Scotland, and not improbably the problematical Gyrichnites of the nearly equivalent deposit of Gaspé are the belongings of these animals. The Mesozoic rocks are singularly deficient in their traces, and may be said to be almost wholly wanting in them; excepting the somewhat doubtful Geophilus proavus from the Jurassic deposits it appears that no chilopodous form is met with before the Tertiary.

II.

Distribution of the Vertebrata.—Fishes.—Amphibians.—Reptiles.—Birds.—
Mammals.

THE geographical distribution of fishes is at the present day, and probably has been for a considerable number of past geological periods, world-wide. Although vastly more abundant, if not individually at least specifically, in the regions of elevated temperature than in those of the opposite extreme, both as regards the marine and fresh-water forms, they are still far from wanting in waters of icy coldness, whether these be in high latitudes, the oceanic abysses, or elevated mountain lakes or streams. The officers of the British Polar Expedition, under command of Sir George Nares, obtained specimens of the charr (Salmo arcturus and S. Naresii) from beyond the eighty-second parallel of north latitude, the highest point at which fresh-water fishes have been observed, and from a still higher latitude, the eighty-third, some half-dozen species of shore fishes, among them a bull-head and cod (Cottus quadricornis, Gadus Fabricii, Icelus hamatus, Cyclopteris spinosus, Liparis Fabricii, Gymnelis viridis).[108] And were it not for the insuperable obstacles that were interposed in the way of fishing, there can be no doubt that many additional forms would have been discovered. The number of forms that descend into, or inhabit, the abyssal waters whose temperature is about that of freezing is very considerable; Günther[109] enumerates thirty-nine species whose range extends to, or passes beyond, the fifteen hundred fathom line, thus penetrating deep into the zone of icy coldness. In the European Alpine region fishes (salmonoids) inhabit the lakes or streams situated at about the level of perpetual snow, and there is very little doubt that the same is the case in nearly all regions of elevated mountains. The loach has been found in the Himalayas at an altitude of eleven thousand feet, and in the South American Andes, Lake Titicaca, at an altitude of thirteen thousand feet, has yielded several species. Indeed, in the latter region, some of the Alpine cyprinodonts, as Trichomycterus, penetrate to heights of fifteen thousand feet, and upwards.

Fresh-Water Fishes.—It will be evident from the present relation existing between land and water—which in its general features unquestionably dates from a very remote geological period—and the resulting barriers opposed to a free migration, that fresh-water fishes will be much more limited in their range than the fishes of an oceanic type, whose distribution is in main part governed by conditions of food-supply and temperature, and to a certain extent by the nature of oceanic currents. This comparative areal restriction among fresh-water forms is exemplified not only in the case of species and genera, but also in that of families, none of which, if we except the cat-fishes (Siluridæ), can be considered to be in any way cosmopolitan. Manifestly, the oceanic basins must prove to the animals of this class an obstruction much in the manner that it does to the higher animals, the reptiles and mammals, for example. Yet, certain peculiar occurrences would seem to indicate that the natural barrier thus formed is not in all cases as effectual in preventing the distribution of fishes as it is with the majority of the animals just mentioned. Thus, several identical species, as the salmon (Salmo salar), perch (Perca fluviatilis), burbot (Lota vulgaris), pike (Esox lucius), and a stickleback (Gasterosteus pungitius), inhabit alike the waters of Europe and Eastern North America; the perch of the Ganges and other East Indian rivers (Lates calcarifer) is found in the waters of Queensland, Australia; and one of the forms of south temperate so-called "trout" (Galaxias attenuatus) inhabits Tasmania, New Zealand, the Falkland Islands, and the southern extremity of the continent of South America. In addition to such divided species, representing, naturally, divided genera as well, there are also several generic types whose limited number of representatives (while distinct specifically) belong to opposite quarters of the globe. The genus Umbra, limited to two species, is represented in the Atlantic States of the American Union by the "dog" or "mud fish" (U. limi), and in the Danubian system of waters by the "Hundsfisch" (U. Krameri); the shovelnose sturgeons (Scaphirhynchus), and the paddle-fishes (Polyodontidæ), both of them restricted to some two or three species, are confined respectively to the river systems of Central Asia and the Mississippi, and the Mississippi and the Yangtse-Kiang; the American suckers (Catostomus) have an outlying representative in Siberia; while the East Indian genus Symbranchus, after skipping Africa, reappears with a single species (S. marmoratus) in the waters of South America.

In what precise manner these equivalent types have found their way to such widely removed portions of the earth's surface it has been thus far impossible to determine. That some transference was effected by way of northern waters over land-surfaces now no longer existing is very nearly certain ; hence, the occurrence of identical or representative specific forms in the streams of Northern Eurasia and North America is not very surprising. But that a similar transference was effected over the broader or equatorial parts of the oceanic basins, a hypothesis necessitating the assumption of the submergence of vast continental land-masses where no traces of their former existence are visible, is more than doubtful. At any rate, it is very unlikely that any such alternation in the relative positions of land and water took place at a time so recent as satisfactorily to explain such anomalies of distribution as are presented by the genera Lates and Symbranchus, already mentioned, and by the genus Pimelodus (Africa and South America) among the cat-fishes. It would appear at first sight far more natural to assume that these fishes were originally of a marine type, spread over the oceanic expanses, and that at a later period they accommodated themselves to fresh-water conditions, and gradually restricted their habitats to regions where we now find them. This is not unlikely, seeing that some of these (Lates, Symbranchus) freely enter brackish water. That such has been the recent origin of many forms of fresh-water fishes is placed beyond question. Several permanent species of the Northern Baltic, where, through an excessive indraught of fresh water from the surrounding streams, the water has lost most of its salinity, are identical with marine types inhabiting the Arctic seas to the north. Yet the accommodation to new conditions of existence was effected since the closing off of the Baltic from the northern ocean, or since the Glacial period. The gobies, blennies, and atherines of the northern lakes of Italy present us with other instances of the permanent establishment of marine types in fresh waters. Indeed, it is scarcely necessary to seek for examples of this kind, seeing how very readily the many marine fishes which periodically ascend the inland streams during the spawning season accommodate themselves to the newly imposed conditions. Of so little importance does a change of medium appear to be in many cases that it is frequently very difficult, or impossible, to indicate whether a given group of fishes is more properly of a marine or fresh-water type. The numerous instances where certain species of a genus are of one habit, and other species of the same genus of the opposite habit, render the determination of this question still more difficult.

An accommodation similar to that which has been noticed in the case of marine fishes also obtains with many of the more strictly fresh-water forms; i. e., they descend without inconvenience into the briny oceanic medium. This we see in the case of the trout, the charr, in several species of Coregonus, and especially among the toothed carps (cyprinodonts) and sticklebacks. How far these may wander out to sea is not exactly known, but there is no special reason for supposing that they might not proceed, at least in some instances, to very considerable distances. A species of the cyprinodont genus Fundulus (F. nigrofasciatus) was obtained by the officers of the "Challenger" from the pelagic fauna of the Atlantic, midway between St. Thomas and Teneriffe. It is manifest, therefore, that an arm of the sea is not an impassable barrier to certain forms of fresh-water fishes; and not impossibly some brackish-water forms may occasionally find their way completely across the oceanic expanse. The irregular distribution of certain types or species thus receives a partial, or, at any rate, a possible solution.

The total number of species of strictly fresh-water fishes recognised by Günther is nearly two thousand three hundred, of which four are lung-fishes, thirty-two ganoids, twelve lampreys, and the remainder teleosts or bony-fishes. Of the last nearly one-third are comprised in the family of the carps (Cyprinidæ) and somewhat more than one-fourth in that of the cat-fishes (Siluridæ). The Characinidæ and Chromides, forms from tropical America and Africa, are represented by somewhat more than two hundred and fifty and one hundred species respectively, the salmonoids by one hundred and thirty-five species, and the toothed carps (cyprinodonts) by one hundred and ten species. As to their geographical distribution, two primary zones might be recognised: the northern, corresponding largely to North America and Temperate Eurasia, characterised by the presence of sturgeons, salmonoids, pikes, and numerous carps, with only a feeble development of the cat-fishes; and the southern or tropical zone, comprising the Indian, Ethiopian, Neotropical, and Australian regions of zoogeographers, characterised by a special development of the cat-fishes. The following scheme for the classification of the southern zone has been proposed by Dr. Günther:

CYPRINOID DIVISION.—Characterised by presence of Cyprinidæ and Labyrinthici.

1. *Indian Region, 625 Species.*—Characterised by Ophiocephalidæ and Mastacembelidæ. Cobitoids numerous.

2. *African Region, 255 Species.*—Characterised by lung-fishes (Protopterus) and ganoids (Polypterus, Calamoichthys). Chromoids and characinoids numerous. Cobitoids absent.

ACYPRINOID DIVISION.—Characterised by absence of Cyprinidæ and Labyrinthici.

1. *Tropical American Region, 672 Species.*—Characterised by lung-fishes (Lepidosiren). Chromoids and characinoids numerous. Gymnotidæ (electric eels).

2. *Tropical Pacific Region, 36 Species.*—Characterised by lung-fishes (Ceratodus Forsteri and C. miolepis, from the waters of Queensland, Australia). Chromoids and characinoids absent.

The eastern and western divisions of the northern zone, with some three hundred and sixty and three hundred and forty species respectively, are very closely related to each other, not only through the preponderance of types that are common to both regions, but in the possession, as has already been seen, of a number of identical species. Two genera of ganoids not found in Eurasia, Lepidosteus and Amia, serve to characterise the American ichthyic fauna, which is further distinguished from its trans-Atlantic correspondent in the special development of the suckers (Catostomidæ) and in the absence of cobitoids and barbels (Barbus).

An Antarctic zone, made to include New Zealand, Tasmania (with a portion of Southeast Australia), the Falkland Islands, Tierra del Fuego, Patagonia, and Chili, whose faunas are very intimately related to one another, is recognised by some authors, but the number of species here represented is too limited to permit of much importance being attached to a negative region of this kind. Galaxias attenuatus, a species of southern "trout," and one of the three species of lamprey are found in New Zealand and the Tasmanian and Fuegian tracts. No fresh-water fishes are thus far known to inhabit any of the islands situated south of the fifty-fifth parallel of south latitude.

The paucity of the fish-fauna of the tropical Pacific region is very remarkable, and only receives a partial explanation through the circumstance that the greater part of the tract belonging to it, the Australian, is deficient in water-courses. The island of Celebes, for example, which, as has been well urged by Günther, would seem to offer favourable conditions for the development of a fresh-water fauna, has thus far offered barely more than a half-dozen species, all of them common Indian forms; nor has New Guinea shown itself to be much more prolific. Long-continued isolation has apparently prevented much of the tract from receiving the necessary supply from the fresh waters of continental areas, although the identity of existing forms with such as are found elsewhere would seem to indicate a recent migration and peopling of the waters. The smaller islands of the Pacific are inhabited principally by such forms, as eels, atherines, gobies, mullets, which can readily exchange fresh water for salt water, and to which, consequently, the oceanic basin constitutes no insuperable barrier to a free migration. A species of Arius, a siluroid, inhabits the Sandwich Islands.

The marine fishes may be conveniently divided into three categories: shore fishes, or such as habitually frequent the coast-lines, and which rarely descend to a greater depth of water than three hundred fathoms; pelagic fishes, which inhabit the waters of the open sea, the majority of them spawning there also; and deep-sea fishes, or the fishes of the greater oceanic depths, where the influence of light and surface temperature is but little felt. No sharp line of separation between these several classes is permissible, however, since the habits normally belonging to the members of one class are occasionally assumed by members of the other classes as well, as must necessarily follow from the different conditions governing their distribution.

Shore Fishes.—The fishes of the first category number upwards of three thousand five hundred species. Their range in the north extends to, or beyond, the eighty-third parallel of latitude, but in the Southern Hemisphere no species have been found to pass beyond the sixtieth parallel, although, doubtless, they exist along some of the more southerly shore-lines. The Arctic fauna, or the fauna occurring north of the sixtieth parallel of latitude, is, as far as we are warranted in believing, a strictly homogeneous one, identical types largely characterising both the Old and the New World divisions. The more extensively represented families are, among the spiny-rayed fishes, the bull-heads (Cottidæ—Cottus, Icelus, Triglops), the Agonidæ, lump-suckers (Discoboli), and blennies (Blenniidæ—Anarrichas, wolf-fish) ; and among the anacanths the cod-fishes, with the cod (Gadus), hake (Merlucius), and ling (Molva). Among the physostomous fishes, or those in which the air-bladder is provided with a pneumatic duct, the herring (Clupea) is represented by a limited number of species. The cartilaginous fishes are very scarce; indeed, thus far only one species, the Greenland shark (Læmargus), is known to penetrate north of the Arctic circle. The chimæra, spiny dog-fish (Acanthias), and ray, are met with along the southern borders of this tract.

The Antarctic shore fauna is in many respects closely related to the Arctic, although nearly one-third of the generic types are peculiar. As in the north, the cartilaginous fishes are scarce, being represented by a single species of shark (Acanthias), and one or more species of ray (Raja, Psammobatis). The Scorpænidæ and Agonidæ among bony-fishes have each one genus, Sebastes and Agonus respectively, which is held in common with the Arctic fauna. The lophobranchs have in addition to the northern pipe-fish (Syngnathus) the remarkable Protocampus, represented by a single species of the Falkland Islands (P. hymenolomus). A most interesting fact connected with the Antarctic fauna is the recurrence of types belonging to the far north, which are wanting in the intermediate region. This we see in the single species of spiny dog-fish (Acanthias vulgaris), which is a member of the Arctic and north temperate faunas, but is absent from the equatorial region. The hakes comprise two species, one of which is restricted to the northern waters and the other to the southern; and a similar separation is found among the species of the Arctic and Antarctic genus Lycodes. The gadoids are in both regions accompanied by the parasitic hag (Myxine).

The shore fishes of the north temperate sea are largely identical in their general types in both the Atlantic and Pacific basins, as well as on opposite sides of these basins. Thus, of the fishes frequenting the British seas we find, among other genera, the following types also represented on the east coast of America : the bass (Labrax), sea-perch (Serranus), porgy (Pagrus), bull-head (Cottus), angler (Lophius), wolf-fish (Anarrichas), Zoarces, cod (Gadus), hake (Merlucius), ling (Molva), and rockling (Motella); and among physostomes the smelt (Osmerus), herring (Clupea), and conger. The surmullets (Mullus), gurnards (Trigla), John Dorys (Zeus), and some of the breams are wanting, or are but rarely met with.* The cartilaginous fishes show as common to the East and West Atlantic the "hounds" (Mustelus), rays, sting-rays (Trygon), and the electric rays (Torpedo) ; the true dog-fishes (Scyllium) appear to be wanting on the East American coast, and Chimæra has thus far been found only in deep water.

Most of the genera of British fishes also frequent the Mediterranean Sea, whose fauna effects a passage to the fauna of the equatorial zone. The number of peculiar genera is very limited. Among the newly appearing forms are the beryces (Beryx), stargazers (Uranoscopus), umbrines (Umbrina), barracudas (Sphyræna), horse-mackerels (Caranx), and sea-horses (Hippocampus), members of the (more southerly) American fauna as well (elements borrowed from the fauna of the West Indies). The flat fishes—turbots, plaices, flounders, soles (Rhombus, Pleuronectes, Solea, &c.)—exhibit an increased development, while the gadoids rapidly diminish in numbers. A most remarkable correspondence exists between the Mediterranean fishes and those of the Japanese province—*i. e.*, the coast of Asia between the thirtieth and thirty-seventh parallels of latitude ; indeed, viewed from a generic standpoint, this correspondence may be almost said to amount to identity. More than one half of all the generic types represented are fishes of the south of Europe, and in many cases even the species are identical with

---

* The occurrence of Trigla cuculus on the American coast is considered very doubtful by Jordan and Gilbert; on the other hand, Mullus barbatus, supposed to be absent, has during late years been reported from Pensacola, Florida, and Wood's Holl, Massachusetts. (Smith's "Misc. Coll.," 1883.)

European forms.[110]   Several of the berycoid genera inhabit the Japanese and Mediterranean waters exclusively, while others, as the
red mullets, John Dorys, and trumpet-fishes (Centriscus), occurring
in these two districts and elsewhere, are wanting on both the East
and West American coasts.*   How the transference of the Mediterranean fauna to the East Asiatic coast, or the converse, was effected,
whether by means of a comparatively recent open water-way between the two regions, as has been supposed by some, our present
knowledge does not permit us to say.

The shore fishes of the south temperate zone, which extends
northward from the Antarctic faunal belt to about the thirtieth
parallel of south latitude, are most intimately related to those of
the north temperate, although very distinct from the forms which
occupy the intermediate or equatorial zone.   Nor is this correspondence restricted to generic types alone, since we find a considerable number of northern species which, skipping the intermediate
tract, reappear here without having undergone even varietal modification.   Such are the chimæra (Chimæra monstrosa), two species
of dog-fish (Acanthias vulgaris and A. Blainvillii), the monk-fish
(Rhina squatina), John Dory (Zeus faber), angler (Lophius piscato-
rius), bellows-fish (Centriscus scolopax), anchovy (Engraulis en-
crasicholus), sprat (Clupea sprattus), and conger (Conger vulgaris).
Among the marked differences separating the two regions may be
mentioned the substitution of the cottoids by the Nototheniæ (family Trachinidæ) and the Discoboli by the Gobiesocidæ.†

Of the four south temperate provinces, that of the Cape of
Good Hope, the South Australian (with New Zealand), the Chilian,
and the Patagonian, the Australian is by far the richest, numbering
in its fauna not less than one hundred and twenty genera and two
hundred species.   Two-thirds of all the genera occurring on the
coasts of Southeastern Australia and Tasmania are also represented
in the shore fauna of New Zealand, which comprises some one
hundred species.   The New Zealand genera not represented in the
Australian coast are about twenty-six in number, with about an
equal number of species.   The most marked difference between
the two nearly contiguous faunas is the absence of gadoids in the
Australian element, while the group is represented by not less than

* Centriscus scolopax is said to be accidental on the American coast.
† The Gobiesocidæ have also northern representatives.

six genera (Gadus, Merlucius, Lotella, Motella, &c.) in the fauna of New Zealand.—The Chilian fauna is generically closely related to the South Australian, and contains but few types that are strictly peculiar to it. One of its more remarkable features is the possession of a species of Polyprion (P. Kneri, from Juan Fernandez), one of the Percidæ, the only other known species of the genus (P. cernium) being a member of the European fauna.* The number of fishes thus far obtained from the Patagonian province is very limited, and scarcely permits us to formulate any definite opinion as to the characters of the entire fauna.

The equatorial shore fishes far exceed those of the temperate zones both in number and diversity of form, agreeing in this respect with the condition presented by equatorial land and fluviatile faunas generally. It is here, too, that we find grotesqueness of outline combined with a most varied and beautiful colouring, an almost infinite arrangement of lively tints, scarlet, yellow, blue, black, &c., imparting to the fishes of this tract an indescribable brilliancy. Throughout the greater part of the equatorial or tropical belt there is manifest a strong faunal identity, rendering the institution of ichthyic provinces practically impossible. Thus, the greater number of the dominant types of the Atlantic Ocean are also represented in the Pacific, and *vice versa*, and in many instances even the species representing these types are identical. The fishes of the tropical waters of the Indian and Pacific oceans are very intimately related, the number of identical species ranging from the Red Sea far into the Polynesian Archipelago being very great. A limit to the eastward extension of the Indo-Pacific fauna appears to have been set, however, by the cold current sweeping northward along the western coast of South America, and as a consequence we find a more or less distinct, or individualised, fauna along the eastern border of the great ocean. Among the more largely represented generic or family groups of the equatorial zone are the sea-perches (Serranus), snappers (Mesoprion), mullet-kings (Apogon), blow-fishes (Chætodon, &c.), Scorpænidæ, horse-mackerels (Caranx), coral-fishes (Pomacentridæ), Julidinæ, flat-fishes (Pleuronectidæ), herrings, and Murænidæ, several of which are more or less distinctive of the zone. In all of these groups the number of species in

* Obtained also in deep water by the United States Fish Commission. (Jordan and Gilbert, " Synopsis Fishes of North America," 1882.)

the Indo-Pacific basin is very much greater than in the Atlantic. The extensive development of reef-structures through the former area, presenting unusually favourable conditions for existence, has, doubtless, much to do with this comparative superabundance.

Pelagic Fishes.—Our knowledge respecting the fishes which spend a considerable, or the greater, part of their existence on the free surface of the oceans, borne resistlessly in the course of the oceanic current, or inhabiting masses of floating sea-weed, although still very meagre, is sufficient to indicate that the number of such types is very limited. As with the other groups of fishes, they are most numerous in the regions of high temperature, diminishing rapidly as we proceed either north or south from the Equator. Most of the tropical genera are also met with in the temperate zones, and probably the converse is also true, although a number of exceptions have been indicated. The warm-water fishes become rare beyond the fortieth parallel, and decline very rapidly with the decline of the temperate fauna itself. Almost the only pelagic fish of the Arctic Ocean is the Greenland shark (Læmargus borealis).

Among the better known bony-fishes that enter into the composition of the pelagic fauna are the flying-gurnards (Dactylopterus), mackerels, tunny, bonitos, dolphins (Coryphæna), skip-jacks, pilot-fishes, sword-fishes, frog-fishes, scopelids, skippers, flying-herrings (Exocœtus), sea-horses, porcupine-fishes, and sun-fishes (Orthagoriscus). The cartilaginous fishes are preeminently pelagic in their habits, and contribute largely to this fauna. The sharks are represented by a number of genera (Carcharias, Galeocerdo, Zygæna, Lamna, Notidanus, &c.), and by forms which are not only the largest of their tribe, but approximately the largest of known fishes. The basking-shark (Selache), the largest shark of the North Atlantic, attains a length of more than thirty feet ; Carcharodon Rondeletii, a tropical or sub-tropical species, the most formidable of all sharks, has been known to measure forty feet, and Rhinodon typicus, a species of the Indian Ocean, fifty to sixty feet. Rivals to these monsters of the deep are the sea-devils or eagle-rays (Myliobatidæ), many of whose forms (Dicerobatis, Ceratoptera) appear to attain a length of twenty feet or more.

The similarity existing between the pelagic faunas of the Atlantic and Indo-Pacific basins is very great, extending not only to genera, but to species. A number of the forms are nocturnal in

14

their habits, living in a deeper zone during the day, and appearing at the surface only at night. The majority of these are provided with phosphorescent organs, a structure largely prevalent among deep-sea fishes, to which these fishes effect a passage.

Deep-Sea Fishes.—Much uncertainty still remains as to the character of the abyssal ichthyic fauna, owing to the difficulty of determining in many or most cases the depths whence specimens caught in the net were obtained. But there can be no doubt that many forms inhabit very nearly the greatest depths that have been reached by the dredge. Günther, in his "Introduction to the Study of Fishes," enumerates upwards of fifty forms which are supposed to have been obtained from depths exceeding 1,000 fathoms; twenty-six from depths exceeding 2,000 fathoms; and nine from depths of 2,500 fathoms or over. Additional forms have since been obtained in the dredgings of the "Talisman" and "Albatross." Bathyophis ferox and Halosaurus rostratus are reported to have been dredged by the "Challenger" in water of 2,750 fathoms (5,019 metres), and Gonostoma microdon at an extreme limit of 2,900 fathoms. Considerable doubt, however, attaches to the last, since the fish is abundant in water of only moderate depth, and may have been taken down in the descent of the dredge, or captured only in its ascent. A similar doubt attaches to many of the other forms with a reputed very broad bathymetrical range, as it is hardly to be supposed that animals, organised specially to meet the conditions (pressure, &c.) of life in the oceanic abyss, should at the same time be so constituted as to endure with impunity the very different conditions governing life near the surface. The "Talisman" obtained Alepocephalus rostratus at depths stated to range between 868 and 3,650 metres, Scopelus Maderensis between 1,090 and 3,655 metres, and Macrurus affinis between 590 and 2,220 metres ; similarly, it is claimed that the "Albatross" obtained Cyclothone (Gonostoma) lusca and Scopelus Mülleri in depths varying from 560 to 5,394 metres (2,949 fathoms—latitude 37° 12′ north; longitude 69° 39′ west), or over a vertical extent of 16,000 feet. Mr. Tarlton H. Bean justly considers the deep catch as doubtful. Less doubt attaches to the position of Aleposomus (?) Copei and Mancalias uranoscopus, both of which were obtained in the 5,394 metre haul.[111] The deepest recorded find of the "Talisman" was in 4,255 metres (Bythites crassus).[112]

The Gadidæ, Ophidiidæ, Macruridæ, and Scopelidæ make up a very large proportion of the deep-sea fauna, which has thus far yielded some ninety to one hundred or more genera. The eels (Murænidæ) are largely represented, and have been dredged from depths extending to 2,500 fathoms. Very few cartilaginous fishes were obtained by the "Challenger," and these (Scyllium, Centrophorus, Raja) were restricted to water not much exceeding six hundred fathoms, although the same group of fishes yielded specimens to the "Talisman" at nearly twice this depth (coast of Portugal).

Whether or not different zones of ichthyic life can be recognised in the oceanic abyss may perhaps still be considered an open question. Dr. Günther maintains, as the result of his studies, that "as far as the observations go at present, no distinct bathymetrical regions which would be characterised by peculiar forms can be defined," and that, "if the vertical range of deep-sea fishes is actually as it appears from the 'Challenger' lists, then there is no more distinct vertical than horizontal distribution of deep-sea fishes." As a result of the explorations of the "Talisman," Filhol arrives at the opposite conclusion, and believes that a series of more or less distinct zones can be indicated. Although in its general features the abyssal fauna ("Bassalian" of Gill) is closely related to that of the superficial tracts, yet a number of distinctive elements, arising in part from certain remarkable abnormalities of structure, or from the presence of types not elsewhere represented, serve in a measure to define it. Professor Gill indicates twenty-eight families of deep-loving fishes.[113]

Of the two lowest orders of fishes, the Pharyngobranchii, as represented by the Amphioxus or lancelet, and the Marsipobranchii, lampreys and hags, no unequivocal fossil remains have as yet been discovered. Possibly, however, some of the singular tooth-like bodies described as conodonts from the Cambrian and Silurian deposits, which by many authors are referred to the dental armature of annelids, may belong to fishes more or less nearly related to the modern hag.

The Elasmobranchii (sharks, rays) have their oldest representatives in the Upper Ludlow horizon of the Upper Silurian formation, being preceded in time by a bucklered ganoid, Pteraspis (Scaphaspis) Ludensis, from the Lower Ludlow. These earliest remains, belonging to the family of the true sharks (Squalidæ), are in the form of teeth (Thelodus) and fin-spines or ichthyodorulites (Onchus, Ctenacanthus). The order is but very scantily represented in the Devonian deposits, but in those of the succeeding Carboniferous age (sub Carboniferous limestone) it acquires a profuse development, the pavement-teeth of forms generally considered to be allied to the modern "Port-Jackson" shark (Cestracion) being particularly abundant—Helodus, Orodus, Chomatodus, Petalodus, Cochliodus, Psammodus. The relationships of these genera are, however, still very obscure, and not impossibly they constitute a group apart by themselves — Psammodontes. Less doubtful representatives (Strophodus, Acrodus, Hybodus) of the type of cestraciont fishes appear in the early and middle Mesozoic periods, Triassic and Jurassic, and in the chalk the teeth of Ptychodus are still very abundant. But in the latter period the more modern type of sabre-toothed sharks and dog-fishes, largely represented by existing genera, Notidanus, Scyllium, Lamna, Carcharias, Hemipristis, Galeocerdo (Corax, Otodus—extinct), appear to have gained the ascendency, which they retained throughout the subsequent Tertiary periods. The most important of the later genera, and especially distinctive of the Miocene deposits, is Carcharodon. The earliest unequivocal traces of rays occur in the deposits of Jurassic age, although not impossibly some of the hypothetically placed forms of the later Paleozoic periods may belong to this group, or, at any rate, effect a union between it and the sharks. Of this nature appear to be the Jurassic Thaumas and Squaloraja, both of them nearly allied to the modern (and Cretaceous) Squatina. The true- and sting-rays are abundantly represented, particularly by fragments of their dental armatures, throughout the Tertiary deposits, the modern genera (Trygon, Myliobatis, Ætobatis, Zygobatis, Raja) predominating. The saw-fish (Pristis) and torpedo-ray date from the Eocene, and not impossibly from the Cretaceous period.

Appearing almost simultaneously with the selachians, but attaining a much earlier considerable development, are the ganoids, a class of fishes which in the earlier geological periods exhibit a remarkable diversity of structure, but at the present day are comprised within a very limited number of genera, whose members, with the exception of the partially marine sturgeon (Accipenser), are all inhabitants of fresh water. Three or more distinct types, based upon characters drawn from the structure of the dermal armor, may be conveniently recognised: the osseous plated ganoids, as represented by the sturgeon ; the enamel rhomb-plated ganoids, typified in the American alligator-gars (Lepidosteus) and the African Polypterus ; and the cycloid scaled ganoids, exemplified in the American Amia, which may in a measure be said to connect these fishes with the herring among the teleosts. None of these types appear to be represented in the Silurian period, unless, indeed, the group of the bucklered ganoids, to which the ancient Scaphaspis Ludensis, and the greater number of the Devonian forms (Cephalaspis, Pteraspis, Pterichthys, Coccosteus) belong, be considered to be nearly related to the sturgeons. This relationship, however, requires further demonstration before it can be accepted as a fact; indeed, it has recently been attempted to show that some of these most ancient ichthyoid forms—e. g., Pterichthys —are not fishes at all, but members of what may, perhaps, be considered to be a degenerated group of the lower Vertebrata, the tunicates. The almost total obliteration of the bucklered type of ganoid with the close of the Devonian period is very remarkable, and has not yet received a satisfactory explanation. It has been conjectured by some that these fishes early withdrew to fresh water, and that, in the absence of fresh-water deposits of any magnitude in the period succeeding, they have necessarily left behind but scanty traces of their existence. It must be confessed, however, that this explanation is more in the nature of an assumption than anything else, and has but little positive to bear it out. In how far the Devonian fishes were of a fresh-water habit still remains to be determined, but it seems more than probable that they were largely of this, or at least of a brackish-water, character. The remains of the sturgeon are not known prior to the Eocene period, although a direct forerunner (Chondrosteus), uniting this family with the Spatularidæ, occurs as low down as the Lias.

The non-bucklered ganoids of the Devonian period (Holoptychius, Glyptolepis, Dipterus, Osteolepis, Diplopterus) belong principally to the type of the fringe-finned or crossopterygian Polypteri, which effect a passage to the lung-fishes (Dipnoi). This series, which is represented by both scaled and plated forms, is continued into the Carboniferous period (Rhizodus, Dendrodus, Megalichthys, Cœlacanthus); but here, as in the succeeding Permian period, they are already largely replaced by the lepidosteoid type

(Palæoniscus, Amblypterus), which has its forerunner in the Devonian Chirolepis. The primitive heterocereal tail which characterises the ganoid fishes of the Paleozoic era, and which is still borne by many of the rhomb-plated fishes of the early Mesozoic period— Palæoniscus, Ischypterus, Catopterus—is converted by gradual modification into the higher homocereal type, which distinguishes the more important genera of Jurassic fishes (Tetragonolepis, Dapedius, Lepidotus, and the teleostoid Leptolepis). In Semionotus, which ranges from the Trias into the Lias, the tail is of a well-marked transitional character. Very few of the Jurassic Ganoidei survive into the Cretaceous period, which practically marks the final collapse of this important order of animals, henceforward succeeded by the more highly constituted bony-fishes. The most important remaining group is that of the pycnodonts, or bean-toothed ganoids, whose numerous closely related forms have been referred to several distinct genera, which range collectively from the Carboniferous period (Platysomus) to the Eocene. Among the Jurassic genera are Microdon, Mesodon, Gyrodus, and Pycnodus, the last, a remarkable example of a persistent type, surviving the close of the Cretaceous period into the Tertiary. Of the group of the cycloid scaled ganoids (Ganoidei cycliferi), represented at the present day by Amia, probably the earliest unequivocal remains are those of the genus Amia itself, which appear in the Cretaceous deposits ; by many naturalists, however, several Paleozoic forms, as Holoptychius, Asterolepis, Bothriolepis, and Cœlacanthus, are referred to this group, and with them also the cœlacanthine Cretaceous genus Macropoma.

The lung-fishes (Dipnoi), which at the present day are represented by the three very isolated genera Lepidosiren (Brazil), Protopterus (Africa), and Ceratodus (Australia), have left undoubted traces of their existence as far back as the Permian period, when the genus Ceratodus itself appears (Bohemia, Texas), presenting us with the most remarkable instance of persistence in the whole range of vertebrate animals. Not unlikely the genus may be found to be of still older date, and to have been nearly contemporaneous with its formidable predecessor, the Dinichthys. Remains of Ceratodus have been found throughout the entire series of Mesozoic deposits, from the Trias to the Cretaceous inclusive. Lepidosiren and Protopterus are not known in a fossil condition.

The osseous fishes proper appear for the first time in Cretaceous strata, being immediately preceded by the teleostoid group of ganoid fishes of the family Leptolepidæ, which effects a passage to them. Indeed, by many authors the genus Leptolepis and its Jurassic allies (Caturus, Thrissops, &c.) are classed with the former, and placed near the herring, with which they appear to have been most nearly related. Although both the physoclist and physostome types, or those in which the swimming-bladder is closed off from, or remains connected with, the gullet, appear very nearly simultaneously in the same deposits, and consequently by their occurrence give no evidence as to their respective antiquity, there can be no question that the physostome is the more ancient type, the severance of the bladder in the physoclists being the result of the disuse of parts. This is further proved by the existence of a connecting air-bladder among the recent (and, doubtless, also among the ancient) ganoids. The Cretaceous teleosts belong largely to existing types—Clupea (herring), Osmerus (smelt), Esox (pike), Beryx; but it is not until the Eocene period that we find a representative modern ichthyic fauna. To this, and the succeeding Tertiary period, most of the more prominent existing types date their first appearance. In addition to a very large representation of both the arthropterous and anarthropterous forms, the Eocene deposits contain remains of the Lophobranchii (Syngnathus, pipefish), Plectognathi (Diodon, porcupine-fish; Ostracion, trunk-fish), and Apodes (Anguilla, eel).

The study of the distribution of fossil fishes renders evident two important facts: First, that there has been a progressive modification and evolution from less to more highly organised types; and, secondly, that among the almost innumerable forms of comparatively recent origin, our existing ichthyic fauna still holds the wreck of a past fauna, whose period of decline belongs already to the earlier part of the earth's history. It, moreover, reveals an extraordinary persistence on the part of some of the individual types. The occurrence of the recent genus Ceratodus in deposits as ancient as the Permian is certainly very remarkable, but it does not argue, as some would lead us to believe, that this particular fish has undergone no modification since the period of its first introduction. The fact that its remains have been found fossil in the deposits of Europe and America, whereas at the present time it is found exclusively in Australia, would seem to imply a different distribution of land and water masses than now exists. But there appears to be no doubt, from their association with a marine fauna, that the early members of the genus were of an oceanic character, and that the fresh-water habit was obtained as the result of later modifications. The peculiar distribution would then be readily accounted for. It appears more than probable that from the order of fishes represented by this genus have been descended at least some of the earlier amphibians; if this be true, it would seem that the lungs of the dipnoans had been developed when the animals were still more or less strictly marine. The genetic relationship of the two remaining genera, the South American Lepidosiren and the African Protopterus, to Ceratodus still remains to be determined; but the development of an additional lung in these forms, coupled with the circumstance of their broad geographical isolation, would indicate an ancient differentiation of the two groups of the Dipneumones and Monopneumones.

In tracing the phylogeny of the class of fishes as a whole we are presented with certain difficulties which in the present state of our knowledge prove an insuperable obstacle to the solution of the problem. Paleontology thus far offers no positive clue as to what might have been the direct ancestors of the animals in question, and until it does so inferences drawn from purely zoological characters will be largely in the nature of pure hypotheses. The earliest fishes that appear, although obviously of a much less perfect structural type, exhibit very nearly the amount of specialisation seen in the modern forms, and are evidently far removed from the period of their first origination. Whether, therefore, the tunicates, which unquestionably possess many points of structural relationship with the fishes, are their true ancestors or not, or whether, instead of representing primitive vertebrate types, they are merely the degenerated remains of a more highly constituted ichthyic stock, must still be considered an open question. Nor would it be safe to affirm that the most ancient representative of the class was a form either closely or remotely allied to amphioxus, the lancelet, or that the latter is an ancient type at all. If the views recently set forth by Professor Cope [114] as to the tunicate affinities of Pterichthys and its allies be proved to be correct, then, indeed, the presumptive evidence would be very great for concluding that what have hitherto been generally considered to constitute an abnormal type of the Mollusca are in reality the true progenitors of the fishes. The facts in the case, however, require further substantiation.

The phylogenetic relationship existing between the selachians and ganoids is equally obscure as that which exists between fishes generally and the other classes of animals. Both types are known to us in their oldest forms from very nearly the same horizon, and consequently give no indication as to priority of birth. Much more positive indication in this direction is afforded with respect to the lung-fishes and teleosts, the former of which appear to be clearly related to, and to have been derived from, the dipteroid ganoids (of the type of Dipterus), and the latter to have held a similar relationship to the rhomb-scaled ganoids, possibly of the type of Leptolepis. From the former, apparently, have descended the amphibians, while the latter have continued to develop as the dominant fish-fauna of existing waters. That the fresh-water forms are modified descendants of types originally inhabiting the seas there can be no reasonable doubt. It is impossible to state when the earliest differentiation of marine and fresh-water forms was effected, but there is every reason for supposing that it dates back far into the Paleozoic era, and that some, if not many, of the Devonian fishes were of a strictly fresh-water habit.

AMPHIBIA.

The most salient facts that present themselves in connection with the geographical distribution of the Amphibia are, first, their almost complete absence from oceanic islands—the Seychelles, New Caledonia, and the Feejee and Solomon Islands forming island groups exceptional to a general rule—and, secondly, the very nearly universal limitation of the tailed forms, sirens, newts, salamanders, &c., to the Northern Hemisphere. The nature of the first condition has already been discussed in treating of the dispersal of animals generally. The total number of known forms are comprised, according to the latest researches of Boulenger, in somewhat more than one hundred and forty genera and nine hundred species, of which only twenty-seven genera and seventy species belong to the urodelous or tailed division, eleven genera and thirty-one species are cœcilians (Apoda), and the remainder, one hundred and five genera and eight hundred species, frogs and toads (Anura).

The cœcilians are tropical forms belonging to the East Indies, Africa (with the Seychelles—Hypogeophis rostratus), and America. The American species, including all of the genus Cœcilia itself, are about twenty in number, and range from Mexico to Peru and Brazil.* Remarkable instances of divided genera are presented by Dermophis, which possesses five American species and one from West Africa (D. Thomensis), and Uræotyphlus, represented by two species in Malabar and likewise one in West Africa.

The urodele amphibians are comprised in four families : The Sirenidæ or sirens, with two or three species, inhabiting the South-eastern United States; the Proteidæ, with two genera, Proteus and Menobranchus (or Necturus), the former confined to the subterranean waters of Carinthia, Carniola, and Dalmatia, and the latter to the streams of Eastern and Central United States and Canada; the Amphiumidæ, with three genera, two of which, Amphiuma and Menopoma, represent North American forms, while the third, Sieboldia (Cryptobranchus or Megalobatrachus), which is closely related to the menopomas, is confined to Japan and China; and the Salamandridæ (newts, salamanders, &c.), comprising upwards of ninety species, very extensively distributed throughout temperate Eurasia and North America, with some fifteen or more species in tropical America (from Mexico southward—Amblystoma, Spelerpes), a limited number in North Africa, and two (Tylotriton) in the Himalayas. The North American forms belong principally to the genera Plethodon, Desmognathus, Diemyctylus, Amblystoma (with Axolotl), and Spelerpes, the first two of which appear to be restricted to the Western Hemisphere.† Spelerpes has one species (S. fuscus) in the south of Europe, and Amblystoma one (A. persimile) in Siam, remarkable instances of separation in genera. The urodele Amphibia of North America, north of the Mexican boundary, number about fifty species. The permanent larval forms of one or more species of Amblystoma (A. tigrinum, A. mavortium), known as

---

* Boulenger gives the range of Chthonerpeton indistinctum as extending to Buenos Ayres; but this is considered doubtful by Peters (" Monatsb. Berl. Akad.," 1879, p. 940).

† American zoologists recognise the Plethodontidæ (with Spelerpes), Desmognathidæ, and Amblystomidæ as distinct families ; Diemyctylus, representing the Pleurodelidæ, is by Boulenger considered to be synonymous with the Eurasiatic Molge (Triton of Laurenti).

axolotls, occur in various parts of Mexico and the Western United States (California, Wyoming).

The greater number of the Old World salamandroids belong to the genus Molge (or Triton), whose range extends from Great Britain (M. cristata; M. palmata) to China and Japan (M. pyrrhogastra; M. Sinensis), and south to Syria and the Mediterranean coast of Africa. The species having the most extended range appear to be M. cristata and M. vulgaris, both of which are distributed throughout the greater part of Europe, and largely also over temperate Asia. The most northerly point reached by any species seems to be about 63° 30′ (M. vulgaris or aquatica, in Norway). The Alpine triton (M. alpestris) ascends the Alps, according to Fatio, to an elevation of about 8,000 feet (2,500 metres), while a Mediterranean species (M. montana) inhabits the Lago d'Argento, on Monte Cinto, in Corsica, at an altitude of 6,000 feet. The genus Salamandra has three species, which collectively inhabit the greater part of Central and Southern Europe, the Caucasus, Asia Minor, and Algeria. Salamandra atra, the black or rain salamander, inhabits the mountain-regions of Savoy, Switzerland, and Austria between altitudes of 2,500 and 10,000 feet.

The anurous, or tailless, amphibians (frogs and toads), which, as has already been seen, comprise not less than eight hundred species, enjoy a much broader distribution than the tailed forms, being absent only from the regions of high northern and southern latitudes, and the remote oceanic islands. The genera Rana and Hyla are each represented by a single species in the Solomon Islands, and Cornufer (Ranidæ) by three species (C. dorsalis, C. Vitianus, C. unilineatus) in the Feejee Islands. The only family that is entitled to be considered in any way cosmopolitan is that of the toads (Bufonidæ), which are only absent, apart from local areas and the strictly oceanic islands, from Madagascar, New Guinea, and New Zealand. The genus Bufo, which in itself comprises nearly eighty out of a total of some ninety species belonging to the family, covers the entire range, with the exception of Australia, where it is replaced by the genera Pseudophryne, Notaden, and Myiobatrachus. The most broadly distributed species of the genus is the common European toad or paddock (B. vulgaris), whose range comprises practically the whole of Europe, Asia as far east as Japan, and Northern Africa; in Switzerland it ascends the Alps to a height of nearly 7,000 feet. Bufo calamita and B. viridis are likewise distributed throughout the greater part of Europe, the latter extending its range eastward to Turkestan.—The greater number of the American species occurring north of the Mexican boundary belong to the Sonoran transition-tract, where some six or seven species are met with. The common form of the Eastern and Southern United States is the Carolina toad (Bufo lentiginosus), of which several distinct varieties are recognised. A number of bufonine species are found in the West Indies, and Bufo (Chilophryne) dialophus is said to inhabit the Sandwich Islands.

Next to the toads the most broadly distributed family is that of the true frogs (Ranidæ), which are most abundantly developed in the Oriental and Ethiopian tracts, but are almost entirely absent from Australia. Of some two hundred species (representing eighteen genera), recognised as belonging to this group, somewhat more than half belong to the genus Rana itself, whose distribution is practically that of the family. The genus is absent from the southern parts of South America—in the whole of which continent there have been determined thus far only three or four species—and from New Zealand, but is represented by a single species (Rana Papua) in North Australia. A solitary species (Rana Krefftii) is also found in the Solomon Islands. The most broadly diffused Old World form is the green or edible frog (Rana esculenta), whose habitat extends from England and Scandinavia to North Africa, and eastward through Central Asia to China and Japan ; the species is wanting in the island of Sardinia.* Somewhat less broadly distributed through Eurasia is the common frog (R. temporaria), which is the most northerly of known species, ranging in Norway (var. platyrhina) to beyond the seventieth parallel of latitude. In the Alps it still frequents the waters at an elevation of 8,000 feet.

The two commonest species of Eurasian frog have their American representatives in the shad- or leopard-frog (R. halecina) and wood-frog (R. sylvatica)—the latter by some authors considered to be identical with R. temporaria—both of which are widely distributed in the United States. The largest American species of the genus, which alone represents the family north of the Mexican frontier,

---

* Schreiber affirms that the species is also wanting in Great Britain ; but the British Museum is in possession of a specimen from Cambridgeshire (Boulenger, "British Museum Catalogue," 1882).

is the common Jull-frog (R. Catesbiana). Among the more important remaining genera of Ranidæ are Rhacophorus (with Polypedates, according to Boulenger), whose thirty or more species inhabit Japan, the Philippines, Southeast Asia, India (with Ceylon), and Madagascar; Ixalus, with about twenty-five species, restricted to the East Indies; and Rappia, with a nearly equal number of species, inhabiting tropical Africa.

The tree-frogs (Hylidæ), with upwards of one hundred and sixty species, find their greatest development in the Neotropical region, which contains somewhat more than one hundred species. The genus Hyla itself is represented by nearly ninety species, or by nearly three-fourths of all the known forms. The species of the North American fauna are comprised in the genera Hyla, Acris, and Chorophilus. Temperate Eurasia has but a solitary representative of the family, the common tree-frog (Hyla arborea), which, in its several varietal forms, is distributed from Great Britain and the Canary Islands to Japan. Hyla Chinensis and H. annectens, the latter from North India, are the only other Asiatic species. The genus Hyla is wanting in the Ethiopian realm, but is represented by several species on the continent of Australia, whose amphibian fauna is made up almost exclusively of the families Cystignathidæ (about twenty species), Bufonidæ (six species), and Hylidæ (eleven species—Hyla and Hylella).

Scarcely inferior in point of specific development to the tree-frogs are the Cystignathidæ, whose one hundred and fifty or more species are almost entirely restricted to Australia (with Tasmania) and South America, a few species penetrating northward into Mexico and the West Indies, and three or four into the Southern and Western United States (Florida, Texas, California). The family may, therefore, be said to be distinctive of the Southern Hemisphere. The most abundantly represented of its numerous genera is Hylodes (forty-five species, tropical America), peeping-frogs, many of whose species partake of the habit of the common tree-frogs. Collectively the species are very broadly distributed, and penetrate far beyond the region of elevated temperatures. Hylodes leptopus, about the most southerly of all known species of frog, descends to the Strait of Magellan, while H. Whymperi was obtained by Mr. Whymper on the slopes of Chimborazo at an altitude of 13,200 feet. Paludicola marmorata (Leiuperus viridis), a member of the same family, was found by Tschudi in the Peruvian Andes at an elevation of nearly 16,000 feet. Among the more distinctive forms of Cystignathidæ are the horned frogs (Ceratophrys), which inhabit tropical America from Guiana to Uruguay.

A remarkable instance of a divided family among the Anura is furnished by the Dendrobatidæ, which comprise two genera and ten species, one genus, Mantella, being confined to Madagascar and Nossi Bé, and the other, Dendrobates, to tropical South America. The intermediate tracts are entirely devoid of representatives of the family. The number of families restricted to a single zoogeographical region is five, of which four characterise the Neotropical realm —Dendrophryniscidæ, Amphignathodontidæ, Hemiphractidæ, and Pipidæ—and one, the Dactylethridæ, the Ethiopian. The most limited of all the families is the Pipidæ, which is restricted to a single species, the Surinam toad (Pipa Americana), an inhabitant of Guiana and Brazil.

Of the four primary groups to which the animals of this class are referable, the Stegocephala (corresponding to the Labyrinthodontia of most authors), Gymnophiona (cœcilians), Urodela (salamanders, tritons), and Anura (frogs and toads), the first acquires special geological importance from the fact that all, or very nearly all, of the older forms are comprised within it. Remains of Urodela are only doubtfully known from the Paleozoic deposits, while the anurous type does not appear before the Tertiary epoch; no fossil cœcilian has as yet been discovered.

The now wholly extinct order Stegocephala, which comprises salamandroid and ophidian forms more or less covered with a protecting armour of bony (ganoid) plates, dates from the Carboniferous period (Hylerpeton, Batrachiderpeton, Pelion, Dolichosoma, Ophiderpeton), when, or at a still considerably earlier era, they appear to have become differentiated from the type of lung-fishes (Dipnoi) or of the dipteroid ganoids. A further development of types, with a partial persistence of Carboniferous genera, is manifest in the Permian deposits, where, as in the older strata, the forms are principally referable to the division Ganocephala (Archægosaurus, Dendrerpeton, Branchiosaurus, Protriton,* Hylonomus,

* Protriton Petrolei is by Deichmüller considered to be identical with Branchiosaurus gracilis.

Limnerpeton, Melanerpeton), in which the peculiar labyrinthine infolding of the teeth, distinctive of the true labyrinthodonts, is largely absent. The apodal and cœcilian-like division Aistopoda is represented among other forms by the Carboniferous genera Dolichosoma and Ophiderpeton, and by Palæosiren and the American Molgophis. Contemporaneously with these types we have also the true labyrinthodonts, whose earliest member appears to be Baphetes, from the Carboniferous deposits of Pictou, Nova Scotia. The full development of this group does not obtain, however, before the Triassic period, at the close of which the entire order of animals seems to have become extinct in most regions.* Among the more distinctive genera of this period are Labyrinthodon, Mastodonsaurus, Trematosaurus, and Metopias, to one or several of which probably belong the foot-prints of the fanciful animal designated Cheirotherium.

Of the perennibranchiate division of the Urodela, in which external gills are retained throughout the entire existence of the animal (Siren, Proteus), we have as yet no positive indications in any of the rock-formations. The Cryptobranchia, which retain a gill-opening after the absorption of the gills—the American Amphiuma and Menopoma, and the giant salamander of Japan, Cryptobranchus Japonicus—seem to have one or more fossil representatives in the genus Andrias (Cryptobranchus of some authors), from the Miocene deposits of Oeningen, Germany, to which are referred the remains presumed by Scheuchzer to be those of earliest man (Homo diluvii testis). It is certainly a very remarkable fact in distribution that the only link uniting the so widely separated, but closely related, genera Menopoma and Cryptobranchus, should be this extraordinary form from the middle Tertiary period. Its existence would seem to indicate a former much broader diffusion of this particular group of animals, and a very different distribution of land and water areas than now obtains.—The caducibranchiate urodeles (salamanders, tritons), which in their transformation to lung-

---

* Brachyops, from the Damuda beds of India, and one or two other genera of labyrinthodonts have been indicated as belonging to the Jurassic period, but it may be questioned whether the age of the deposits in which these remains occur has been as yet satisfactorily determined. Excepting these somewhat doubtfully placed forms, no amphibians are known from Jurassic strata.

breathers pass one stage beyond the cryptobranchs in the obliteration of the gill-aperture, date from the Eocene period, when forms
more or less nearly allied to recent types appear. Both Triton and
Salamandra are represented. Remains of tailless amphibians (Anura), more or less nearly allied to modern forms, have been obtained
from the Tertiary lignitic and fresh-water strata (Oligocene, Miocene) of Western and Central Europe. Palæobatrachus diluvianus,
one of the oldest known forms, is from the lignitic strata of Orsberg,
near Bonn, Germany. The lacustrine deposits of Oeningen have
yielded several extinct genera, among which are Latonia (related to
the Brazilian horned toad, Ceratophrys), Palæophrynus (a bufonine
type), and Pelophilus, the last not impossibly a true Bombinator.
Among recent genera, Rana, Bufo, and Pipa have also Tertiary
representatives.

The paucity of remains of existing types of amphibians, combined with the circumstance of their very late appearance, renders
impracticable the determination of the phylogenetic relationships
which bind together the various groups. Equally uncertain are
the stages which mark the differentiation of the modern fauna
from that of the Palæozoic and the early Mesozoic periods, nor is
it likely that any progress towards the solution of this problem
will be effected until the void which is caused by the almost total
absence of amphibian remains from the deposits of Jurassic and
Cretaceous age will have been in great part filled.

That the animals in question are derived either in whole or in
part from the dipnoan type of fishes there is very little doubt, but
the immediate connecting link or links between their ichthyic progenitors, whatever these may have been, and the earliest stego-
cephalic forms are still wanting. The apparently sudden disappearance with the Triassic period of the largely represented order
which contained all, or very nearly all, the earlier forms of amphibians, without leaving in the modern fauna any positive indications
of its former existence, is not a little surprising, but it appears not
unlikely that the cœcilians, which in many points of structure
resemble the ophidian labyrinthodonts, represent at least a part of
this ancient stock. Again, by many geologists the crocodiles are
assumed to be the modified descendants of the true labyrinthodonts.

REPTILES.

Chelonia.—The total number of known species of chelonians is estimated by Hoffmann (1880) to be somewhat more than two hundred and fifty. Of these only five are marine forms, the rest being inhabitants of the land and its fresh waters. The former, comprised in the genera Dermatochelys (or Sphargis), Chelone, and Thalassochelys, are very broadly distributed throughout the tropical and sub-tropical or temperate waters of both the Old and the New World, most of the species being cosmopolitan, or nearly so. Dermatochelys coriacea, the leathery turtle, is found along the American border from Brazil to South Carolina and Massachusetts, exceptionally on the European coast, and in the Indian and Pacific oceans, from Africa to Chili. The green turtle (Chelone viridis), which is held in such high estimation as an article of food, has an equally extended range, although it is but very rarely found on the European coast (England to the Mediterranean). Of still rarer occurrence in the European seas is Chelone imbricata, the hawk's bill, which yields the tortoise-shell of commerce, and whose habitat embraces nearly the whole circumference of the globe. The loggerhead (Thalassochelys corticata) is abundant on both sides of the Atlantic, and in the Mediterranean, and is at rarer intervals also met with in the Indo-Pacific basin.

The land and fresh-water chelonians have a very unequal distribution, being most abundant in the region of the tropics, and rapidly diminishing as we pass either north or south into the temperate zones. The greatest number of forms belong to tropical and sub-tropical America, and the smallest number to Australia and temperate Eurasia, each of which possesses some fifteen species. The northern limit reached by these animals in the Western Hemisphere is about the fiftieth parallel of north latitude (Chelydra serpentina), and not improbably the same parallel marks the corresponding general limit in the Eastern Hemisphere, although in Europe Cistudo lutaria or Europæa, the most wide-spread species of the continent, is found as far north as the fifty-fourth parallel (Mecklenburg), and possibly still farther. The total number of European species is five, most of which more properly belong to the region about the Mediterranean. No species is known from Great Britain, the Scandinavian Peninsula, Denmark, Holland, or

Belgium. Cistudo lutaria and Testudo Græca, the latter introduced, inhabit the waters of Southern France; the first of these
also inhabits Switzerland, but it is only doubtfully indigenous to
that country. The number of species occurring in North America
north of the Mexican boundary is about forty, nearly one-half
of which properly belong to the Southern United States. Among
the commoner or better known forms are the box-turtles (Cistudo,
Cinosternum), wood-turtles (Chelopus), painted-turtles (Chrysemys),
marsh-turtles (Malacoclemmys), terrapins (Pseudemys), musk-turtles
(Aromochelys), snappers (Chelydra), and soft-shells (Aspidonectes),
all of which are very broadly distributed, especially in the Eastern
and Southern United States. The species having the most extended range is the common snapper (Chelydra serpentina), which
is found from Canada to Ecuador. Two other species, the common
wood-turtle (Chelopus insculptus) and the painted-turtle (Chrysemys
picta), range as far north as Canada.

Most of the species of Chelonia are restricted to a single faunal
region, and where identical species are found in more than one
continent, the range of the species on the continent not properly its
home is, as a rule, very limited. Two species are known to be
common to Europe and (North) Africa—Testudo nemoralis and
Cistudo lutaria (C. Europæa); one species, Pyxis arachnoides, is
common to the continent of Africa (with Madagascar, and some of
the neighbouring islands) and India; and likewise one, Manauria
fusca, common to the East Indies (Java, &c.) and Australia. Turtles are wanting in the true oceanic islands, but they are sufficiently
abundant in many of the continental islands, even where these are
distant several hundred miles from the nearest mainland. Two of
the most ponderous representatives of the order belong to such isl-
and groups: the Galapagos turtle (Testudo nigra) and the elephant
turtle (T. elephantina), the latter, whose weight is known to reach
five hundred pounds, inhabiting the Seychelles and some of minor
island groups of the Mozambique Channel.

Of the four more generally recognised families of land and
fresh-water turtles, the Testudinidæ, Emydæ, Chelydæ, and Trionychidæ, only the first has representatives in all the major divisions
of the earth's surface. Australia is lacking in both the Emydæ and
Trionychidæ, the latter being also absent from South America,
while the three southern continents are almost the sole possessors

of the Chelydæ. Remarkable instances of discontinuous genera are seen in Hydromedusa, one species of which inhabits the Oriental realm and the remainder the continent of South America, and in Podocnemis, whose species are divided between South America and Australia.

The earliest chelonian remains occur in deposits of Jurassic age (Switzerland, Germany, France), in which a well-marked differentiation of the modern families Emydæ (as seen in the genera Thalassemys, Eurysternum, Tropidemys, Helemys, the last supposed to have been closely related to the American snapper) and Chelydæ (Plesiochelys, Idiochelys, Craspedochelys) already appears. The number of forms is materially increased in the succeeding Cretaceous deposits, where, in addition to the representatives of the two families already indicated (e. g., Platemys, Pleurosternum, Adocus, Euclastes, Osteopygis), we have those of the Trionychidæ and Cheloniidæ (Trionyx, New Jersey; Chelone, Maestricht chalk, and greensand of New Jersey). Protostega gigas, a marine turtle from the deposits of this age of Kansas, attained a length of upwards of twelve feet. Many of the recent genera, as Testudo, Chelydra, Emys, Cistudo, &c., appear as fossils in the early or middle Tertiary deposits. The most extraordinary of all extinct forms is the giant land-tortoise of the Siwalik Hills of India, Colossochelys atlas, which measured apparently not less than fifteen to twenty feet in length. Of somewhat less than one-half these dimensions was the Macrochelys mira, from the *molasse* of Southern Germany (Oberkirchberg, near Ulm), whose modern representative is the Mississippi snapper (Macrochelys lacertina).

Lacertilia.—The number of known species of lizard is estimated by Günther to be about seventeen hundred, of which by far the largest part is confined to the warmer regions of the earth's surface. But comparatively few forms are found to pass beyond the fortieth parallel of latitude, and at about the sixtieth parallel (north) the order practically disappears. The most northerly species is Lacerta vivipara, whose range comprises nearly the whole of Europe, and extends northward to the seventieth parallel (in Norway); it is accompanied as far as Lapland by the no less broadly distributed blind-worm (Anguis fragilis). In the Western Hemisphere the northward extension of the order is much more limited than in the Eastern, and it would appear that only one species, a

Gerrhonotus, passes beyond the fiftieth parallel ; in the Middle United States the northern skink (Eumeces septentrionalis) penetrates into Minnesota, and along the Atlantic border Eumeces fasciatus, a species, singularly enough, also found in Japan, forms part of the Massachusetts fauna. The most southerly range of any species is that of Liolæmus Magellanicus, which reaches the Strait of Magellan.

In the whole of Europe north of the forty-fifth parallel of latitude, or what might be considered to be Central and Northern Europe, there are scarcely more than a dozen species of lizard, of which nearly one-half belong to the genus Lacerta, or common lizard. Scandinavia, Great Britain, and Denmark have each three (and the same) species: Lacerta vivipara, L. agilis, and Anguis fragilis. An additional species, the wall-lizard (L. muralis), belongs to Belgium and Holland, and a fifth one, the green lizard (L. viridis), which has also found a congenial home on the island of Guernsey, to Germany. All of these species form part of the southern or Mediterranean fauna, which in Europe comprises some thirty-five or more species, many, or most of them, of a distinctively African type. The affinities with the tropical faunas are seen in the development of the geckotine type (Hemidactylus verruculatus, the common gecko of the houses of Southern Europe ; Gymnodactylus, Phyllodactylus, Platydactylus) and the agamas (Agama, Stellio—South Russia and the Balkan Peninsula), the Old World representatives of the American iguanas. One species of chamæleon (Chamæleo vulgaris) is found in Andalusia.

In temperate North America lizards are even more scarce than in the equivalent region of the Old World. Indeed, in the whole of the continent north of a line that might be considered to unite San Francisco with Galveston in Texas there are probably less than twenty species, of which more than one-half belong to the Old World genus of skinks, Eumeces. A distinctive feature separating the saurian fauna of this tract from the European is the absence of the group to which all the commoner European forms (Lacerta) belong, although the genus Xantusia, from the Pacific coast, is by some authors doubtfully referred to the Lacertidæ. On the other hand, a distinct Old World relationship is established in the glass-snake (Ophiosaurus—from Tennessee southward and westward), a near ally of which is the glass-snake (Pseudopus) of Southern Europe (Dalmatia, Hungary, Russia) and West-Central Asia.* In the region lying south of the San Francisco-Galveston line, which is largely in the form of parched or desert tracts, the prevalence of a considerable number of tropical or South American types imparts a distinct individuality, or non-North American character, to the fauna, which is best expressed in the family of iguanas (Iguanidæ). This group is represented by not less than forty species, the greater number of which belong to the genera Sceloporus and Phrynosoma ("horned-toad"), one species of the latter genus penetrating as far north as Dakota.† Inhabiting the same tracts, but extending its range to Tehuantepec, is the venomous Heloderma. A single species of amphisbænian, the "thunder-worm" (Rhineura Floridana), is known from Florida. The so-called chamæleon of the Southern United States is the green goitred lizard Anolis.

The more distinctive or most largely represented tropical families of lizards are the iguanas, agamas, monitors, geckos, amphisbænians, and chamæleons. The first of these is almost exclusively American, and is represented by probably not less than three hundred species, of which nearly, or fully, one-third belong to the genus Anolis, whose members especially abound in the West India islands. The genus Iguana is more properly South American, although also found in some of the West Indies, and penetrating northward into Mexico. Basiliscus, the basilisk, likewise ranges into Mexico. In addition to the forms that have already been indicated as belonging to the United States, may be mentioned Uta, Callisaurus, and Holbrookia, the last of which is sufficiently abundant in certain parts of Texas and the transition-region to the northwest. The most remarkable member of the family is the Galapagos leguan (Amblyrhynchus), which is partially marine in its habits. Brachylophus inhabits the Feejee Islands. No iguanian

---

* A variety of this species also occurs in Morocco. M. Boulenger has recently attempted to show ("Ann. and Mag. Nat. Hist.," Aug., 1885) that the North and South American lacertilian faunas are, strictly speaking, one, the Neogean, a conclusion which is not borne out by the facts of distribution. The misconception arises from the incorporation of the tract lying south of the line indicated above with the North American faunal region proper, while in reality it is a transition-tract more nearly Neotropical in character than " Nearctic."

† Phrynosoma orbiculare was found by Mr. Geddes on the plateau of Mexico at an altitude of 7,500 feet.

is found on any of the continental divisions of the Old World, but two genera of terrestrial habits, Hoplurus and Chalarodon, appear in Madagascar.

The true Old World representatives of the iguanas are the agamas (Agamidæ), which might be said to present a parallel series of forms to the iguanian types of the New World. Their distribution covers the greater part of the continent of Africa, the warmer tracts of Asia, especially the islands of the East Indian Archipelago, and much of Australia. No species has thus far been obtained from New Zealand. A limited number of species is found in Asia north of the Himalayas (Trapelus, Phrynocephalus — Tartary to Afghanistan), and their occurrence in Southern Europe (Agama, Stellio *) has already been noted. Several genera have representatives in the Andaman and Feejee Islands groups. Among the more remarkable forms of the family are the flying-lizards (Draco)—inhabitants of the East Indies (except Ceylon)—which are provided with a tegumentary expansion specially adapted for sailing through the air; the frilled lizard of Queensland, Australia (Chlamydosaurus Kingii), which is ornamented with a broad fan-like collar nearly encircling the head and neck; and the spine-covered Moloch horridus of Southern and Western Australia. The agamas proper range throughout Africa, and eastward to India.

The geckos (Geckotidæ), which, with the exception of the cosmopolitan skinks, have the broadest distribution of all the lacertilian families, number about two hundred species. They occur in the hotter parts of all the continental regions, and are largely represented even in the more distant oceanic islands—Madeira, Ascension, the Seychelles, New Zealand; the Solomon, Andaman, and Sandwich Islands groups, &c.—evidently possessing some special means for dispersion which is wanting in other reptiles. Several of the more largely represented genera, as Gymnodactylus, Phyllodactylus, and Hemidactylus, have practically the range of the entire family; Gonatodes is found in tropical America and East India, but is wanting in Africa. The genus Gecko, as restricted, has about seven species, which are confined to China, Japan, the Papuan Islands, and the islands of the East Indian Archipelago. Most of the geckos are nocturnal in their habits,

---

* Boulenger ("Catalogue of Lizards," British Museum, 1885) considers Trapelus and Stellio as synonyms of Agama.

and it would appear that the different species intentionally keep apart from each other. Colonel Tytler observes that "although several species of geckos may inhabit the same locality, yet, as a general rule, they keep separate and aloof from each other; for instance, in a house the dark cellars may be the resort of one species, the roof of another, and crevices in the walls may be exclusively occupied by a third species. However, at night they issue forth in quest of insects, and may be found mixed up together in the same spot; but on the slightest disturbance, or when they have done feeding, they return hurriedly to their particular hiding-places." [115] Remarkable instances of broad specific range are presented by Hemidactylus mabouia, which inhabits Brazil, San Domingo, Eastern Africa, and Madagascar, and Gehyra mutilata, whose range extends from the Mascarene Islands through India, the Malay Peninsula, and New Guinea to Mexico. With the exception of the common chamæleon (Chamæleo vulgaris), whose range extends from Andalusia through North Africa eastward to India and Ceylon, all the species of the family are restricted to the African continent and the neighbouring islands (Madagascar, Bourbon, and Fernando Po, the first with nearly one-half the total number of species). The monitors, or water-lizards (Varanidæ), which range over the greater part of Africa, East India, Australia, and the Austro-Malayan islands, comprise the largest Old World members of the class, some of the species measuring, or exceeding, six feet in length. The common monitor of the Nile (Monitor Niloticus) is found in the neighbourhood of all the more important streams of tropical Africa. Psammosaurus scincus, a North African species, is strictly terrestrial in its habits.

The amphisbænians, or footless lizards, which by Dr. Gray are elevated to the rank of a distinct order, are principally tropical American forms, although a considerable number of species are known from the African continent, and a few, of the genus Blanus, from the Mediterranean districts of Europe and Asia. In America the species range from the Argentine Republic through the West Indies to Florida (Rhineura [Lepidosternon] Floridana). Chirotes lumbricoides, which is provided with the anterior pair of appendages, is a Mexican species. A distinctively American family of lizards is the Teiidæ, or teguexins, which may be said to replace the Old World Lacertidæ, and whose range extends from Patagonia to Montana and Virginia. Upwards of a hundred species have been described.

New Zealand (or rather the small islands off the northeast coast) possesses a remarkable lizard in the genus Hatteria or Sphenodon, which in many points of structure departs from the type of true lizards, and approximates it to an ancient lost form from the Trias, the genus Hyperodapedon.

The earliest known representative of the Lacertilia is Protero-saurus, from the Permian deposits of Germany and England, which appears to be most nearly related to the monitors, from which, however, it differs in its thecodont dentition.* Hyperodapedon (which, with the contemporaneous Rhynchosaurus, and the recent Hatteria, is by some authors constituted into a distinct order, Rhynchocephala) and the acrodont genus Telerpeton (Elgin lime-stones of Scotland) appear in the Trias, and are succeeded in the deposits of Jurassic age by a number of more or less obscurely defined genera (Geosaurus, Homœosaurus, Acrosaurus, Anguisau-rus), whose relationships with modern forms are in most cases not clearly indicated. Lacertilian remains are not abundant in the Cre-taceous deposits, and such as have been preserved are mainly in a fragmentary condition; the recent genus Hydrosaurus, one of the monitors, is indicated. In Tertiary strata the remains become nu-merous, and belong in considerable part to modern types. Frag-mentary skeletons from the European Miocene deposits have been referred to Iguana and Lacerta, and, doubtfully, also to Scincus and Anguis.—No true lacertilians are known from American de-posits older than the Eocene. The western lake-basins of this age have yielded numerous remains, which are referable to a number of distinct genera—Glyptosaurus, Iguanavus, Oreosaurus, Tino-saurus, Saniva—and some of which appear to have survived into the Miocene. Among the very limited number of forms of this period may be mentioned Peltosaurus, doubtfully referred to the Gerrhonotidæ, and Cremastosaurus, the latter of about the size of the horned-toad.

Ophidia.—The distribution of the Ophidia is very similar to that of the Lacertilia, the order being most numerously represented

* Professor Seeley believes it probable that Proterosaurus is a dinosaur. (Phillips, "Manual of Geology," edited by Etheridge and Seeley, 1885.)

in the tropical regions of the earth's surface, and rapidly diminishing toward either pole. Excepting, however, members of the family of water-snakes (Hydrophidæ), which are especially abundant in the Australian and Indian seas—ranging westward to Madagascar, and eastward to Panama—the order is only exceptionally represented in the strictly oceanic islands, in this respect differing from the lizards and agreeing with the amphibians. Evidently, the animals of this class, like the Amphibia, possess no facilities for traversing broad arms of the sea.

With our deficient knowledge of many of the more favoured regions of the globe it is impossible to arrive at any estimate of the numerical extent of the order, but it may be safely assumed that there are considerably more than one thousand clearly defined species known to naturalists, of which very nearly one-half are found in British and Farther India, and the East Indian Archipelago. Mr. Blanford places the number of species from British India and its dependencies alone at two hundred and seventy-four.* In Europe, north (and inclusive) of the Alps, there are some fifteen or more species, of which three, the common viper or adder (Vipera [Pelias] berus), the grass or ringed snake (Tropidonotus natrix), and the Coronella Austriaca (lævis), penetrate beyond the fifty-fifth parallel of latitude. These are the only species found in Scandinavia, the British Isles, Denmark, Holland, and Belgium. The most northerly of all serpents is the common viper, whose range embraces the whole of Europe and Northern Asia, and which in Scandinavia extends to the Arctic circle; in the Alps it is occasionally met with at an altitude of nine thousand feet. The northern limit of the ringed snake appears to be the sixty-fifth parallel. Germany has in all six or seven species,[116] the three above mentioned, and Tropidonotus tessellatus, Elaphis flavescens (Æsculapii), Zamenis viridiflavus (doubtful), and Vipera aspis (the asp), the last very largely distributed throughout the whole of France and Switzerland, and the commonest of the venomous serpents of Italy. It does not appear to ascend the Alps to elevations much exceeding

* The census of the other Reptilia is as follows: Chelonia fifty-four, Crocodilia four, Lacertilia one hundred and eighty-two. The Amphibia comprise about one hundred species, of which one only belongs to the tailed division, and five to the Pseudophidia (Cœcilia). "Journ. Asiatic Soc. Bengal," Dec., 1881.

five thousand feet. The species, like many of the other Mediterranean forms, is found also in Algeria, but seems to be absent from Morocco.[117] All of the German species occur in Switzerland, which, however, numbers one additional form, Tropidonotus viperrinus.

The Mediterranean fauna, or what might be considered to be the fauna of Southern Europe, comprises about thirty species, in which are included probably all the forms that occur elsewhere in Europe; the Iberian Peninsula numbers twelve species,[118] Italy about fifteen, and Greece fourteen.[119] In the western half of this region the similarity existing between the ophidian faunas of the several countries amounts almost to identity, but eastward, from the Balkan Peninsula to the Crimea, a gradual exchange of species is effected, so that in both Turkey and Russia nearly, or fully, one-half of the species (about fifteen in each country) are distinct. Somewhat more than one-half of the Italian and Iberian species are also found in the region south of the Mediterranean—Algeria and Morocco.

The North American serpents, or those found north of the Mexican boundary, belong in the main to two families, the colubers (Colubridæ) and pit-vipers or rattlesnakes (Crotalidæ), the former numbering some one hundred and ten or more species, and the latter about twenty. In addition to these there are a limited number of representatives of three or four other families. Thus, the worm or burrowing snakes (Typhlopidæ), whose species are abundantly distributed over the tropical regions of both hemispheres, occur sparingly in California and Texas (Stenostoma); the Erycidæ, a limited family of Old and New World serpents allied to the boas, are represented on the west coast by two species of Charina (California to Puget Sound); and the venomous Elapidæ, to which very nearly two-thirds of all the Australian snakes, and the deadly cobra (Naja), Bungarus, and Ophiophagus of India belong, are represented by the harlequin-snake (Elaps fulvius) in the Southern United States (east of the Mississippi), and by Elaps euryxanthus in Arizona. The greater number of these forms can scarcely be said to constitute a part of the North American ophidian fauna proper, inasmuch as they occur principally in a border tract whose general faunal relationship is more nearly with the region lying to the south than the north.

The North American colubrine snakes are comprised principally in five or six groups or genera: 1. Tropidonotus, water-snakes, whose

range is coextensive with the whole United States, and whose best
known exponents are the ribbon-snake (T. [Eutænia] saurita), water-
snake or adder (T. sipedon), and garter (T. [Eutænia] sirtalis), the
first two abundant in the region east of the Mississippi, and the last
found almost everywhere from Canada and Nova Scotia to Mexico
and Panama. 2. Coluber, whose range is no less extensive than
that of the water-snakes, and which embraces among other forms
the most broadly distributed black-snake or constrictor (C. [Basca-
nium] constrictor) and the coachwhip-snake (C. flagelliformis) of
the Southern States. 3. Pityophis, pine-snakes. 4. Elaphis, to
which the spotted racer (E. [Scotophis] guttatus), chicken-snake (E.
quadrivittatus), and pilot (E. obsoletus) belong, the first two principally from the Southern States, and the last generally distributed
over the Atlantic border, from New England to Alabama. 5. Ophi-
bolus, king-snakes, whose species are widely diffused throughout
the United States, and whose best known representatives are the
southern chain-snake (O. getulus)—with a western variety known
as the king-snake (O. Sayi)—the red-snake (O. doliatus), and the
very common milk-snake or spotted adder (O. triangulus), whose
range extends from the Atlantic border to the Mississippi, and
northward to Canada; and, 6. Diadophis, ring-necked snakes, ranging nearly through the entire continent south of the Canadian line.
The genus Cyclophis comprises two common species of green-
or grass-snake, the summer-snake (C. æstivus) and spring-snake
(C. vernalis), both of which have a very extensive distribution.
Two species of hog-nose snake (Heterodon) occupy a considerable
part of the United States, and are locally known as blowing-vipers
or adders.

The remaining colubrine forms are embraced in genera largely
limited as to the number of species, and which in many cases, as in
Contia, Tantilla, Sonora, &c., are confined to the transition-tract
which unites with the Neotropical realm. The North American
crotaloids are comprised in three or more genera: Crotalus, the
rattlesnakes proper, Sistrurus (or Crotalophorus), the prairie or
grass rattlesnakes, which are confined principally to the Central
and Southern United States, and Ancistrodon, the copperheads and
moccasins. Of the last there are three species: A. contortrix, the
copperhead, whose habitat is the greater part of the region east of
the Mississippi; A. piscivorus, the true or water moccasin, which

inhabits the waters of the Southern States from South Carolina to Texas; and A. atrofuscus, the highland moccasin, found in the mountain-region south of Virginia, and by many authors considered to be only a variety of the last. The rattlesnakes proper are represented by some ten or more species, most of which are found in the region of the Southwestern United States. Five species (or varieties) are known east of the Mississippi, of which the common or banded rattlesnake (C. horridus), which is still abundantly distributed between Texas and New England, has the most extended range. The diamond-rattlesnake (C. adamanteus) inhabits the Southern States.

Although the greater number of species of North American non-venomous Ophidia belong to genera or groups which are also largely developed in, and are equally characteristic of, the Old World, as Tropidonotus (Eutænia), Coluber (Bascanium), and Elaphis (Scotophis), types but barely represented in the Neotropical realm—thus clearly indicating the Old World affinities of the so-called "Nearctic" fauna, it appears that all the species are distinct.* This is not very surprising in view of the limited northern range, especially in the Western Hemisphere,† of the majority of the species, which are incapable, and have been incapable for a long period past, of traversing the chilled northern tracts by which at one time, doubtless, a union was effected between the two hemispheres. As a result of this isolation new species have been formed. It is more remarkable that the most northern of all ophidian genera, Viperus, the viper, whose appearance on the American continent might have been confidently looked for as a result of its extended range, is completely wanting. Other anomalies of distribution are presented by the distinctively American genera Heterodon and Dromicus, both of which have representatives in the island of Madagascar, and the family of pit-vipers (Crotalidæ), which is largely developed in the Oriental realm, but is wanting in Africa.

Of the more important families of tropical and sub-tropical

---

* By most American herpetologists Eutænia, Bascanium, and Scotophis are considered to be distinct from the Old World genera with which they have been united by the greater number of European naturalists.

† Several species are found in British Columbia along the Canadian boundary-line, but it is doubtful whether any penetrate much beyond the fiftieth parallel of latitude,

snakes—indeed, of all snakes—the colubers take first rank, numbering probably fully one-fourth of all known species of Ophidia. They are, strictly speaking, the most cosmopolitan of all the various groups, and are represented, in addition to genera whose distribution embraces several of the zoogeographical regions, by a number of distinct genera in each of the great zoogeographical regions except Australia, where the family is but feebly developed (Tropidonotus, Coronella). Next in importance, and more strictly tropical, are the venomous colubrine snakes (Elapidæ), with probably upwards of one hundred species, about one-half of which are confined to Australia and the neighbouring islands. The family, which is almost wholly wanting in the north temperate region—represented by the genus Callophis in Japan and by the harlequin-snakes (Elaps) in the United States—comprises many of the most deadly of the Thanatophidia, as the cobra (Naja tripudians), Bungarus, and Ophiophagus of India and some of the eastern islands. Callophis bilineatus appears to be the only poisonous snake of the Philippines. The genus Elaps embraces all or most of the American species of the family, including the much-dreaded Brazilian coral-snake (Elaps corallinus).*

Partaking very nearly of the distribution of the last family are the burrowing-snakes (Typhlopidæ), whose numerous members, belonging chiefly to the genus Typhlops, are found in nearly all the warmer regions of the earth's surface. One species of the genus, Typhlops lumbricalis, is found in Greece and on some of the Grecian islands. The tree-snakes proper (Dendrophidæ) are found in all the tropical regions; the nocturnal tree-snakes (Dipsadidæ) and the arboreal whip-snakes (Dryiophidæ) are also essentially tropical, but they are either wholly, or almost wholly, wanting in Australia.

The boas or pythons (Boidæ; Pythonidæ) are one of the most distinctively tropical families, comprising some fifty or more species. The pythons proper (genus Python) are distributed throughout nearly the whole of the Oriental region—the islands as well as the

---

* Many travellers and naturalists, and notably Maximilian, Prince of Wied, have denied the venomous nature of this animal. The researches of Ihering, however, conclusively demonstrate this nature in Elaps Marcgravii, and would seem, consequently, to uphold the common notion concerning E. corallinus ("Zoologischer Anzeiger," August, 1881).

mainland—and over the greater part of the continent of Africa, although by some naturalists the Ethiopian species are placed in a distinct genus, Hortulia. The netted python (P. reticulatus) inhabits nearly all the islands of the Malay Archipelago, besides portions of the mainland (Farther India), where it shares in part the habitat of the common Indian species, P. molurus. Among the African species are the royal python of the western forests (P. regia), Seba's python, or the fetich-snake (P. Sebæ), whose distribution is much more general, and the Natal rock-snake (P. [Hortulia] Natalensis). The Australian Pythonidæ are included in the genera Morelia, Aspidiotes, Liasis (islands of the Arafura Sea), and Nardoa, to the first of which belong the diamond-snake (M. spilotes) and the carpet-snake (M. variegata) of the colonists. In the New World the pythons are replaced by the boas and anacondas, which by many naturalists have been constituted into a distinct family, Boidæ, and whose habitat is principally the warmer parts of the South American continent. Boa constrictor, whose home is more properly the equatorial forest region, is represented by several closely allied forms in Central America and Mexico, as B. isthmica, B. imperator, and B. Mexicana, which are by some authorities considered to be mere varieties of the common southern constrictor, and by others as distinct species. A fourth species, the yellow boa (Chilabothrus inornatus), whose home is the West Indies, is doubtfully said to inhabit Central America and Mexico as well.* The anaconda (Eunectes murinus) is found in the tropical waters.

The most remarkable instance of a localised family of any extent is presented by the earth-snakes, or rough-tailed burrowing-snakes, as they are sometimes called, the Uropeltidæ, whose thirty-five or more species are confined almost entirely to Ceylon and the southern part of the Indian Peninsula, or to the tract constituting the Cingalese sub-region of the Oriental realm. Their headquarters on the peninsula are the western mountain-ranges between Canara and Cape Comorin, only one species, according to Beddome,[120] being found in the mountains of the east coast, and but three on the west, whose range extends northward beyond Kudra Mukh in South Canara. Several species of Silybura ascend the Neilgherries to an

---

* The naturalists of the United States Fish Commission steamer "Albatross" found a species of boæform serpent on the island of New Providence, Bahamas ("Science," June, 1886).

elevation of seven thousand feet, and Plectrurus Perrotetii is found between five thousand and eight thousand feet.

Fossil remains of serpents are not numerous, and only one species, the Simoliophis Rochebruni, from the Upper Cretaceous deposits of the Charente, France, is known to antedate the Tertiary period. Several species of Palæophis, considered by some authors to have been closely related to the boas, which they rivalled in size, and by others to constitute the type of a distinct family, have been found in the Lower Eocene deposits (Londonian) of England, France, and Italy; two or three species have been likewise described from the nearly equivalent deposits of the State of New Jersey.* Boæform serpents appear to be indicated by the Python Eubœicus, from Kumi, in the island of Eubœa, and by the remains from the Eocene fresh-water deposits of the Western United States which have been referred to the genera Boavus, Lithophis, and Limnophis. The genus Coluber is represented by several species from the Miocene fresh-water deposits of the continent of Europe (Oeningen, &c.). Fossil Toxicophidia, or venomous serpents, appear to be still less abundantly represented than the non-venomous types. A form supposed to be related to the rattlesnakes has been described from Salonica as Laophis crotaloides, and one, related to the cobra, from Steinheim, as Naja Suevica. The most ancient remains of Ophidia in the New World appear to be those of Helagras prisciformis, from the Puerco Eocene, which was of about the size of the black constrictor (Coluber constrictor).

The paucity of ophidian remains leaves very uncertain any speculations as to the origin or evolution of this order of animals. Whether or not they are in part the modified descendants of the lacertilian pythonomorphs, which they seem to approximate in certain points of structure, still remains to be determined.

Crocodilia.—Of the four orders of existing reptiles the Crocodilia are numerically the least important, and at the same time the most restricted in their distribution. Some twenty-five more or less well-defined species, inhabiting the tropical and sub-tropical regions of the earth's surface, are known to naturalists, by whom three distinct groups or families are recognised: the gavials, crocodiles proper, and alligators. The gavials are exclusively Old World forms, and the alligators forms belonging to the New World. The

* Palæophis littoralis, P. Halidanus, P. (Dinophis) grandis.

former, as understood by most systematists, are comprised in two genera, Gavialis (with a single species, G. Gangeticus), restricted to the waters of the Indian Peninsula, and Tomistoma, a Bornean form, whose range probably extends to North Australia.

The true crocodiles, of which some authors recognise two genera, Crocodilus and Mecistops,* inhabit nearly all the larger streams (and many of the lakes) of Africa, India, and the north coast of Australia. Although for a long time supposed to be entirely wanting in the New World, they are now known to inhabit the waters of tropical America on both sides of the Andes (Ecuador, Colombia, the Orinoco, &c.), extending their range to Mexico and the West India Islands (Cuba, San Domingo, Jamaica). Crocodilus Americanus enters some of the streams of Florida. The species having the broadest distribution appear to be Crocodilus porosus, whose range embraces the area included between the North Australian coast, the Indian Peninsula, and China, and C. vulgaris, the common African form, which is found throughout the greater part of the continent, and which has been reported, although doubtfully, also from Palestine. Two species of crocodile, C. robustus and C. Madagascariensis, the one related to the common Indian form and the other to the African, are found on the island of Madagascar.

The alligators (Alligator), also known as caymans and jacarés, and comprising, according to some authors, not less than ten distinct species, are confined to the waters of tropical and sub-tropical America, ranging from the Argentine Republic to Tennessee. The single species of the United States is the Alligator Mississippiensis. It is not a little surprising, seeing the presence there of crocodiles, that alligators should be almost wholly absent from the West Indies; one species (A. latirostris) is said to inhabit the island of Guadeloupe.

Geologically the crocodiles represent an ancient group, dating their first appearance, as far as is yet known, from the Triassic period. Three genera of this age are recognised: Stagonolepis, from the Elgin sandstones of Scotland, Belodon, from Würtemberg, the

---

* Dr. Gray, in his " Catalogue of the Shield Reptiles of the British Museum " (1872), makes seven genera, of which Oopholis is Asiatic and Australian, Bombifrons Asiatic, Palinia and Molinia American, and the remainder, Crocodilus, Halcrosia, and Mecistops, African. It is questionable whether any of these forms is entitled to generic distinction.

Eastern United States, and India, and Parasuchus, from India. In the deposits of the succeeding Jurassic age the number of distinct types and species is very largely increased. No less than forty species, belonging in the main to the genera Mystriosaurus, Teleosaurus, Steneosaurus, Metriorhynchus, and Dakosaurus, are known from British strata alone.[121] Many of these are also found in the deposits of the continent of Europe, which comprise a considerable number of additional types. The amphicœlous, or biconcave, type of vertebra, distinctive of the Triassic and Jurassic crocodilians, is retained in a measure by the Cretaceous forms, as in Goniopholis and the American Hyposaurus,* but we now also meet, and for the first time, with the type of the modern procœlian crocodile. Gavialis and Crocodilus, abundantly developed as Tertiary forms, both occur in the Upper Cretaceous beds of Europe, and are represented in the nearly equivalent American deposits by the gavialine genera Holops and Thoracosaurus. Tomistoma is found in the Miocene of Malta and Lower Austria. Alligator does not appear before the Tertiary (Eocene) period (Europe and America). A gavialine form from the Siwalik deposits of India, Rhamphosuchus crassidens, is supposed to have attained a length of from fifty to sixty feet.

The origin of the crocodilian line is involved in much obscurity. Whether or not the animals of this group stand in direct genetic relation with some of the earlier labyrinthodonts, as is maintained by some paleontologists, our present knowledge does not permit us to determine. Among themselves, however, the different crocodilian types exhibit a remarkable gradational series of structural peculiarities, which connect the most ancient and the modern forms, and place them in an almost unbroken sequence. Professor Huxley has indicated the line of succession as passing from the Parasuchia —the Triassic forms, in which neither the palatine nor pterygoid bones enter into the formation of secondary posterior nares—through the Jurassic and Cretaceous Mesosuchia, in which the palatines alone are produced to form these nares, to the modern and Upper Cretaceous (procœlian) Eusuchia, in which both bones are similarly produced. M. Dollo recognises in Bernissartia, a recently discov-

---

* Hyposaurus Rogersi, from the "greensands" of the Eastern United States, was until recently the only known species of the genus; a second species, H. Derbianus, has been described by Professor Cope from the Province of Pernambuco, Brazil ("Trans. Am. Phil. Soc.," Jan., 1886).

ered form from the Cretaceous deposits of Belgium, the ancestral type of the short nosed modern crocodilians—*i. e.*, the crocodile and alligator.

BIRDS.

The principal features connected with the geographical distribution of birds having been discussed in the early part of this work, only the geological distribution of the class will be considered here.

The earliest known birds are the Archæopteryx, whose remains have thus far been found only in the Solenhofen limestone (Upper Oolite) of Bavaria, and the Laopteryx priscus, from a nearly equivalent horizon of the Western United States (Wyoming Territory). The latter, which was of about the size of the great blue heron (Ardea Herodias), is apparently a member of the heterogeneous group designated by Marsh the Odontornithes, or toothed-birds, to which the more remarkable of the American (Middle) Cretaceous birds, Ichthyornis, Hesperornis, and Apatornis, belong. Ornithic remains, with somewhat doubtful relationships, and referred to the genera Graculavus, Laornis, Palæotringa, and Telmatornis, have also been obtained from a somewhat higher horizon (Upper Cretaceous) in the Eastern United States (New Jersey). Almost the only clearly determined bird-remains of this period occurring in Europe are those of Enaliornis (Pelagornis; Upper Greensand of Cambridge), which appears to have had some resemblance to a penguin.

In the Tertiary deposits remains of this class are very much more numerous, and there is a close approximation to modern type-structures. Thus, in the Eocene deposits of the Paris Basin and elsewhere in France (Auvergne, Provence, Languedoc) we find the remains of the true quail (Coturnix), grouse (Tetrao), cormorant, godwit, rail, sandpiper, nuthatch, and falcon, associated with which are a number of forms whose relationships have not in all cases as yet been absolutely determined. The most remarkable of these is probably Gastornis Parisiensis, a bird of about the stature of the African ostrich, but possessing so many well-marked anatine characters as to have induced some naturalists to class it with the ducks and geese.* Agnopterus and Elornis appear to have repre-

---

* Gastornis Klaasseni, a bird apparently exceeding the ostrich in size, has recently been described by Mr. Newton from the Lower Eocene strata of Croydon, England ("Proc. Geol. Assoc.," Feb., 1886).

sented the flamingoes, and Palæocircus and Palæortyx, as is indicated in their names, the raptorial and gallinaceous birds respectively. It is not a little remarkable that Leptosomus, the type of
a small family now absolutely restricted to the island of Madagascar,
should constitute a part of this fauna. The deposits of the Swabian
Alps have yielded a limited number of bird-remains (harrier, cormorant), and so have those of Glarus, Switzerland, whence was
obtained the nearly perfect skeleton of the passerine form known
as Protornis or Osteornis.

The equivalent, or nearly equivalent, deposits of the London
Basin, the island of Sheppey, and of Hempstead, in the Isle of
Wight, have also yielded a number of avian forms, some of which
appear to have been most intimately related to types now living,
such as the herons, gulls, and kingfishers (Halcyon or Halcyornis).
But here, as in the Paris Basin, there occur several distinct types
whose position among living forms it is very difficult or impossible
to establish. Such are the Megalornis, a bird somewhat smaller in
size than the emu ; Dasornis, which apparently combines true
struthious characters with those of the recently exterminated moas
of New Zealand; Macrornis, also with struthious characters; and
the very singular anatine Odontopteryx toliapicus, recalling in its
dental armature the Cretaceous toothed-birds of America. All the
older Tertiary bird-remains that have thus far been described
from the American continent are from the Western United States,
and belong in principal part to the gruiform genus Aletornis (Wyoming), some of whose species appear to have attained to nearly
the stature of the sand-hill crane. A true owl (Bubo leptosteus),
about two-thirds as large as the great horned-owl (B. Virginianus),
represents the birds of prey, and the passerine Palæospiza bella the
songsters; the former is from Wyoming (Eocene), and the latter
from the insect-bearing shales of Florissant, Colorado (Oligocene?).
A giant struthious bird, combining some of the characters of the
extinct moas, has been described by Professor Cope from the Eocene
deposits of New Mexico, as Diatryma gigantea, a form not unlikely generically identical with the European Gastornis.

Ornithic remains are much more abundant in the Miocene deposits than in the Eocene, and there is a corresponding further
approximation to modern type structures. From the lacustrine
deposits of Central and Southern France, whence the greatest

number of distinct types has been obtained, upwards of fifty species have been described, the greater number of which are referable to the modern genera Aquila (eagle), Haliaetus (fishing eagle), Milvus (kite), Bubo (owl), Picus (woodpecker), Corvus (crow), Motacilla (wagtail), Passer (sparrow), Columba (pigeon), Rallus (rail), Phœnicopterus (flamingo), Grus (crane), Ardea (heron), Ibis, Totanus (tattler), Numenius (curlew), Tringa (sandpiper), Larus (gull), Phalacrocorax (cormorant), Sula (gannet), Pelecanus, and Anas (duck). The occurrence of a parrot (Psittacus) and of several species of pheasant (Phasianus; also in Greece) is rather re· markable, since the former is no longer an inhabitant of the European continent, or of any adjoining tract, and the latter is generally conceived to have been a modern introduction from Asia. Several of the generic types found in France have also been recognised in the deposits of South Germany (Steinheim, &c.). The Siwalik Hills formation of India has yielded the remains of two species of pelican, a cormorant, stork (Leptoptilus), merganser, ostrich (Struthio Asiaticus) and emu (Dromæus? Sivalensis). With reference to the occurrence in India of the last named bird, whose relationship with its living Australian congener, Dromæus Novæ-Hollandiæ, is very intimate, Mr. Lydekker says : "The former occurrence in India of a large struthioid closely allied to the emu is one more instance of the originally wide distribution of the struthioid birds; and it not improbably indicates that the home of the group of which the cassowaries, emus, and moas are diverging branches, was originally somewhere in the neighbourhood of the Indian region, whence a migration took place during some part of the Tertiary period towards the southeast, where the group, in regions more or less completely free from the larger mammals, subsequently attained its greatest development." [122] The American Miocene birds are limited to some four or five species, a turkey (Meleagris antiquus ; Colorado), nearly as large as the common wild species (M. gallopavo), gannet, shearwater, and guillemot.

The Pliocene and Post-Pliocene birds of the continent of Europe are much less numerous than the Miocene, and in the greater number of cases do not admit of absolute determination. Several species of waders, swimmers, and gallinaceous birds (Gallus, Scolopax, Anas, Anser), more or less intimately related to existing forms, have been described from England, France, and Germany. The mallard (Anas boschas) and grey lag-goose (Anser cinereus) both appear to be represented in the later deposits. Among the cave deposits of France have been discovered the remains of the snowy-owl (Nyctea Scandiaca) and willow-grouse (Lagopus albus), northern forms which appear to have followed the southward migration of the reindeer, and of a large extinct species of crane (Grus primigenia). Two or more species of swan have been found in the ossiferous cavern of Zebbug, in the island of Malta, one of which, Cygnus Falconeri, an extinct form, exceeded by about one-third the dimensions of the common C. olor.

Of the group of sub-fossil birds, or those whose remains belong to a comparatively very recent period, are the giant struthious birds of New Zealand, known as "moas" (Dinornis and Mionornis, with some seven or more species), and the palapteryxes (Palapteryx and Euryapteryx); the Æpyornis maximus of Madagascar; and the Australian Dromæornis australis, the precursor of the modern emu. A giant goose (Cnemiornis), associated with which are the remains of several remarkable ralline forms (Aptornis and Notornis—the latter surviving up to our own period), and a number of other birds, also occur in the newer deposits of New Zealand. In this connection may be mentioned, although not strictly falling under the category of fossils or sub-fossils, the recently exterminated didine birds of the Mascarene Islands—the dodo (Didus ineptus) of Mauritius, and the solitaire (Pezophaps solitarius) of Rodriguez; the crested parrot of Mauritius (Lophopsittacus Mauritianus), and the Aphanapteryx, an abnormal ralline species, from the same islands.

MAMMALIA.

Monotremata.—This, the most limited, order of terrestrial Mammalia, forming the sub-class Ornithodelphia of most naturalists, comprises two families, the Ornithorhynchidæ, or duck-bills, and Echidnidæ, or Australian hedgehogs, the former of which is restricted to the continent of Australia and Tasmania, and the latter to the same region with the addition of New Guinea. The duck-bills are represented by a single species, the platypus or water-mole of the colonists (Ornithorhynchus paradoxus or anatinus). No fossil remains referable to this genus have as yet been discovered.

The Echidnidæ comprise two recent genera : Echidna, with

three or four more or less clearly defined species (E. hystrix or aculeata, E. cetosa, E. acanthion *) inhabiting Australia and Tasmania, and a single one (E. Lawesii) New Guinea; and Acanthoglossus, represented by A. Bruijnii, from Northern New Guinea. Fossil remains of this family are not numerous, and belong exclusively to the Post-Pliocene deposits of the Australian continent. Echidna Oweni is founded upon a portion of a humerus from Darling Downs, and indicates an animal considerably larger than the common recent species. E. Ramsayi, from a breccia cave in Wellington Valley, is likewise founded upon a humerus.

Marsupialia.—All the existing members of this order, if we except the single family of American opossums (Didelphidæ), are restricted to the Old World, and are in the main confined to the Australian continent and New Guinea, a limited number of forms finding a habitat in the debatable tract between the Australian and Oriental realms. The general features of their distribution are discussed in the chapter treating of the Australian realm. The order is not represented in either of the continents of Eurasia or Africa.

The opossums comprise a considerable number of species, the majority of which are confined to South and Central America; two species, Didelphys Virginiana and D. Californica, are found in the United States, the former, the common American species, ranging from the Gulf border to the State of New York. An aberrant web-footed form, the yapock (Chironectes), inhabits South and Central America.

Marsupial remains in deposits older than the Tertiary are not abundant, and are in the main comprised in a number of genera whose exact relationships have not as yet been absolutely determined. Indeed, it is not a little doubtful whether the earliest forms usually referred to this order—those from the Trias—actually belong here, or represent an even more primitive type of mammal. To this category of uncertain forms may be referred the Microlestes antiquus, from the Keuper of Germany, M. Moorei and Hypsiprym-

---

* Described by Collett from North Queensland ("Forh. Selsk. Christiania," 1884). Lütken indicates the possible existence of a fourth Australian species ("Proc. Zool. Soc.," London, 1884, p. 150), while Dubois defines a supposed new species, named Proechidna villosissima, from New Guinea ("Bull. Mus. Belg.," iii., p. 109).

nopsis Rhæticus, from the Rhætic deposits of Somersetshire, England, and Dromatherium sylvestre, from the Chatham coal-fields of North Carolina. Of less doubtful affinity are Tritylodon, from the Triassic deposits of South Africa, and the numerous forms whose fragments have been obtained from the British Oolites and the island of Purbeck—Amphitherium, Phascolotherium, Stereognathus, Spalacotherium, Amblotherium, Stylodon, Triconodon, Triacanthodon, Plagiaulax—and from the nearly equivalent deposits of the Western United States (Diplocynodon, Stylacodon, Tinodon, Triconodon, Dryolestes, Ctenacodon). Many, or most, of these forms appear to depart to a certain extent from the normal type of marsupial structure—approximating the Insectivora — hence, by some naturalists, as Professor Marsh, they are relegated to distinct groups—Pantotheria and Allotheria—supposed to have no living representatives.* The Marsupialia are not represented in the Cretaceous deposits ; Meniscoessus, a form whose closest relationship appears to be with the Jurassic Stereognathus, occurs (in association with dinosaurian remains) in the Laramie formation of the Western United States, the position of which in the geological scale, as has already been intimated, is more properly with the Cainozoic than with the Mesozoic series.

The Tertiary marsupial remains of the Northern Hemisphere belong principally to the earlier periods, beginning with the oldest Eocene; in Europe they have not been recognised higher than the middle Miocene, and in North America, if we except the pygmy opossum (Didelphys pygmæa) from the Miocene of Chalk Bluffs, Colorado, no representative is known from deposits newer than the Oligocene (White River beds). Barring the opossums,† whose earliest remains have been found in the Eocene deposits of both

---

* According to Professor Seeley, Hypsiprymnopsis, which appears to be most intimately related to the modern kangaroo-rat (Hypsiprymnus), is founded on the premolar teeth of Microlestes. The same authority recognises in Amphitherium and Phascolotherium a strong combination of marsupial and insectivore characters, and the indications of a "generalised insectivorous type, modified from a monotreme stock in the direction of the marsupial plan" (Phillips's "Manual of Geology," i., p. 520, 1885).

† Separated by some authors from the genus Didelphys as Peratherium and Amphiperatherium ; Peratherium, whose range extends into the Miocene, is represented by five or more species in the White River deposits of Colorado, the largest of which about equals in size the mole (Scalops aquaticus).

France and England, none of the recent families are indicated. The herbivorous type seems to be entirely wanting in the European deposits, but in America, the forms that have been described from the Puerco formation of New Mexico as Polymastodon are referred to this type by Professor Cope. Neoplagiaulax, from the basal Eocene beds of France and New Mexico, and its near ally Ptilodus, represent the Jurassic Plagiaulacidæ, and appear to effect a partial transition from these to the Pliocene or Post-Pliocene Thylacoleo of Australia.

The remains of true kangaroos (Macropodidæ), some of them, as Palorchestes, considerably exceeding in size the largest of the modern representatives of the family, occur in the newer Pliocene or Post-Pliocene deposits of the Australian continent. Associated with these are a number of remarkable forms whose precise affinities still remain to be determined, although in the general character of their dentition they approximate the kangaroos and phalangers. Diprotodon australis, with less disproportionate limbs than in the kangaroos, appears to have exceeded the rhinoceros in size. Of somewhat smaller dimensions are the species of Nototherium. Thylacoleo carnifex, described as "one of the fellest and most destructive of predatory beasts," is held by many naturalists to have been an herbivore.

Edentata.—The animals of this order are at the present day confined almost wholly to the southern continents—indeed, it might be said principally to the continent of South America (with Central America), which possesses more than three-fourths of all the known species. Of the five recognised families, the sloths (Bradypodidæ), ant-eaters (Myrmecophagidæ), and armadillos (Dasypodidæ), are exclusively American; the aard-varks (with two or three species— Orycteropus Capensis, the Cape ant-eater, O. Æthiopicus, from Northeast Africa, and a possible third species from Senegal) are African; and the scaly ant-eaters or pangolins (Manididæ), both African and Asiatic. The species of the last, some eight or more, are properly referable to a single genus, Manis, although several sections, by some authors considered to be of generic value, have been constituted to receive certain well-marked, but unimportant, peculiarities of structure. The common pangolin (Manis penta-dactyla) inhabits the Indian peninsula and the island of Ceylon, sharing in part the distributional area of the Chinese species (M.

aurita), whose range extends from North India to the island of Formosa. The Javan pangolin is a native of Burmah, the Malay Peninsula, and the larger islands—Java, Borneo, Sumatra—of the Eastern Archipelago. A limited number of species are known from Western Africa, one of which, M. (Pholidotus) gigantea, measures about five feet in length to the tip of the tail. The most aberrant form of the family is M. (Smutsia) Temminckii, from the southern and eastern portions of the African continent.

Of the American groups the most restricted in point of numbers are the ant-eaters, whose range, collectively, embraces the greater portion of the Neotropical realm included between Mexico and Paraguay, east of the Cordilleras. The better known species—indeed, the only species admitted by most authors—are the great ant-eater (Myrmecophaga jubata), the tamandua (Tamandua tetradactyla), and the little or two-toed ant-eater (Cycloturus didactylus), whose individual ranges coincide largely with the range of the entire family. The sloths, of which some authors recognise not less than a dozen fairly well-marked species or varieties, occupy much the same area as the ant-eaters, although they do not appear to enter Paraguay. They are inhabitants of the forest region, which limits their distribution. Two genera, founded upon the number of toes on the fore-feet, are generally admitted : Bradypus, the three-toed sloths, and Cholœpus, two-toed sloths, both of which are very extensively distributed.*

The armadillos (Dasypodidæ), which comprise nearly twenty clearly defined species, are the most broadly distributed of the American edentates, their range extending from the most northern limits of the Neotropical realm to the fiftieth parallel in Patagonia. A single species, the peba or seven-banded armadillo (Tatusia septemcincta), which is found in South America as far south as Paraguay, enters the United States in Texas. Among the better known species are the six-banded armadillo or encoubert (Dasypus sexcinctus), an inhabitant of Brazil and Paraguay; the tatouay or cabassou (Xenurus unicinctus), with much the same range as the last, but extending into Guiana ; the three-banded armadillo or

---

* Arctopithecus appears to have no distinctive generic characters. A specimen of Bradypus tridactylus in the museum of the Royal College of Surgeons, of London, corresponds, according to Professor Flower, with Gray's Arctopithecus gularis.

apar (Tolypeutes tricinctus), which, with the remaining members of the generic group to which it belongs, has the power of rolling itself into a complete ball; and the great armadillo (Priodon gigas), an inhabitant of the forests of Brazil and Guiana, the largest living representative of the family, measuring upwards of three feet from the tip of the nose to the root of the tail.

Fossil remains of edentate animals are not numerous, and are in the main confined to the Pliocene and Post-Pliocene deposits of the New World, especially South America (Pampean formation of the Argentine Republic; bone-caves of Brazil). The oldest known form is Ancylotherium priscum, from the phosphorites of Quercy, France (Oligocene), a generalised type of animal, considered by some authors to stand intermediate between the Edentata and Ungulata; the same genus (A. Pentelici) occurs in the Miocene deposits of Pikermi, Greece. An apparently allied form, Macrotherium, whose remains indicate a possible climber of gigantic proportions, with comparatively feebly developed hinder extremities, is represented by several species in the Miocene deposits of both France and Germany. No New World forms are known to antedate the middle Tertiary period. Moropus, from the Miocene and Pliocene deposits of the Western United States, comprises animals ranging in size between the tapir and rhinoceros, but with uncertain affinities; equally uncertain is the position to be assigned to the Pliocene Morotherium, whose remains have been found at various localities in Idaho and California.

The South American edentate fauna (Pliocene and Post-Pliocene) comprises, according to Gervais and Ameghino, some eighty or more species, the greater number of which belong to genera now no longer living. The better known forms are referable to the families Megatheriidæ and Glyptodontidæ or Hoplophoridæ, the former of which appear to hold an intermediate position between the modern sloths and ant-eaters—combining the head and dentition of the one with the trunk and appendages of the other—while the latter, in the presence of a carapace, approach the armadillos. Included in the family Megatheriidæ, besides other forms, are the genera Megatherium, with animals of the size of the rhinoceros,*

---

* Megatherium Americanum, from the Argentine Republic and Paraguay, was only inferior in size to the elephant, far surpassing all other land animals. A mounted skeleton measures eighteen feet in length from the fore part of the head to the tip of the tail.

Cœlodon, Mylodon, Lestodon, Scelidotherium, Plationyx, and Megalonyx, the majority of which embrace species of very robust dimensions. The species are mainly found in the bone-caves of Brazil, and in the Pampean and diluvial deposits of the Argentine Republic and Patagonia. Megalochnus rodens, a diminutive species, is from the island of Cuba. Nothropus priscus, from the Argentine Republic, appears to have possessed arboreal habits, in this respect agreeing more closely with the modern sloths than any of the other forms. Of the North American members of this family the best known species are Megatherium mirabile, a somewhat smaller form than the M. Americanum, from the superficial deposits of the Southern United States ; Megalonyx Jeffersoni, originally described from a cave in Virginia; and Mylodon Harlani, from the Western and Southern United States. A peculiar genus from the deposits of Natchez, on the Mississippi, has been described as Ereptodon.

The glyptodons embrace a considerable number of Pampean species, which by Burmeister and other authors are referred to several distinct genera—Hoplophorus, Panochthus, Dœdicurus, Euryurus, Glyptodon, and Schistopleurum. The best known species are Glyptodon typus and clavipes.

Sirenia (Sea-cows).—This order is at the present day limited to some half-dozen species, referable to two genera: Manatus, the manatees, and Halicore, the dugongs, the former of which is common to both the Eastern and Western Hemispheres, while the latter is strictly confined to the Old World. Of the two American species of Manatus the West Indian sea-cow (Manatus latirostris) inhabits the creeks, lagoons, and estuaries of the north of South America, the West Indies, and Florida, and the Brazilian sea-cow (M. Americanus or inunguis) the South American coast-line to about the twentieth parallel of south latitude, and the more important Brazilian rivers, very nearly to their sources.* The only Old World form (M. Senegalensis) inhabits the West African coast for about ten degrees on either side of the Equator, and the interior as far as, or farther than, Lake Tchad.

Of the three species of Halicore, one (Halicore tabernaculi) is

---

* The identity of the coast species and that of the Upper Amazon and Orinoco Rivers has not yet been absolutely established, but is considered highly probable by Hartlaub (Spengel's " Zool. Jahrb.," 1886).

restricted to the East African coast and the Red Sea, another (Hali-core dugong) inhabits the Indian and Pacific oceans, eastward from the home of the last to the Philippines, and the third (Halicore australis) the waters of Eastern and Northern Australia.

Fossil remains of sirenians, although not very numerous, occur throughout all the Tertiary formations, from the Eocene to the Pliocene, inclusive. The earliest forms are the Eotherium Ægyptia-cum and Manatus Coulombi, from the Mokattam limestone of Egypt, Hemicaulodon effodiens, from the basal Tertiary beds of Shark River, New Jersey, and the somewhat doubtful Halitherium du-bium, from the deposits of the Gironde, France. In the last-named genus are included a considerable number of species from the Mio-cene deposits of Germany, France, Belgium, and Italy, and a single undetermined (?) form from the Isthmus of Suez. Felsinotherium and Chirotherium are Pliocene forms from Central and Northern Italy, and Rhytiodus Capgrandi a species from the nearly equivalent deposits of the Garonne, France. The American fossil sirenians comprise, in addition to the Eocene Hemicaulodon, two or more species from the Miocene deposits of South Carolina, Manatus an-tiquus, M. inornatus, and Dioplotherium Manigaulti; the first also occurs in New Jersey and Virginia. Of the Post-Pliocene forms the best known is the Rhytina gigas or Stelleri, "Steller's sea-cow," an animal which appears to have been fairly abundant about Behring and Copper Islands as late as the second half of the last century, but which is now apparently entirely extinct. The im-bedded remains occur principally in the raised beaches and peat-mosses of Behring Island.

Woodward calls attention to the significant fact that, if we "take the belt of the tropics, that is, $23\frac{1}{2}°$ N. and $23\frac{1}{2}°$ S. of the Equator (or, better still, say 30° N. and S. of the Equator), we shall cover the geographical distribution of all the living sirenians. If we take another belt of 30° north beyond the Tropic of Cancer, we shall embrace the whole geographical area in which fossil remains of sirenians have been met with. Assuming, as I think we may, that the Sirenia at the present day belong exclusively to the tropi-cal regions of the earth, and that Rhytina, in its boreal home, was simply a surviving relic from the past (a sort of geological 'out-lier,' as of a stratum elsewhere entirely denuded away), we must conclude that the presence of about twelve genera and twenty-seven species of fossil Sirenia, as widely distributed then as the recent forms are at the present day, but with a range from the Tropic of Cancer up to 60° of north latitude, affords a most valuable piece of evidence (if such were needed) attesting the former northern extension of subtropical conditions of climate which must have prevailed over Europe, Asia, and North America, in Eocene and Miocene times, and in the older Pliocene also." [123]

Cetacea (Whales, &c.).—The animals of this order are distributed throughout almost the entire oceanic expanse, and a limited number of forms, the members of the family Platanistidæ, and some delphinoids, are also found in fresh or estuarine waters. Platanista Gangetica, an inhabitant of the waters of Northern India —Ganges, Brahmaputra, and Indus, and their tributaries—is entirely fluviatile, never being known to pass out to sea. The only absolutely fluviatile form occurring in America is Inia Geoffrensis, from the Upper Amazonian water system; Pontoporia Blainvillii inhabits the estuary of the Rio de la Plata, but is not positively known to ascend that stream into fresh water.

Of the marine cetaceans two distinct types are usually recognised by naturalists: the whalebone or toothless whales (Mystacoceti), as represented by the right-whales (Balæna), rorquals, or finwhales (Balænoptera), and the humpbacks (Megaptera), and the toothed-whales (Odontoceti), which comprise the sperm-whales, dolphin, porpoise, grampus, &c. The right-whales, which are confined principally to the northern and southern seas, have been divided into some half-dozen species, which, however, so closely resemble one another that not improbably they represent only varietal forms of one and the same species. The best known is the Greenland right-whale (Balæna mysticetus) of the Arctic seas; other northern forms are B. Biscayensis and B. Japonica. The southern species or varieties are B. australis, of the South Atlantic, and B. antipodarum and B. Novæ-Zelandiæ, of the South Pacific. A nearly equal uncertainty attaches to the different varieties or species of rorquals and humpbacks, especially the latter, some forms of which are found in almost every sea. Four or more apparently distinct types of rorqual inhabit the northern seas, but whether these are absolutely separable from their antipodal congeners still remains to be determined. Much confusion exists as to the synonymy of the species; hence, the great difficulty of their identification. The members of this genus are probably the largest of all living animals, some of the forms, as Balænoptera Sibbaldii and B. sulfurea, attaining a length of eighty or perhaps even a hundred feet. The smallest of the known species of whalebone whales is the rare Neobalæna marginata, from the Australian and New Zealand seas, which attains a greatest length of about twenty feet.

Of the toothed-whales, other than the members of the family Delphinidæ (dolphins, porpoises, &c.), the best known and probably most widely distributed species is the cachalot or sperm-whale (Physeter macrocephalus), a giant form measuring upwards of sixty feet in length, whose habitat is more properly the tropical and subtropical seas, the animal but rarely appearing in the polar waters. More or less closely related forms of the same family (Physeteridæ) are Kogia, Ziphius, and Mesoplodon, the species in each group of which have either individually or collectively a very broad extension.* Two species of bottle-nose whale (Hyperoodon rostratus and H. latifrons) inhabit the North Atlantic.

Of the delphinoid type of cetaceans the most numerously represented genus is Delphinus, the dolphin, or, as it is frequently miscalled, porpoise, the numerous species of which are distributed throughout most seas, a limited number even habitually ascending some of the larger streams, as the Amazon. The type-form of the genus is the common or Mediterranean dolphin, the *hieros ichthys* or sacred fish of the ancients (D. delphis), which is also abundant in the Atlantic, and of which closely allied, if not identical, forms are found in the Australian seas (D. Forsteri) and in the North Pacific (D. Bairdii).[124] One of the most northerly species of dolphin is the tursio, or nesarnak of the Greenlanders (D. tursio), which inhabits the Atlantic between Greenland and the European shores. Modifications of the ordinary delphinoid type are seen in the long-beaked forms of the group Steno, and in a South Sea species, Leucorhamphus (Delphinapterus) Peronii, in which there is no dorsal fin. The bottle-heads (Globicephalus) are inhabitants of nearly all

* Much diversity of opinion exists as to the number of species belonging to the different genera. Thus, while Gray recognised not less than six species of Kogia, founded upon about as many individual specimens, only one (Kogia breviceps, found in the South and North Pacific oceans) is admitted by Flower (" Encycl. Brit.," 9th ed., xv., p. 396). The same authority likewise considers the species of the other genera as being in great part founded upon insufficient characters.

seas, especially the north and south temperate, and exhibit a marked specific identity between the most widely removed forms (Australia and North Atlantic). The type of the genus is the pilot-whale (G. melas; Delphinus globiceps), the grindhval of the Faroe-Islanders, whose distribution is practically coextensive with the northern seas. Related to the preceding are the so-called grampuses of the genus Grampus, of which only one species (G. griseus), remarkable for the variability of its colouring, has been thus far clearly determined; it inhabits the northern ocean, more rarely descending into the Mediterranean. A second form, from the Cape of Good Hope, has been described as G. Richardsoni.

The true grampuses, also known as "killers," from their predacious habits, constitute the genus Orca, and are more nearly related to the true porpoises than to the dolphins proper. They are distributed over the greater portion of the oceanic expanse, from Greenland to Tasmania; but neither the relationships of the different so-called species, nor the limitations of their respective habitats, have as yet been determined. Orca gladiator, the common grampus, is more properly a northern form, and is fairly abundant in the polar seas. Pseudorca crassidens, a much rarer form of grampus, found on the Danish coast, appears to be identical with a species from the Australian waters. The genus Orcella is represented by two species, one of which (O. brevirostris) inhabits the Bay of Bengal, and the other (O. fluminalis), a fluviatile form, the Irrawaddy River, at a distance of several hundreds of miles from its mouth.

Of the true porpoises of the genus Phocæna, whose limited species are confined principally to the waters of the Northern Hemisphere,* the best known and probably most widely distributed form is P. communis, the common porpoise, which inhabits in shoals or schools the North Atlantic, from Britain to Greenland and the American coast, frequently ascending the outflowing streams for a considerable distance above their mouths. They have been observed on the Thames, at London, and appear to have occasionally penetrated up the Seine as far as Paris. The animal does not seem to enter the Mediterranean. By some naturalists the common porpoise of the Atlantic coast of the United States is considered to be a distinct species, to which the name P.

---

* Phocæna spinipennis has been described from the mouth of the Rio de la Plata.

Americana has been applied. A peculiar form of porpoise, from the Indian Ocean (?) and the Japanese coast, destitute of the dorsal fin, has been described as Neomeris phocæniformis.

Of the remaining delphinoids the narwhal, or sea-unicorn (Monodon monoceros), inhabits the Arctic Ocean, rarely passing south of the sixty-fifth parallel; the beluga, or white whale (Delphinapterus leucas), closely related to the last, is also an inhabitant of the Arctic Ocean, descending on the American coast to the St. Lawrence River, and more rarely, in European waters, to the shores of Scotland.

Excepting the Palæocetus Sedgwicki, from the boulder clay of Roswell Pit, Ely, England, whose remains were encased in a matrix supposed to be Upper Jurassic (Kimmeridgian), the earliest cetaceans of whose organisation we know anything are the zeuglodons, which apparently represent a type intermediate between the toothed and toothless forms of the present day. They occur in the Upper Eocene deposits (Jacksonian) of the Southern United States; one species (Zeuglodon Wanklyni) has also been discovered in the equivalent Barton sands of England, and two, corresponding to the American forms—Z. macrospondylus and Z. brachyspondylus—in the deposits (Eocene or Oligocene) of Birket-el-Keroun, Egypt.*
Closely related in dental characters to the zeuglodons, but differing in well-marked cranial features, are the squalodons, whose remains are abundantly scattered throughout the Miocene and Pliocene deposits of many parts of continental and insular Europe (Vienna Basin, France, Antwerp and Suffolk Crags, &c.). They have also been noted from nearly contemporaneous strata in North America (Squalodon Atlanticus, from Shiloh, New Jersey †) and Australia.

The oldest known form of modern-type cetacean is Balænoptera Juddi, from the Oligocene (Headon) beds of the Hampshire basin, England; no other representative of an existing genus has thus far been found to antedate the Miocene period. Both the toothed and

---

* Dames suggests that the two forms of Müller may only represent the male and female of a single species, which would then be the Zeuglodon cetoides of Owen (" Sitzungsb. Berl. Ak.," 1883).

† The forms described from the Eocene deposits of the Ashley River, South Carolina, are considered by Van Beneden and Gervais to be only doubtfully referable to this genus.

toothless whales are represented in the deposits of this age (Miocene), the latter, however, apparently only by the rorquals or their immediate allies (Balænopteræ); Cetotherium seems to have occupied a position intermediate between the Mystacoceti and Odontoceti. Of the latter the earliest representatives appear to have been the genera Ziphius and Mesoplodon, although not improbably the true dolphin was an immediate contemporary. The baleen whales proper are not known before the Pliocene, when, in addition to Balæna, and possibly Neobalæna, we meet with a number of extinct types more or less closely related to these—Balænula, Idiocetus, Plesiocetus. Balænotus, from the Antwerp Crag, is a connecting form between the baleen whales and the rorquals, and Bartinopsis between the rorquals and humpbacks. Cetacean remains are abundant in the Post-Pliocene deposits, and comprise a variety of recent types; the narwhal has been indicated from the deposits of England and Siberia. The Miocene deposits of the Eastern United States have yielded a number of delphinoid remains to which the generic names Priscodelphinus, Tretosphys, Zarachis, Lophocetus, Rhabdosteus, and Ixacanthus have been applied.

Insectivora.—The animals of this order, which comprises barely more than one hundred and thirty to one hundred and forty living species, of which about one-half are true shrews (Soricidæ), are distributed over the greater part of the earth's surface, but are absent from both South America and Australia. The greater number of the nine generally recognised families are limited to comparatively narrowly circumscribed distributional areas, which in some cases are only co-extensive with the sub-regions or provinces of the main zoogeographical divisions. The Galeopithecidæ, or flying-lemurs, which were formerly referred to the true lemurs, and by some naturalists to the bats, are the most aberrant forms of the order, and constitute the type of a distinct sub-order, Insectivora dermoptera. But two species are known, Galeopithecus volans and G. Philippinensis—the former an inhabitant of the forests of the Malay Peninsula, Sumatra, and Borneo, and the latter of the similar districts of the Philippine Islands. The squirrel- or tree-shrews (Tupaiidæ), small arboreal insectivores resembling squirrels in outline and habits, are restricted to the Oriental region, where they range from the Khasia Hills, in India, to Java and Borneo. Of the two genera, Tupaia and Ptilocercus, the latter contains but

16

a single species, the very interesting Bornean pentail (P. Lowii), remarkable for its long quill-shaped tail.—The elephant-shrews (Macroscelidæ), small leaping animals furnished with a trunk-like snout, are confined principally to South Africa; a single species of Macroscelides, to which genus the greater number of species belong, is found north of the Atlas Mountains. Several fossil forms, apparently referable to genera belonging to this family (Oxygomphus, Parasorex, Echinogale), have been described from the Miocene deposits of France and Germany. Likewise restricted to the African continent are the potamogales and golden-moles (Chrysochloridæ) —the former, represented by a single species (Potamogale velox), inhabiting the territory about the Gaboon River, and the latter the region south of the Equator, but more particularly the Cape of Good Hope districts. Neither family is known by fossil representatives. The Centetidæ occupy two widely separated regions of the earth's surface, namely, Madagascar, with the adjoining islands (possibly introduced into Mauritius and Réunion) and the two larger islands of the Antilles, Cuba and Hayti. Most of the forms, with the genus Centetes itself, the Madagascar hedge-hog, occupy the former region; the American species (two) belong to the genus Solenodon, which in certain anatomical points differs so essentially from its nearest allies as to constitute in the opinions of some systematists[125] the type of a distinct family, Solenodontidæ.

Of the three remaining families of insectivores, the shrews (Soricidæ), moles (Talpidæ), and hedge-hogs (Erinaceidæ), the first, which, as has already been stated, comprise about one-half of all the species of the order, have the broadest distribution, embracing, in fact, the entire tract covered by the Insectivora generally. By some naturalists but a single genus (Sorex), apart from the very remarkable web-footed Nectogale from Thibet, is recognised, which is differentiated into a number of more or less well-marked sub-genera, founded upon the number and colour of the teeth, and the bristles on the tail—Crocidura (Old World), Sorex (the entire range), Blarina (Canada to Costa Rica), Neosorex, and Crossopus, the last two amphibious forms of the New and the Old World respectively. Several species referable to the genus Sorex (and Crocidura) have been described from the Miocene formations of France, and similar remains occur in the Quaternary cavern deposits and breccias of both Europe and Asia. One or two extinct genera or sub-genera, Mysaracihne and Plesiosorex,* have also been indicated (Miocene).

The hedge-hogs comprise two genera, Gymnura and Erinaceus (the hedge-hog proper)—the former of which inhabits the Malay Peninsula and some of the islands of the Indian Archipelago, and the latter, with about nineteen species, the greater part of Europe, Asia, and Africa, although wanting in Madagascar, Ceylon, Burmah, Siam, the Malay Peninsula, and the Archipelago. The range of the common European species, Erinaceus Europæus, extends from Ireland and the Shetland Isles (possibly introduced) to Eastern China, and from the sixty-third parallel of latitude in the Scandinavian Peninsula to Southern Italy, Asia Minor, and Syria, ascending the Alps and Caucasus to elevations of 6,000 and 8,000 feet respectively. In view of this very remarkable range, its absence from the New World is not a little surprising. Remains of several species of hedge-hog have been found in the Miocene deposits of France and Germany, but none of the recent species, except E. Europæus, which forms part of the Quaternary cave fauna, are known as fossils. The genus Neurogymnurus, which is probably closely related to Gymnura, is represented by a single species (G. Cayluxi) in the French Eocene.

The moles are generically the most numerous of the Insectivora, although the number of species is comparatively limited. The greater number of the ten or twelve recognised genera are represented by but one or two species. The family belongs almost exclusively to the Holarctic region, through which it is very generally distributed, only a very limited number of species passing beyond the confines of that region into the Oriental tract. The moles proper (Talpa), with about four species, are found throughout nearly the whole of the Eurasiatic tract, the range of the common species, Talpa Europæa, extending from Britain to Japan, and from Scandinavia and Siberia to Italy and the southern slopes of the Himalayas. The water-moles (Myogale) are comprised in two species, one of which, M. Pyrenaica, inhabits the northern valleys of the Pyrenees, and the other, M. Muscovitica, the region of the Don and Volga rivers. The American moles belong in principal part to three genera: Condylura (the star-nosed mole, which in-

* Referred by Trouessart to the Talpidæ ("Catalogue Mamm. Viv. et Foss.," 1881).

habits the Atlantic slope from Nova Scotia to South Carolina, and westward to Oregon), Scapanus, and Scalops, the last (Scalops aquaticus) the common American form, whose range covers the greater part of the North American continent east of the Rocky Mountains. A somewhat aberrant form, Neurotrichus Gibbsii, found in the Western United States (Cascade Mountains to Texas), has its analogue in the Urotrichus of Japan, to which genus it has generally been referred, and from which it differs mainly in the dental formula.—Fossil remains of Talpidæ date back to the Eocene period (Prototalpa, Quercy, France ; Talpavus, Wyoming), but the genus Talpa itself is not known prior to the Miocene; the common European species is found in the Quaternary deposits. Myogale (with Palæospalax and Galeospalax) occurs in the Miocene and Pliocene deposits of France and England respectively. None of the existing American genera have as yet been found in a fossil state. Other insectivorous forms, however, known principally in a fragmentary condition, and not impossibly referable, at least in part, to the type of insectivorous Marsupialia, have been described from the Eocene of Wyoming and New Mexico (Passalacodon, Centetodon, Entomodon, Entomacodon, Triacodon, Esthonyx), and, together with a number of Miocene forms from Dakota and Colorado (Leptictis, Isacis, Ictops), constitute a distinct family, Leptictidæ.

Insectivorous Forms of Doubtful Position.—Numerous insectivorous animals, known largely by portions of their dental armature alone, are found in the older (principally Eocene) Tertiary deposits of France and the western territory of the United States (Wyoming, New Mexico). They are not improbably, as Professor Cope suggests, referable in part to the lemurs, although from their imperfect state of preservation it is in many or most cases impossible to determine their true relationship, whether with the class of animals just mentioned or with the true insectivores. By Trouessart they are all ranged with the Insectivora as the group of the protolemurs. Among the better known of these forms are Microsyops, Palæacodon, Hyopsodus, Sarcolemur, Tomitherium, Notharctus, Necrolemur (supposed by Filhol to have its nearest ally in the galago of Africa), Adapis (Palæolemur), and Protoadapis, the last three from the Eocene of France, and representing distinctively lemuroid types. Galerix is from the Miocene deposits of the same country. In the

North American lower Eocene genus Anaptomorphus (A. æmulus and A. homunculus), which comprises animals of about the size of the ground-squirrel (Tamias), and whose dentition approximates that of the higher apes and man, Professor Cope recognises the most simian type of lemur yet discovered, and believes that it "represents the family from which the anthropoid monkeys and men were derived. Its discovery is an important addition to our knowledge of the phylogeny of man." [126]

Cheiroptera (Bats).—Bats are practically of world-wide distribution, being found almost everywhere over the continental tracts where there is a sufficient supply of insect food. The number of species is very much greater in the region of the tropics than in the temperate zone—probably three times as great—the specific and individual diminution corresponding to a marked elevation of latitude being very rapid. Vesperugo noctivagans alone among the American bats appears to reach the fifty-fifth parallel, but in Eurasia one or more forms (Vesperugo borealis) penetrate to the Arctic circle. Most of the species inhabiting the region of elevated mountain-chains do not seem to ascend to any very great altitude, preferring to remain in the basal zone of from 4,000 to 6,000 feet; a few instances are noted of habitation at nearly twice this height. Vesperugo montanus and Molossus rufus have been observed on the Peruvian Andes (Huasampilla) at an elevation of 9,000 feet; Vespertilio muricola on the Himalayas at 8,000 feet; and Vespertilio oxyotus on the slopes of Chimborazo at nearly 10,000 feet. Vesperugo maurus is in Europe found chiefly in the region of the higher Alps.

Bats, differing from all other mammals, are found in most of the oceanic islands, but none have so far been observed in either Iceland, St. Helena, the Galapagos, Kerguelen Land, or the islands of the Low Archipelago. [127] Three species inhabit the Bermudas— Vesperugo noctivagans, Atalapha cinerea, and Trachyops cirrhosus (the last a vampyre *)—but only a single one the Azores—Vesperugo Leisleri. It would thus appear that the members of this order of animals were specially endowed with the power of crossing broad arms of the sea, standing, in this respect, next to the birds. It is more than probable, however, that, in the case of the species in-

---

* An individual of Molossus rufus, var. obscurus, is catalogued by Dobson as coming from Bermuda.

habiting the last two named island groups, their appearance there is an accidental circumstance, depending upon the prevalence of certain storm-winds by means of which an unlooked-for transport has been effected; and even in the case of the species found in the Pacific islands, it is not exactly unlikely that the broad distribution has been brought about in a manner not necessarily indicating sustained flight, or at a period when the physical configuration of that portion of the earth's surface—the relation of land to water —was different from what it is at the present day. For, as Mr. Dobson suggestively points out, if the "Cheiroptera possess great powers of dispersal, it is certain that quite nine-tenths of the species avail themselves of them in a very limited degree indeed, and it is significant that the distribution of the species is limited by barriers similar to those which govern it in the case of other species of mammals."* Thus, it is shown that out of a total of somewhat more than four hundred recognised species only about twenty-five pass beyond the confines of the regions to which they properly belong—i. e., about ninety-five per cent. are characteristic regional forms; and of this number more than two-thirds belong to the pre-eminently wandering family Vespertilionidæ, which has by far the broadest geographical distribution of any family of the order. If, however, it be urged that this restriction is not so much due to any inability on the part of the animals to migrate, but to considerations connected with altered conditions of food and climate—the influence of which must be very marked—we have the still more salient fact presented that of the numerous flying-foxes (Pteropus) which inhabit Madagascar and the Comoro Islands, not a single species is found on the east coast of Africa, the narrow channel of one hundred and eighty to two hundred miles which intervenes between the continent and Great Comoro Island seemingly being sufficient to form an effectual barrier to a westward migration. Still, in this special instance we are, perhaps, not presented with a just criterion of the actual powers of dispersion possessed by this class of animals, since the slow and laboured flight of the large flying-foxes can scarcely be compared with that of the smaller insectivorous species; indeed, there is a striking similarity between the insectivorous bat fauna of Africa and Madagascar.

Instances of very broad specific distribution among the Cheirop-

* "Rept. Brit. Assoc.," 1878.

tera are numerous, and perhaps most notably so in the case of the genus Vesperugo, of the family Vespertilionidæ. Vesperugo noctula, the noctule, is distributed throughout the greater part of the Old World, from England to Japan, and from the Scandinavian peninsula to Southern Africa; it extends through India to Ceylon and the islands of the Malay Archipelago. Vesperugo abramus, whose home is primarily the Oriental region, extending from Japan to Northern Australia, is found during the summer months throughout Middle Europe, and even as far north as Sweden; the species furnishes us with a remarkable example of a true migrant. The range of V. maurus extends from the Canary Islands (Palma, Teneriffe) through Central Europe (Switzerland, the Tyrol) to China and Java, and that of Miniopterus Schreibersii from Southern Europe (Spain, Italy) to Japan and the Philippine Islands, and throughout the whole of Africa (with Madagascar) eastward to Australia. The last is probably the most widely distributed of all known species of bats, with the exception of the little serotine (Vesperugo serotinus), whose distributional area covers nearly the whole of Eurasia, Northern and Central Africa, and, in the New World, the American continent from Lake Winnipeg to Central America, and the West Indian islands. This is the only species that has been thus far positively identified as being common to both the Eastern and Western Hemispheres.* Of the strictly American species Atalapha Noveboracensis, in its several varietal forms, appears to be the most widely distributed, ranging from the Aleutian Islands to Chili.

Of the six families into which the Cheiroptera have been divided only two, the Vespertilionidæ and the Emballonuridæ, are common to both the Eastern and Western Hemispheres; the former comprise some one hundred and sixty or more species, fully three-quarters of which are confined to the Old World, over which they are very extensively distributed. This is the most broadly distributed of all the families, and is that which has the most northerly range. Of its sixteen or more genera, at least five of which —Antrozous, Nycticejus, Atalapha, Natalus, and Thryoptera—are peculiar to America, only two, Vesperugo and Vespertilio, the

---

* Vesperugo abramus is thought by Dobson to be possibly identical with a species (Scotophilus hesperus of Allen) from Vancouver's Island ("Cat. Cheir. Brit. Mus.," p. 229).

former with about fifty species, and the latter with about forty, approach cosmopolitanism. Of the Emballonuridæ but a single genus, Nyctinomus, is common to both hemispheres, and its early differentiation is shown in the fact that, while all the American forms are closely related to one another, they depart widely from their European representatives.

The Phyllostomidæ, or simple leaf-nosed bats, which comprise the vampyres, number upwards of fifty species, all of them very closely related, and, with one exception—Trachyops cirrhosus, which has been noted from the Bermudas (and also doubtfully recorded from South Carolina)—confined to the Neotropical realm, over the forest-covered tracts of which they range from Mexico to about the thirtieth parallel of south latitude.* Vampyrus spectrum, the best known species of vampyre — whose habits appear to be mainly frugivorous—and the largest of all the American bats, is distributed over the greater portion of the tract covered by the entire family.

Of the strictly Old World families of bats the Pteropodidæ, fruit-eating bats or flying-foxes, are specifically the most numerous, comprising about seventy species, distributed between the Australian, Oriental, and Ethiopian realms, and some of the intervening tracts, with a preponderance of species in the first-named region. They are restricted almost wholly to the region of the tropics, where a continuous supply of tree-fruits might be obtained; no species has thus far been noted from either New Zealand or Tasmania. Cynonycteris, alone of the genera, has the distribution of the entire family. The most largely represented genus is Pteropus, which includes more than one-half of all the recognised species belonging to the family; its range extends from the Comoro Islands on the west to the Navigators' Islands, in the Pacific, on the east, and through much the greater portion of the Oriental and Australian regions ; but few of the island groups of the Pacific—Sandwich Islands, Low Archipelago, Gilbert's and Ellice's groups—are deficient in the members of this genus, to which the largest known forms of bats belong. Pteropus edulis, which inhabits the islands of the Malay Archipelago, measures five feet in expanse of wing. Only one species, Pteropus medius, the common flying-fox,

---

* Macrotus Californicus or Waterhousii just enters the United States (Fort Yuma, California), but at a point which more properly belongs to the Neotropical than to the Holarctic tract.

has thus far been obtained from the Indian peninsula, and this, singularly enough, is more nearly related to the Madagascan P. Edwardsii than to any of the more eastern species, although separated from its habitat by a continuous water-way of upwards of one thousand miles. The rarity of species in the Indian peninsula is not readily accounted for, seeing how numerous are the individuals belonging to the single species of the genus.*

Of the two remaining families of bats, the Nycteridæ and Rhinolophidæ, the former, comprising some twelve species, are almost wholly restricted to the Oriental and Ethiopian realms, while the latter, numbering about fifty species, are spread over the greater portion of the Old World, from Ireland eastward to Japan and New Ireland, and southward to the Cape of Good Hope. No species appears to have been thus far positively identified from any of the Polynesian islands. Nearly all the species are included in the genera Rhinolophus and Phyllorhina, the former of which has practically the range of the whole family ; Rhinolophus ferrum-equinum, the common horseshoe bat, is distributed over almost the entire tract included between the south of England, Japan, and the Cape of Good Hope. The species of Phyllorhina are confined principally to the tropical and sub-tropical parts of the Old World ; Phyllorhina armigera, the most northerly species, has been found at Amoy, China, and at Mussoree, on the Himalayas, at an elevation of five thousand feet.

* Dr. J. Anderson thus describes the appearance of these animals (" Cat. Mamm. Ind. Mus.," 1881, p. 101, Part I): " This species has been flying for the last few days from the north to the south of the city [Calcutta] in immense numbers, immediately after sunset. The sky, from east to west, has been covered with them as far as the eye could reach, and all were flying with an evident purpose, and making for some common feeding-ground. Over a transverse area of two hundred and fifty yards as many as seventy bats passed overhead in one minute ; and as they were spread over an area of great breadth and could be detected in the sky on both sides as far as could be seen, their numbers were very great, but yet they continued to pass overhead for about half an hour. This is not the first time I have observed this habit in this species ; indeed, it was more markedly seen in August, 1864, while I was residing in the Botanical Gardens, Calcutta. The sky, immediately after sunset, was covered with this bat, travelling in a steady manner from west to east, and spread over a great expanse, all evidently making for one goal, and travelling, as it were, like birds of passage, with a steady purpose."

Fossil remains of Cheiroptera, although of rare occurrence, are found as far back as the upper Eocene, from the deposits of which period a limited number of forms, representing in the main modern genera, have been described. The oldest known forms are Vesperugo Parisiensis, much resembling the broadly distributed serotine, from the gypsum of Montmartre, Vespertilio Morloti from the nearly equivalent deposits of Switzerland, and Vesperugo velox, V. priscus, and Nyctilestes serotinus from the United States. In this family of placental mammals alone do representatives of existing genera extend to such an ancient epoch as the Eocene. A species of Rhinolophus, R. antiquus, has been described from the phosphorites of Quercy, France—usually referred to the Oligocene period—while a generically related form, Palæonycteris, occurs in the Miocene. The Miocene deposits of both France and Germany also contain several species of the genus Vespertilio. Post-Pliocene or late Pliocene cave-remains closely approximate the forms now living in the equivalent or adjoining region. The early specialisation of this class of animals, concerning whose differentiation practically nothing is known, indicates a most ancient line of ancestry, which must be traced far into the Mesozoic era.

Rodentia.—With the exception of the bats the rodents are the only order of terrestrial mammals which can be said to have a nearly universal distribution, being found in all the primary zoogeographical regions but the Polynesian. On the continent of Australia, however, only a single family, that of the mice (Muridæ), is represented, and that by a comparatively limited number of species; the squirrels have a few representatives in the Austro-Malaysian transition-tract.

Of the four great divisions into which the rodents are divided, the myomorphs, or mouse-forms, are numerically much the most important, the family of mice alone comprising considerably more than one-third of the total number of species—some eight hundred or more—belonging to the order. The geographical distribution of this family is practically coextensive with that of the order. The mice proper (Mus), of which there are upwards of one hundred species known, are restricted exclusively to the Old World, over which they are very extensively distributed; they are almost wholly wanting in the Pacific and the greater number of the Austro-Malaysian islands, but are found sparingly in both Australia and New

Guinea. Tasmania has several species, and one or more forms (M. Novæ-Zelandiæ, M. Maorium) appear also to be indigenous to New Zealand; M. Salamonis inhabits the island of Ugi, in the Solomon group. The better known members of the group, the Norway or brown rat (Mus decumanus), and black rat (Mus rattus), whose original home seems to have been Southern or Central Asia, and the common mouse (Mus musculus), probably a native of India, have been spread through man's intervention over the greater portion of the inhabited globe, rapidly displacing in many quarters the indigenous races of similar or allied forms that originally occupied the conquered territory. The black rat, which appears to have been the earliest intruder, is now largely supplanted by the brown species; in England there would appear to be but a single colony left.* The wood-mouse (Mus sylvaticus) and harvest-mouse (M. minutus), the latter the smallest of the European species of mice, are distributed over the greater portion of Europe and Northern Asia. The largest member of the rat tribe is the great bandicoot or pig-rat of the Indian peninsula (Nesokia bandicota), which frequently exceeds one foot in length. Other distinctive rat-like forms of the Old World are the spiny-mice (Acanthomys), which are confined principally to Syria (Palestine), and the east coast of Africa; the jumping-mice (Meriones or Gerbillus) of the continent of Africa, the warmer tracts of Southern and Southwestern Asia, and the steppe region about the Caspian Sea ; and the jumping-rats of Australia (Hapalotis), which recall in general appearance the jerboas.

The Old World forms of the genus Mus are represented in the New World by the vesper-mice (Hesperomys), which very closely resemble them in general character, but differ in certain peculiarities of dental structure, which likewise serve to distinguish most of the American rats; some seventy or more species and varieties have been described, ranging collectively from the Arctic regions to the Strait of Magellan. Hesperomys leucopus, the white-footed or deer mouse, inhabits the greater portion of the North American

---

* While in most regions where the black and the brown rat have been introduced the latter has been rapidly driving out or exterminating the former, it would appear that in some parts of Central Germany the reverse phenomenon is presented—that is to say, the black rat is regaining its ascendancy over the brown species (Magnus, "Sitzungsber. d. Gesell. naturf. Freunde," 1883, p. 47).

continent. The wood-rats, constituting the genus Neotoma, are the largest of the American murine forms, and inhabit the greater portion of the region included between Guatemala and Canada. The cotton-rat (Sigmodon hispidus) is confined principally to the Southern United States, Mexico, and Central America, occasionally, it appears, wandering into the northern portions of South America. In the genus Reithrodon are included a limited number of remarkable leporine forms, which, though differing very essentially in general appearance, do not seem to be distantly removed from the North American harvest-mice of the genus Ochetodon; they are confined principally to the extremity of the South American continent and to Tierra del Fuego.—Interesting modifications of structure or habit are seen in some of the murine forms, as in the partially web-footed water-rats of the Australian region (Hydromys), and in their Brazilian analogues of the genus Holocheilus; again, in the arboreal dormouse-like forms that have been referred to the genera Dendromys (Ethiopian) and Rhipidomys (American).

Of the less murine or rat-like forms of the mouse family the voles or meadow-mice (Arvicolæ), which in a measure replace the true mice in the far north, and on elevated mountain-summits, and whose distribution embraces nearly the whole of temperate and Arctic Eurasia and North America, are probably the most numerous specifically, while in point of individual numbers they far exceed any other mammal, with the possible exception of the closely related lemming. Many of the species enjoy a very broad distribution, but none are known to be common to both the Eastern and Western Hemispheres. Arvicola arvalis or agrestis, the common meadow-mouse, which ascends the Alps to a height of 7,000 feet, is distributed over nearly the whole of Europe (including Italy) and Siberia, its range corresponding approximately with that of A. amphibia (water-vole); A. alpina or nivalis inhabits the region of the higher Alps, between 5,000 and 12,000 feet elevation; on the Finster-Aarhorn it has been observed at an altitude of 4,000 metres. The most broadly distributed of the American species is the common meadow-mouse (A. riparia), whose range extends from the Atlantic to the Pacific, and from the Carolinas to the Hudson Bay territory. A single species, A. quasiater, is known from Mexico. Evotomys rutilus, a form very closely related to the arvicoles, inhabits the circumpolar regions of both hemispheres.

The hamsters (Cricetus) inhabit the greater portion of Europe and Central and Northern Asia, the range of the common species (C. frumentarius or vulgaris) extending from the Rhine to the Obi, and from the Obi and Irtish southward to Persia and the Caucasus; other species inhabit the elevated steppes of Mongolia. Pouched rats allied to the hamsters (Cricetomys, Saccostomus) are also found in various parts of Africa. The lemmings, which are readily distinguished from the field-mice by the hairy covering on the soles of the feet, and their sickle-shaped claws, are the most strictly northern forms of all Rodentia. The better known species are the Scandinavian or Norwegian lemming (Myodes lemmus), the Siberian lemming (M. Obensis), which inhabits the boreal regions of both hemispheres, and the Hudson Bay lemming (Cuniculus torquatus or Hudsonius), an inhabitant of Arctic America, Greenland, and corresponding latitudes in the Eastern Hemisphere; it is also found in Nova Zembla.

A strictly American genus of Muridæ is Fiber, of which the only recognised species is the musk-rat.(F. zibethicus), whose range embraces practically the whole of North America. A closely related, but considerably smaller, form is the recently described Neofiber Alleni, from Brevard County, Florida. Of the non-murine families of myomorphs the dormice (Myoxidæ) and mole-rats (Spalacidæ) belong to the Old World exclusively, the pouched rats (Saccomyidæ) are American, and the true jumping-mice or jerboas (Dipodidæ) both Old and New World forms. The dormice are scattered over the greater portion of temperate Eurasia, from Britain to Japan, and southward over almost the whole of Africa; they appear to be wanting in the warmer parts of Asia (India). The common northern species, Muscardinus avellanarius, is more generally replaced in the south by Glis vulgaris, the "seven-sleeper" of the Germans, whose range extends eastward to the Volga River and Georgia. Of the mole-rats, which are confined almost wholly to the African continent and the tracts comprised in the Oriental region, only a very limited number of species (Spalax) pass within the European boundaries, and these are restricted largely to the southeastern districts (Southeast Russia, Greece, Hungary). Bathyergus maritimus, the "great rodent-mole," inhabits the sand dunes of the Cape coast of Africa. The distribution of the distinctively North American family of pouched rats has already been adverted to in the general treatment of the North American fauna. The jerboas or true jumping-mice are distributed from the eastern confines of the Mediterranean to India, and southward over nearly the whole of Africa. Alactaga jaculus extends its range westward from the Altai Mountains to the Danube, and Dipus sagitta from the far east of Mongolia to the Volga. The largest representative of the family is the Cape jumping-hare (Pedetes Caffer), whose range extends from Mozambique and Angola to the southern extremity of the African continent. Only one trans-Atlantic species is known, the American jumping-mouse (Zapus Hudsonius), which is distributed over almost the whole of the North American continent between Labrador and Mexico.

Of the squirrel forms, or sciuromorphs, the family of squirrels (Sciuridæ), which comprises, in addition to the ordinary and flying-squirrels, also the marmots and prairie-dogs, embraces very nearly all the species belonging to the group. The true squirrels (Sciurus), of which there are probably not less than one hundred species, are extensively distributed over all of the continental divisions of the globe with the exception of the Australian, being limited only or primarily by a deficiency in the forest growth. The headquarters of the genus might be said to be the Oriental region, which holds nearly one-half of all the recognised species; no species are known from either Madagascar or the West Indian islands, although several forms inhabit the larger islands of the Malay Archipelago—Java, Borneo, Sumatra, and Celebes.

In the whole of Europe, excluding the Caucasus, there is but a single species of squirrel, Sciurus vulgaris, whose range extends from the extreme north to the Mediterranean, and eastward throughout Siberia. In the Engadine it ascends the Alps to an elevation of 7,000 feet. North America north of the Mexican boundary possesses six species (and about an equal number of well-marked varieties depending upon size and colouration), of which the most familiar form is the common chickaree (Sciurus Hudsonius), in its several varieties —eastern chickaree, western chickaree, Fremont's chickaree, and Richardson's chickaree—whose range extends from the northern limit of forest vegetation to the highlands of Georgia and Alabama; on the Atlantic border its southern limit appears to be Delaware Bay. This is the only species of squirrel found north of the Canadian boundary. The flying-squirrels are usually separated into two dis-

tinct groups—the flat-tailed forms (Sciuropterus), which are abundantly distributed over the northern parts of the North American
continent, and whose range in Eurasia extends from Lapland to
China and Java; and the round-tailed forms (Pteromys), constituting a more southerly group, whose home is the wooded districts of
tropical Southeast Asia, Japan, and some of the Malaysian islands.
By some authors the separation into two generic groups is not
recognised. The spermophiles or pouched marmots (Spermophilus)
and ground-squirrels (Tamias) are spread over the greater portion
of the temperate and boreal regions of the Northern Hemisphere,
but find their greatest numerical development in the New World.
The former, which in a measure connect the true marmots with
the squirrels, although sufficiently abundant in the far north. of
Siberia and on the most elevated slopes of the Caucasus (S. musi-
cus), appear to be wanting in the Alps. The American species
occupy the western portion of the continent, ranging from the
Arctic seas to the plains of Mexico; none are found east of the
central plains or prairies. Of the ground-squirrels there are some
four or five recognised species, all of which are represented in North
America; the northern ground-squirrel or chipmunk (Tamias Asiaticus) is common to both the Eastern and Western Hemispheres,
ranging in America from Lake Superior and Arizona to the Barren
Lands, and in Eurasia from Saghalien and Japan through Siberia
to the Dwina River. In the eastern portions of the American continent this species is replaced by the common chipmunk or striped
squirrel (Tamias striatus).

The marmots (Arctomys) are restricted to the middle and northern portions of the Northern Hemisphere, and comprise some ten
or more species, three of which are American. Of the last the
best known form is the woodchuck (A. monax), whose habitat extends from the Carolinas to the sixty-second parallel of latitude,
and from the Atlantic border to Minnesota. Of the two European
species the bobac (A. bobac) is more properly an Asiatic form,
ranging from Kamtchatka to the German frontier. The true marmot, the Murmelthier of the Germans (A. marmota or Alpina),
inhabits the mountain-tracts of Southern Europe—Pyrenees, Alps,
Carpathians—between elevations of 5,000 and 10,000 feet. The
American "prairie" or "barking dogs," more properly marmots
(Cynomys), of which there are two species known, appear to be

restricted to the parks and plains of the Rocky Mountain plateau region.

The remaining sciuromorphs comprise the sewellels (Haplodon), leporine rodents, somewhat of the habit of the musk-rat, inhabiting the Northwestern United States; the singular anomalures from Western Africa, which, in the possession of a lateral cutaneous expansion adapted to aerial sailing, recall the flying-squirrels; and the beavers (Castoridæ), of which the only species (Castor fiber or Canadensis) inhabits the northern parts of both the Eastern and Western Hemispheres. In its American home the beaver is still met with in tolerable abundance west of the Mississippi in the entire tract included between Alaska and Mexico; to what extent its range extends into the last-named country has not yet been ascertained. East of the Mississippi it is now but sparingly found south of the Great Lakes, a limited number being still harboured in the Maine and Adirondack wildernesses, and a still smaller number probably finding their way along the thinly settled districts southward to Alabama and Mississippi. In some portions of Virginia and Pennsylvania they appear to be still fairly numerous. The present range of the beaver in Europe is even more restricted than in America, the animal being almost wholly confined to Russia (and Poland), specially the streams of the Ural Mountains and those emptying into the Caspian Sea, and, in isolated colonies, to the Rhone, Weser, Elbe, and Danube rivers. The animal is now extinct in Great Britain, and appears, also, to have completely disappeared from Scandinavia.

The hystricomorphs embrace a number of families whose representatives depart widely from one another in many essential characters ; their greatest development is in the Neotropical realm, which alone possesses the chinchillas, the agoutis, and the cavies, besides the greater number of the partially Ethiopian family of spiny-rats (Echiomyidæ). The best-known representative of the last is the coypu (Myopotamus coypu), a large beaver-like animal found only in Chili, measuring nearly two feet in length. Of scarcely smaller dimensions is the arboreal Capromys pilorides, indigenous to Cuba, where it constitutes the largest native mammalian. Plagiodontia ædium, a member of the same family, also found in San Domingo, appears to be the only indigenous mammal of the island of Jamaica, except the bats and mice (the latter probably introduced).

The chinchillas comprise a limited number of species which are restricted to the Alpine zones of the Peruvian and Chilian Andes (the true chinchillas, C. lanigera, 8,000 to 12,000 feet; Alpine viscachas, Lagidium Peruanum, 10,000 to 16,000 feet), and the pampas between the Rio Negro and the Uruguay (true viscachas, Lagostomus trichodactylus). All the agoutis and cavies (or Guinea-pigs) are, as has already been stated, restricted to the Neotropical realm, over which (principally east of the Andes) they are very extensively distributed. The range of the agoutis proper (Dasyprocta) extends from Mexico to Paraguay, one species (D. cristata) finding its way into some of the smaller West Indian islands—St. Vincent, Santa Lucia, Grenada; the paca (Cælogenys paca), the largest member of the family, inhabits the river-bottom forests over almost the entire tract covered by the remainder of the species. The cavies proper (Cavia) are spread throughout nearly the whole of the South American continent, from Guiana to the Strait of Magellan, and from the lowlands to the plateau region of perpetual snow; one or more doubtful species are said to occur west of the Andes. Brazil is most favoured as to number of species, from one of which (Cavia aperea) appears to have descended the domestic Guinea-pig. The Patagonian cavy (Dolichotis Patagonica), an animal measuring nearly three feet in length, inhabits the plains between Mendoza and the forty-ninth parallel. The capybara (Hydrochœrus capybara), the largest of all living rodents, inhabits the whole of South America east of the Andes and north of the Rio de la Plata, wherever water is found; its range at one time appears to have extended as far south as the Salado, or even farther.

Of the remaining hystricomorphs the porcupines, of which some authors recognise two distinct families, the true porcupines or porcupines of the Old World (Hystricidæ) and the tree-porcupines, or the species of the New World (Cercolabidæ), comprise a considerable number of forms, which though closely related to one another in point of anatomical structure, affect most diverse conditions of habit. The American species range from the northern limits of trees to Paraguay, but the South American forms are generically distinct from those inhabiting North America (except Mexico). Two well-marked varieties of a single species, the Canada porcupine (Erethizon dorsatus), inhabit the forest region of North America. The eastern porcupine, or Canada porcupine proper.

whose range formerly extended southward to Virginia (and possibly also to Kentucky), is now largely restricted to its northern habitat, although it is still found in certain portions of the State of Pennsylvania. According to Dr. Allen it is but rarely met with in New England south of Central Maine and Northern New Hampshire. The western variety (E. dorsatus, var. epixanthus) occupies the western half of the continent, ranging between Alaska and the Mexican frontier. The greater number of the South American species are included in the genus Cercolabes, tree-porcupine, whose combined range extends from Paraguay to Mexico; Chætomys subspinosus inhabits the warmer parts of Brazil. Of the Old World forms the best known is the crested or common porcupine (Hystrix cristata), whose home is the Mediterranean districts of both Europe and Africa, with some considerable reaches of territory also in the western part of the latter continent. In South Africa this species is replaced by Hystrix Africæ-australis, and in India by the hairy-nosed porcupine (H. leucura), which is found from the Himalayas to the extreme south of the peninsula. Other species of the genus Hystrix and of the brush-tailed porcupines (Atherura) are distributed over the Oriental realm from Nepaul to Borneo; a species of atherure is also found in Western Africa.

The last division of the rodents, the rabbit-forms or lagomorphs, embraces the rabbits or hares, and the pikas (Lagomys), small Guinea-pig-like animals which are restricted almost wholly to the elevated mountain districts (11,000 to 15,000 feet) of Northern and Central Asia, with a single species found in Southeastern Europe, and another, Lagomys princeps, in the Rocky Mountain region of the Western United States and Canada. The rabbits and hares (Lepus) include some twenty or more species, which are almost entirely confined to the Northern Hemisphere, where they occupy very nearly the whole of North America and Eurasia, and also Northern Africa. A single species (Lepus Brasiliensis) is found in South America, while several are known from South Africa, although in the vast interior of the last-named continent the genus does not appear to be represented. The distribution of the principal American species has already been discussed in the general consideration of the North American fauna. Only one of these, the polar or Arctic hare (Lepus glacialis or timidus), whose range extends over Greenland and the Barren Grounds to the Arctic coast, is common to the Old World, where its range extends over the greater portion of Europe, from Scotland to the Ural Mountains and the Caucasus. Singularly enough, the species appears to be wanting in Scandinavia. The variable hare (L. variabilis), which is by many authors considered to be identical with the last, or at best only a varietal race, inhabits Eurasia (from Ireland to Japan) north of the fifty-fifth parallel of latitude, but reappears on the more elevated mountain regions of the south where the climatic conditions are approximately those of the northern lowlands. Thus, we find the animal in the Swiss, Bavarian, and Austrian Alps, in the Pyrenees, and in the Caucasus, although in much or most of the intervening lowland it is completely wanting. Unlike the last species, which in the Alpine region occupies principally the basal tracts, rarely ascending above 5,500 feet,* the variable hare more properly frequents the elevated summits, up to 10,000 feet or more, and only occasionally descends below the level of 4,000 feet. The species is wanting in the Jura Mountains. Much uncertainty attaches to the true home of the semi-domesticated rabbit (Lepus cuniculus), which is at the present time so extensively distributed throughout Europe, and the contiguous parts of Asia and Africa. Until recently supposed to have been introduced from Spain, the discovery of its remains in the Quaternary deposits north of the Alps would seem to throw considerable doubt upon the accuracy of this hypothesis.

The most complete analysis of the extinct rodent fauna of the Northern Hemisphere is furnished by Schlosser ("Palæontographica," 1884), who recognises about seventy well-characterised species from the Tertiary deposits of Europe alone. Of these the most ancient, or those of the Eocene and Oligocene periods, belong in principal part to types that are either entirely extinct or have their nearest analogues among forms living at the present day in tropical America. Such are the genera Nesokerodon, Theridomys, and Protechimys, from the French phosphorites, the first of which appears to be closely related to the South American cavies, and the last two to the spiny-rats, and more distantly to the chinchillas. The squirrels and dormice are represented by the modern gen-

---

* Théobald affirms the presence of the common hare at an altitude of 7,000 feet in the Grisons (Fatio, "Faune des Vertébrés de la Suisse," p. 250).

era Sciurus and Myoxus respectively, while Plesiarctomys presents us with an ancestral form of the true marmots. Among the extinct genera we find some remarkable suggestions of marsupial structure; thus, Pseudosciurus and Sciurodon, in the character of their dentition, approach the Australian koala (Phascolarctos cinereus), and Sciuroides recalls the phalangers and the kangaroo-rats (Hypsiprymnus).

The Miocene rodents, principally represented in France and Germany, although still retaining a number of the older types, as Theridomys, show a much closer approximation to the modern fauna. Among living genera we find (in the newer deposits) representatives of the true squirrels, dormice, porcupines, and pikas or tailless hares (Lagomys). The rabbits or hares (Lepus) are entirely wanting, as, indeed, they are from the whole of the European Tertiary series, and likewise the true mice (Mus), if we except the single species, Mus (Acomys) Gaudryi, from Pikermi, Greece. The genus Cricetodon, whose earliest appearance is in the Oligocene deposits of France, was without doubt the near ally of the hamsters (Cricetus). One or more species of marmot (Arctomys) have been indicated as belonging to the Miocene deposits of both France and Germany, but it is a little doubtful whether the horizons whence the remains were obtained have been correctly identified. Arctomys primigenia, from Eppelshcim, is not improbably a comparatively recent species, and not impossibly identical with either A. marmotta or A. bobac. The most abundant forms of this period are Myolagus Meyeri and Steneofiber Jägeri, both from the Upper Miocene of Germany and France, the last replacing the more modern true beavers (Castor). Of the older Miocene genera Archæomys stands intermediate between the still earlier Protechimys and the chinchillas, and Issiodoromys between Nesokerodon and the cavies. The Pliocene rodent fauna does not differ essentially from the Upper Miocene, of which it may be considered to be a mere amplification. The recent genera occurring here are already in the main represented in the period preceding, although a limited number of new types, such as the beaver (Castor Issiodorensis, possibly identical with the recent Castor fiber; Puy-de-Dôme, France) and the Arvicolidæ, are for the first time introduced. The forms related to the South American fauna have, on the other hand, completely disappeared from the continent. The remains of the existing species of porcupine and beaver (cave of Gailenreuth) are found in the Quaternary deposits.

The rodent fauna of the American (western) Tertiaries is very closely related to the European, a large number of identical, or analogous, genera being represented. This is especially the case with the forms belonging to the Miocene period, where, in addition to a considerable number of extinct types, we find such forms as Steneofiber, true beavers (Castor—several species), squirrels (Sciurus), vesper-mice (Hesperomys), and not impossibly also the true porcupine (Hystrix). Eumys does not appear to differ essentially from Cricetodon, while Ischyromys represents Sciuromys. A distinctive feature separating the American from the European fauna is the introduction of the hares, which are not only represented by forms now no longer living (Palæolagus), but also by the modern genus Lepus. In the Pliocene fauna there is a still further approximation to the fauna of the present day in the appearance of an additional number of recent genera—Erethizon (Canada porcupine), Geomys (gopher). The last genus is also found in the Quaternary deposits, as also other members of the same family (Saccomys), and the vole, musk-rat, wood-chuck, ground-squirrel, wood-hare (Lepus sylvaticus), beaver, and a form of capybara (Hydrochærus Æsopi). Castoroides Ohioensis, the largest of all known rodents, recent or fossil, appears to have been of the dimensions of the black bear.

Proboscidea (Elephants).—At the present day there are but two living species of this order known—the one being the Asiatic elephant, Elephas Indicus, which inhabits the forest lands of India and Southeast Asia generally, with the islands of Ceylon, Sumatra, and (?) Borneo, and the other the African elephant, E. (Loxodon) Africanus, a native of the greater part of the continent of Africa south of the Sahara. The insular Asiatic form is by some authors considered to represent a distinct species, to which the name E. Sumatranus has been applied. Although now restricted in a general way to the warmer parts of the earth's surface, there can be no doubt that the range of the species was very much greater at an earlier period of the earth's history than it is at present, seeing how very broad was the distribution of the genus. The remains of elephants undistinguishable from the African form have been discovered in the Post-Pliocene deposits of Algeria, Sicily, and Spain.

The extinct species of elephants are numerous, and their remains are largely distributed over the continents of North America (Alaska to Mexico) and Eurasia. None date back in time beyond the Pliocene period, if we except the forms from the Siwalik deposits of the peninsula of India and the island of Perim, whose horizon is somewhat doubtfully placed by geologists as Mio-Pliocene. The best-known species is the mammoth (Elephas primigenius), which was very closely related to the Indian elephant, and whose range covered the greater part of Northern Eurasia (extending as far south as Santander in Spain and Rome in Italy) and Northwest America. Its remains are found most abundantly along almost the entire Arctic shore of Siberia. The species belongs exclusively to the Post-Pliocene period, and was doubtless contemporaneous with man in many of the regions inhabited by it. Other well-known and somewhat earlier species are the European E. antiquus and E. meridionalis, and the American E. Americanus. Elephas Melitensis, from the island of Malta, is the smallest known species, barely exceeding three feet in height when adult ; by Pohlig it is considered to represent only a diminutive variety of E. antiquus.

The closely related genus Mastodon antedates the true elephants by one period, appearing in Europe in the Middle Miocene (Pliocene in America).* Its extinction in the Old World appears to have been effected in the Pliocene period, although in America several species, and more particularly the commonest form, M. Ohioticus or M. giganteus, survived into the late Post-Pliocene.

Remains of true elephants have been found in China and Japan,† and it appears not unlikely, from a fragment of tusk recently described by Professor Owen as Notelephas, that a proboscidean,

---

* The Loup Fork beds, in which several of the American species occur, are by Professor Cope considered to be of Upper Miocene age. This authority recognises nine species of North American mastodon, to which a tenth one, M. Floridanus, has recently been added by Dr. Leidy.

† At least two of the Indian forms—E. Cliftii and E. insignis—referred to the group Stegodon, which in dental characters stands intermediate between the mastodons and true elephants, have been identified by Koken as occurring in the Pliocene deposits of China, and by Naumann in the nearly equivalent series of Japan.

whether elephant or mastodon, also existed in Australia. The
mastodon is also known from South America (M. Andium, M. Humboldtii).

Of the recognised true proboscideans the genus Dinotherium
may be considered to represent the earliest type, inasmuch as its
remains are thus far known only from the Miocene and Mio-Pliocene deposits (Europe and Asia).

Extinct Animals related to the Proboscidea. — Numerous
extinct animal forms, of both small and gigantic dimensions, exhibiting more or less intimate relationship with the Proboscidea,
have been described from the Eocene deposits of both Europe and
America. Among the best known of these is the genus Coryphodon
(order Amblypoda of Cope), with about fourteen species, vegetable
feeders, ranging in size from the dimensions of a tapir to that of an
ox, and, judging from the skeleton, most nearly resembling among
living animals the bear in outward appearance. The structure of
the foot was largely that of the elephant. All the species are
Lower Eocene (England, France, America). Belonging to the same
order are the American Dinocerata, animals equalling or surpassing
in size the modern elephants, to which they bore many points of
structural resemblance. The upper jaw was provided with a pair
of vertically descending canine tusks. The type genus of this
group is Uintatherium, and seemingly the other forms which have
been described under Loxolophodon, Eobasileus, Dinoceras, and
Tinoceras also belong here. Some twenty-nine species are known,
all of them of the Middle Eocene period.—A form uniting the
coryphodons with the Dinocerata has recently been discovered by
Professor Scott in the Bridger (Middle Eocene) beds of Wyoming,
and named Elachoceras parvum.[128]

The members of the amblypod order of Mammalia, as well as
the more recent Proboscidea and Hyracoidea (conies), are traced
back by Professor Cope to a type of ungulate animals which largely
preceded these in the order of their development, and which are
by that naturalist considered to represent the primitive hoofed
forms whence the modern even- and odd-toed ungulates, the Artiodactyla (deer, ox, camel, &c.) and Perissodactyla (horse, rhinoceros,
tapir), ultimately descended. The teeth were tuberculated (of the
bunodont or hog type), and the feet largely plantigrade, provided
with five toes, both front and rear, constituting the generalised

mammalian foot. The best-known genus of the order (Condylarthra), which comprised animals intermediate in size between the opossum and tapir, is Phenacodus, from the Puerco and Wasatch Eocene.

Ungulata Perissodactyla (Odd-toed hoofed-animals).—The only modern representatives of the odd-toed ungulates are the rhinoceros, horse (including the zebra and ass), and tapir. Of the first some five or six species are known, which are generally referred to a single genus Rhinoceros, although by some authors several distinct genera are recognised. The African species (R. bicornis, R. keitloa(?), and R. simus, the so-called white rhinoceros), all two-horned, occupy the greater part of the continent south of the desert, while in Asia the species, both single- and double-horned, range from the forest-covered foot-hills of the Himalayas through Farther India and the Malay Peninsula to Borneo, Java, and Sumatra. The common Indian species, R. unicornis or Indicus, is now restricted in its range almost wholly to the terai region of Nepaul and Bhotan, and to the upper valley of the Brahmaputra. Many species of rhinoceros, in part referable to the genus or genera which contain the modern forms, are found fossil in Europe and India in deposits dating from the Upper Miocene; Rhinoceros (Ceratorhinus) Schleiermacheri is Middle Miocene. The hornless genus Aceratherium, which, on the American continent, is preceded in the Upper Eocene by the somewhat rhinocerotic Amynodon, and which may be considered as the first true rhinoceros, appears in the Lower Miocene deposits of both the Old and the New World, and is looked upon by many as the ancestral type whence, through migration and subsequent development, the existing and Post-Pliocene (R. tichorhinus, &c.) species have been derived. The American genus Aphelops, likewise hornless, belongs to a series of deposits (Loup Fork) which by some authors are referred to the Upper Miocene, and by others to the Lower Pliocene. Hyracodon, a genus in many respects allied to the rhinoceros, but possessing only three toes to the foot, is Lower Miocene or Oligocene (North American). Singularly enough, no rhinocerotic form is found in the New World above the Pliocene.—None of the existing species of rhinoceros antedate the Post-Pliocene period, although the African bicorn type is actually represented in the earlier deposits of Greece (R. pachygnathus) and the Siwalik Hills of India, which last also contain the remains of several forms closely allied to the existing Indian species (R. Sivalensis, R. palæindicus).

Of the tapirs (Tapiridæ) there are at present five or six recog-nised species, one of which, the Malay tapir (Tapirus Malayanus), inhabits the Malay Peninsula and the islands of Borneo and Suma-tra, and the remainder the forest regions of South and Central America, one or more of the species ascending the Andean slopes to heights of from 10,000 to 12,000 feet. The Central American forms (T. Bairdi, T. Dowi) have been referred by Gill to a distinct genus, Elasmognathus. The genus Tapirus itself, which is not known in North America previous to the Post-Pliocene period, extends back in Europe to the Upper or Middle Miocene, and has continued with but slight modification of form from that time up to the present day. Its precursor appears to have been the Listriodon (Middle Miocene), which united it with the somewhat tapiroid group of the lophiodons (Lophiodontidæ—Eocene), the earliest group of known perissodactyls, comprising animals ranging in size from the rabbit to the ox. It is difficult, or impossible, to determine just whether the tapirs constitute a primarily Old World or New World group of animals, for, despite the intimate relationship which is established between them and the European Lophiodon through Listriodon, an equally close connection unites them on the western side of the Atlantic with genera—Helaletes (Tapirulus?), Desmato-therium, and Hyrachyus—which appear to have been contempora-neous with Lophiodon, and, indeed, may have preceded it. Nor is it exactly impossible, as Professor Vogt has suggested, that a parallel development on opposite sides of the Atlantic may have evolved similar forms from slightly different ancestors. The scanty remains of tapirs in the American Miocene formation are referred by Professor Marsh to the genus Tapiravus ; between these and those of Post-Pliocene age there intervenes a complete hi-atus.

Of the horses (Equidæ) there are usually recognised three groups: the horses proper, the asses, and the zebras. By most zoologists these are all placed in the one genus Equus, the characters defining Asinus (the asses) not being considered to be of generic value. Until Poliakof quite recently made known the existence of a new species of horse (E. Przevalskii) from the desert wilds of Cen-tral Asia it was generally supposed that the domestic animal (E.

17

caballus) was the only living representative of the caballine section of the family, and that no truly wild stock any longer existed on the surface of the earth. Whether Przevalski's horse proves to be a good species or not, there can be little or no question as to the normally wild state in which it occurs. The domestic animal has been spread through the agency of man over the greater part of the globe, where in nearly all localities it has flourished to a remarkable degree. That America was wholly, or in great part, deficient in horses at the time of the Spanish conquest, is proved beyond doubt, but at the same time it is equally proved, from the number of fossil remains that have been found between Escholz Bay in the north and Patagonia in the south, that the animal not only inhabited, but abounded on, the continent during a period comparatively recent preceding. There is, further, very little question as to the contemporaneity of the horse and man on the American continent, and, indeed, it would appear not exactly improbable, from certain references contained in old narratives, that at least in South America the animal still lingered on even after the advent of the Europeans. What led to its general extermination, when under apparently similar physical conditions the introduced animal has been able to thrive to such a wonderful degree, is a problem which still awaits solution.

The species of ass appear to be more numerous than those of the horse, although not unlikely one or more of the forms usually considered distinct will have to be classed as mere varietal types. Zoologists are practically agreed that the domestic animal (E. asinus) is either identical with, or only a feebly modified derivative from, the wild ass of Abyssinia (E. tæniopus), the only African species, which it very closely resembles. Three generally recognised species of ass roam over the wilds of West-Central Asia, the Syrian ass (E. hemippus), the onager (E. onager), from Persia and Northwest India, and the kiang or dziggetai (E. hemionus), the most horse-like in appearance, which inhabits the high table-lands of Thibet, at elevations of 15,000 feet and upwards. Two species of zebra—the quagga (E. quagga) and dauw or Burchell's zebra (E. Burchellii)—inhabit the plains of South Africa, while a third species, the mountain zebra (E. zebra), frequents the mountainous districts of the same region. A fourth form (E. Grévyi) has recently been described by Milne-Edwards from the land of Shoa, and is probably identical with the form seen by Speke and Grant during their journey to the lake regions.

Of the fossil forms of Equus there have been thus far described some twenty or more species, which date back in both hemispheres to the Pliocene period. Only two of the recent species are positively known in the fossil state—the E. asinus, or ass, which is Post-Pliocene, and the horse (E. caballus), which appears to be first known from the Upper Pliocene ; it is not improbable, however, that some of the Post-Pliocene equine remains of Central Europe belong to the dziggetai. The Pliocene species seemingly most nearly related to the modern horse are the E. major or E. Americanus, from the deposits of North America, and the E. Stenonis, from the Val d'Arno, Italy. The only other genus of Equidæ besides Equus is Hippidium (Pliohippus ?), which occurs fossil in the Pliocene deposits of both North and South America.

Nehring, from a careful study of the numerous fossil remains of horses found in Germany and elsewhere, arrives at the conclusion that the present European animal—at least, as representing some of the races—instead of being, as is commonly supposed, a recent introduction from Asia, is in reality indigenous to the region which it now inhabits in a domesticated state, and that it has been a continuous inhabitant of Central Europe, then largely in the form of a steppe country, supporting a steppe fauna, ever since the early part of the Quaternary epoch.

Extinct Animals related to the Horse and Tapir.—In no group of mammals, probably, is the difficulty of drawing family boundaries greater than among the perissodactyl ungulates, a circumstance due chiefly to the perfection with which many of the lines of descent have been traced out, and to the intimate relationship which the animals of one line bear to the animals of one or more other lines. Thus the primitive ancestors of the horse— the four-toed and three-toed (fore and aft) Eohippus and Orohippus (Hyracotherium) from the Eocene, the three-toed Mesohippus and Miohippus (Anchitherium) from the Miocene, and the Pliocene Hipparion (also Miocene) and Protohippus—which form an almost continuous chain connecting Eohippus and the Equidæ, belong to two or three families, Lophiodontidæ, Palæotheridæ,* Anchithe-

* To the type genus of this family, the Eocene Palæotherium, many paleontologists have traced the ancestral line of the European horse—Equus Stenonis,

ridæ, one of which, at least, lies as much in the direct line of descent of the tapir as it does of the horse. It will thus be seen that from the evolutionary standpoint, or from the point which views relationship by descent as of equal importance with that which unites forms solely through a community of general characters, family lines could be traced among these earlier ungulates at points other than where they have actually been drawn, and with fully as much reason. For it can scarcely be gainsaid that the direct ancestors of the horse, for example, would form as natural a group among themselves as they now form three or four groups in the way they have been scattered about.

Other tapiroid forms, more or less closely related to the lophiodons, are Limnohyus and Palæosyops, both from the Eocene; Chalicotherium, whose remains have been found in Oregon, and in Eurasia from France to China, appears in the Oligocene and Miocene. Approximately contemporaneous with the last, and embracing animals of the dimensions of the elephant, or even larger, with certain resemblances to the rhinoceros, were the Menodontidæ, whose remains have been found in both hemispheres. Several genera of this family—Menodus, Titanotherium, Symborodon, Brontotherium—have been described, but it would appear that not all of these are entitled to generic recognition.

A group of highly specialised and abnormal forms of Perissodactyla, concerning which there has been much diversity of opinion expressed, and whose position in the zoölogical scale has not yet been definitely established, is that of the Macrauchenidæ, with the

Hipparion, Anchitherium, Palæotherium.—Dr. Max Schlosser, in an elaborate review of the phylogenetic relationships of the Ungulata ("Morphologisches Jahrbuch," 1886), considers the equine line of descent to pass through Phenacodus (as earliest form), Hyracotherium, Anchitherium, Merychippus, Hipparion, and Pliohippus. Hyracotherium and Hipparion, and possibly also Anchitherium, are assumed to be by origination American forms, which subsequently wandered over to Europe, but the relative appearance of these forms on the two continents scarcely warrants this supposition. The same authority identifies the American Mesohippus and Miohippus with Anchitherium, and Protohippus with Hipparion; Eohippus is considered to be probably identical with Hyracotherium. By Mr. Wortman, on the other hand, who first recognised in Phenacodus the earliest ancestor of the horse, and who unequivocally identifies Orohippus with Hyracotherium, both Mesohippus and Protohippus are considered to represent types of distinct generic value ("Revue Scientifique," June, 1888).

two species Macrauchenia Patachonica and M. Boliviensis, remnants
of the remarkable South American Pliocene fauna.   With certain
characters approximating it to the camel and horse, it is claimed
by Burmeister that the animal was provided with a proboscidiform
trunk.

Ungulata Artiodactyla (Even-toed hoofed-animals). — The
members of this sub-order, both recent and fossil, are conveniently
divided into two groups—those which, like the hog, have the
grinding surfaces of the molar teeth tuberculated (Bunodonta),
and those, in which these surfaces are crescentically ridged, as in
the sheep, ox, deer (Selenodonta).   The first of these groups, the
Bunodonta, comprises but two families, the swine (Suidæ) and
the hippopotami (Hippopotamidæ).

Of the hippopotami there are, as generally recognised, only
two species, the common form (H. amphibius), which until recently inhabited most of the larger streams of the continent of
Africa, from the Congo, Senegal, and Zambesi to the Nile, but
whose domain has of late been rapidly narrowing (completely ex-
cluded from the Egyptian Nile), and the West African Chœropsis
Liberiensis, a comparatively small animal, differing primarily from
the first in the possession of only a single pair of incisors in the lower
jaw instead of two pairs.   A third species, whose remains have
been found sub-fossil in the swamp deposits of the island of Madagascar (in association with the giant Æpyornis), and which may
consequently be classed with the recent period, has been described
by Goldberg (1883) under the name of H. Madagascariensis.   The
common species of hippopotamus represents one of the most ancient of the mammalian types entering into the formation of the
modern fauna, it being one of the very few forms which survived
the Pliocene period up to the present day.   Its range was formerly
very much greater than it now is, and even as late as the Post-Pliocene it appears to have inhabited Europe as far north as Northern
Wales, where its remains have been found associated with human
implements.   Seven other species of the genus have been described from the Pliocene and Post-Pliocene deposits of Europe
and India, but none have so far been recorded from America.—The
pigmy hippopotamus of Malta (H. minutus) appears to have been
closely allied to, if not identical with, the living Liberian species.

The recent hogs are commonly divided into three more or less distinct groups : the peccaries (genus Dicotyles), whose range comprises the region included between Arkansas and Paraguay ; the wart-hogs (Phacochœrus) of East and South-Central Africa; and the true swine, under which are ranged the hog proper (Sus), the babyrousa (Babirusa), and river-hog (Potamochœrus). The last is exclusively West African, while the babyrousa is confined to Celebes, and some of the smaller islands of the Eastern Archipelago (Bouro, &c.). Much diversity of opinion still exists as to the number of species that are comprised in the genus Sus. Some fifteen or more have at various times been recognised by zoologists, but by Forsyth Major,[129] who has probably enjoyed better opportunities for making a critical study of the group than any of his predecessors, the number is restricted to four, with a probable fifth one, concerning which we know little, but which appears to occupy a considerable part of the Ethiopian region: Sus vittatus, whose distributional area extends from Sardinia to New Guinea, and from Japan to Damara-Land (Southwest Africa) ; S. verrucosus, from Java and Celebes; S. barbatus, from Borneo; and S. scrofa, the boar, or common hog, whose domain extends, or did extend before man had greatly narrowed its limits, over the greater part of temperate Europe and Asia. This species, which was an early inhabitant of Britain, as is indicated by the remains found in the forest-bed (Post-Pliocene) of Norfolk, was completely exterminated in that region a number of centuries ago. Several species of the genus have been found fossil in the Miocene and Pliocene deposits of France, Italy, Germany, and Greece, and five are described from the Siwalik Hills of India, one of which, S. Titan, is considered by Lydekker to have attained in extreme specimens a height to the shoulder of forty-nine inches or more. None of the genera of recent bunodont Artiodactyla are found fossil in America with the exception of Dicotyles (peccary), which, in association with a nearly related genus, Platygonus (also Pliocene), occurs in the Post-Pliocene deposits. Closely related to the last are the Miocene Thinohyus, Chœnohyus, and Hyotherium (Palæochœrus ?), the last a genus also abundantly represented in Europe in deposits of both Upper Eocene and Miocene age. The oldest representatives of the suilline tribe appear to be the forms that have been described as Eohyus and Achænodon (Parahyus ?), of Lower and Middle Eocene age, the former of which is stated by Marsh to have had at least four functional toes, while the latter united with its predominant suilline characters certain peculiar carnivore modifications of the skull. Some very remarkable bunodonts of a hog-like character, found in the phosphorite deposits of Quercy, France, and named by Filhol the Pachysimia, are considered by that author to possess some striking structural features allying them with the Primates, and rendering it not exactly improbable that the last may find their earliest ancestors in these ancient types.

Of somewhat doubtful position, but with distinctively suilline affinities, are Chœropotamus (Eocene) and Anthracotherium and Hyopotamus (Eocene—Miocene).

Artiodactyla Selenodonta.—Among recent forms this section comprises the camels (Tylopoda), chevrotains (Tragulina), and true ruminants (Ruminantia), with such well-known forms as the ox, goat, deer, and antelope.

The camels, constituting the family Camelidæ, comprise, as generally recognised, two genera—the Old World Camelus, the camel proper, and the New World Auchenia, the llamas (guanaco, vicuña, alpaca). Of the former there are two species, the dromedary, or one-humped animal (C. dromedarius), a native of the deserts of Arabia, whence it has spread eastward to India, and the Bactrian, or two-humped camel (C. Bactrianus), which occupies the region of Central Asia from the Black Sea to China, and from the Himalayas to beyond the Siberian boundaries. Although the camel is generally considered to be a belonging of the hotter regions of the earth's surface, it is well known that the Bactrian species thrives admirably in the northern districts of Mongolia, and that even as far north as the southern extremity of Lake Baikal, on the fifty-third parallel of latitude, it passes the rigours of a Siberian winter without apparent discomfort. That the dromedary, which is now one of the distinctive animals of North Africa, was unknown to the ancient Egyptians is proved by the absence of representations of it from all monumental inscriptions. The American representatives of the Camelidæ are the llama, alpaca, guanaco, and vicuña, constituting the genus Auchenia, the first two of which, inhabitants of the Peruvian and Bolivian Andes, exist now only in a state of domestication, while the vicuña inhabits the Andean slopes of Peru and Chili, and the guanaco the plains of Patagonia and Tierra del Fuego.

The Tertiary deposits of Europe have thus far yielded no traces of the Camelidæ, and if we except the Camelus Sivalensis and C. antiquus, from the Siwalik Hills of India, and C. Thomasi, from the Quaternary deposits of Algeria, the same may be said of the Eastern Hemisphere generally. On the other hand, animals refer-able in part to this family, and, again, others closely related to these, are abundant in America, where they form a connecting series or chain almost as complete as that which has been established for the horse. The cameline line of descent has been traced by Professor Cope from the Oligocene or Miocene Poebrotherium, in which, as well as in the succeeding genus, the metapodial bones were dis-tinct, and the mouth was furnished with a complete series of in-cisor teeth, through Protolabis, Procamelus, and Pliauchenia (Plio-cene), the last standing in the relation of its dentition intermedi-ately between Procamelus and the camel. Auchenia, the llama, which may be considered to terminate the series, is late Pliocene and Post-Pliocene ; with it occur associated several related forms, as Protauchenia, Palæolama, &c. Professor Marsh indicates the Eocene Parameryx as the probable most ancient ancestor of the camels, whereas by Scott, Osborn, and Speir this place is given to a contemporaneous genus Ithygrammodon.

It would appear, therefore, that the camels are a New World family, but this is by no means proved to be the case ; the absence of the true camel in America and its occurrence in India in deposits as ancient as the older Pliocene, render it very probable that an-cestral cameline forms will be found in the Old World as well as in the New.

The chevrotains, or mouse-deer (Tragulidæ), which comprise some of the smallest of known ungulates, and which in structure stand in a measure intermediate between the deer and hog, are ranged under two genera, Tragulus and Hyæmoschus, the former restricted to Southern and Southeast Asia, and the larger islands of the Eastern Archipelago, and the latter to West Africa. The family dates from the Miocene period, of which the genus Hyæmoschus is a belonging.

Of the true ruminants, the Camelopardalidæ, or giraffes, con-stitute perhaps the most peculiar group. Only one species, Camelopardalis giraffa, is known, which ranges over the greater part of the grass-covered plains of East-Central and South Africa. An extinct species of the genus, C. Attica, which occurs in the Miocene deposits of Greece, appears to have rivalled or fully equalled the modern form in size. Other allied species have been described from the Siwalik Hills of India. The Helladotherium, an animal of less elevated proportions than the giraffe, but closely related to it, roamed during the Miocene, or early Pliocene, epoch over the south of Europe, from France to Greece, and across to India. With the same family are possibly to be placed also the Siwalik genera Brahmatherium, Vislinutherium, and Sivatherium, the last a huge antelopine form, referred by most zoologists to the true antelopes.

The antelopes, whose special distribution has been considered in connection with the several zoogeographical regions of the earth's surface, constitute by far the most extensive group of the Ungulata, there being probably not less than one hundred distinct species. The greater number of these belong to the continent of Africa, where they inhabit as well the desert tracts as the open plains and forests, from the Sahara to the Cape, and from Senegal to the Nile. Among the better known of these forms are the springbok, blesbok, bontebok, hartebeest, buschbok, waterbok, koodoo, oryx, gemsbok, klipspringer, gnu, and eland, the last a bubaline form equalling in size a large ox. The opposite extreme in the series is presented by the western guevi, which barely exceeds the dimensions of a rabbit. The fifteen (?) or more Asiatic species, whose combined range comprises very nearly the entire extent of the continent, with the islands of Japan, Formosa, and Sumatra, are nearly all distinct from the African, and even the generic types that are held in common are limited almost exclusively to such forms, as Oryx, Addax, and Gazella, whose domain embraces the almost contiguous desert tracts of Northeast Africa and Arabia.

Europe has but two antelopine species, the Alpine chamois and the saiga, or steppe antelope, the latter of which may perhaps with more propriety be considered an Asiatic species, whose range extends over Russia to the confines of Poland. North America is equally deficient with Europe, holding likewise but two species—the prong-horn (Antilocapra) and Rocky Mountain goat (Aplocerus) —while in South America the family or group is entirely wanting.

The remains of antelopes, if we except certain doubtful forms from the cave deposits of Brazil, which have been referred to Antilope and Leptotherium, are completely absent from America. In Southern and Western Europe they date from the Miocene period, continuing through the Pliocene and Post-Pliocene. The richest antelopine deposit is that of Pikermi, Greece, whence Professor Gaudry has described the genera Gazella, Palæotragus, Tragoceros, Palæoryx, and Palæoreas, the last two most intimately related to the recent Oryx (gemsbok) and Oreas (koodoo) respectively. Several, or most, of these types have also been recognised in the more or less equivalent deposits of France, Spain, Italy, and the Vienna Basin ; Antilope cristata of the Middle Miocene of Switzerland appears to have had its nearest ally among recent forms in the chamois. African-type antelopes are still met with in the late Pliocene, or Post-Pliocene, volcanic deposits of the Auvergne (France). The only unequivocal antelopine remains from Britain are those of Gazella Anglica, recently described by Mr. Newton from the newer Pliocene beds of Norwich, England ; not impossibly, however, the saiga is also represented in the English fauna. Some eight or more species of the family, principally referable to modern genera—Oreas, Palæoryx, Portax (nylghau), Gazella, Antilope, and Alcelaphus—have been described from the early Pliocene of the Siwalik Hills.

The bubaline or bovine ruminants proper (Bovina) comprise about thirteen recent species, which are distributed over the greater part of Eurasia, Africa, and North America. The buffaloes are represented by four species, two African, Bubalus Caffer (Cape buffalo) and B. brachyceros, the former of which roams over the greater part of Southern and Central Africa, and two Asiatic, the Buffelus Sondaicus and B. Indicus, from the last of which has descended the domesticated variety which has been so extensively acclimatised in North Africa, Italy, Greece, and Hungary. A form related to the buffaloes, but differing in certain important essentials, is the dwarf wild-cow of the island of Celebes (Anoa or Probubalus Celebensis), whose early representatives occur fossil in the Pliocene deposits of the Siwalik Hills of India. Two species of buffalo, referable to the genus Buffelus, likewise occur fossil in the Indian deposits (Pliocene and Post-Pliocene), and one species, B. Pallasii, in the Quaternary of Danzig, Germany. Associated with the former are the bubaline forms Amphibos (Hemibos) and Leptobos. The African buffalo is thus far known only from the Quaternary of that continent (Bubalus antiquus, from Setif, Algeria).

The wild cattle (ghaurs) of India, including, according to some authorities, the Thibetan yak * (Poephaga grunniens), and constituting the genus Bibos, range from Southern India through the Malay Peninsula to Java and Borneo. The earliest representative of this group appears to be the Etruscan bull (Bos or Bibos Etruscus), from the Pliocene deposits of France and Italy.

The bisons (genus Bison) are at present comprised in two species, the American (B. Americanus), which until recently inhabited the greater part of the continent of North America, but which is now restricted to a few hundred individuals, and the European (B. Europæus), also known as the aurochs, which up to the period of the Roman Empire appears to have been sufficiently abundant in South-Central Europe, but which is now limited to the imperial preserves of Lithuania and the wilds of the Ural and Caucasus. Its immediate precursor was the Bison priscus, whose remains are distributed throughout the Quaternary deposits of almost every country in Europe and of Siberia; they have also been found at Escholtz Bay, Alaska. The Pliocene B. Sivalensis would seem to be a closely allied form, and, according to Rütimeyer, nearer to it than to the living American species, or to its Post-Pliocene predecessors, the B. latifrons and B. antiquus.

Of the taurine genus Bos, which comprises the domestic cattle, several well-marked varieties are recognised, all of which are by most authorities referred back in their descent to the urus (Bos primigenius), which was abundant in Central Europe in the time of the early Roman emperors, but is now wholly extinct. By Wilckens, on the other hand, it is claimed that at least some of the breeds of cattle are the descendants of the European bison.

The sheep and goats constitute an almost exclusively Old World group of hollow-horned ruminants, of which there are but two indigenous representatives in the Western Hemisphere (North America). One of these is the Rocky Mountain sheep (Ovis montana), which is very closely related to the argali of East-Central Asia, and the musk-ox (Ovibos moschatus), which inhabits the region of Arctic America north of the sixtieth parallel of latitude, or thereabouts, but whose fossil remains are met with as well in the

* Przevalski describes a second species of yak as P. mutus.

Quaternary deposits of Eurasia—England, France, Germany, Siberia, &c.—as in America. The range of this species, or of forms closely allied to it (O. bombifrons, O. cavifrons), on the American continent at one time extended to the confines of Arkansas.

The Old World Caprina, whose special distribution has already been discussed in the zoogeographical part of this work, comprises some twenty or more species, which, with two exceptions—a Neilgherry goat and an Abyssinian ibex—are confined to the Hólarctic region, or to this and the Mediterranean transition tract. It is a somewhat surprising circumstance, in view of the broad distribution of the goats, that the sheep, which have obtained such a firm foothold in the mountainous and intermountainous regions of Asia, should be almost entirely wanting from the continent of Europe—indeed, from the entire Eurafrican region, if we except the islands of Corsica, Sardinia, and Crete, the Balkan Peninsula, and some isolated spots on the Atlas Mountains. No unequivocal remains of the goat or sheep, except Ovibos, have thus far been discovered in any American formation, and in Europe such remains are confined almost exclusively to the Quaternary cave and breccia deposits (France, Italy). In India, however, they have been traced back to an older period; Capra Sivalensis, a form closely related to the recent Iharal of the Neilgherries, has been described from the Pliocene of the Siwalik Hills, and Capra Perimensis, whose remains would seem to have been associated with those of Dinotherium and Aceratherium, animals indicative of the Miocene period, from the island of Perim. A remarkable hornless form, to which Rütimeyer has given the name of Bucapra Daviesii, also belongs to the Siwalik fauna.

The most important group of ungulates after the antelopes is constituted by the deer (Cervidæ), which comprises some sixty or more species—excluding the Central Asiatic musk (Moschus) and the giraffe which are referred here by some authors (Rütimeyer)—distributed over the greater portion of both the Old and the New World. Australia, as in nearly all other mammalian groups, is entirely deficient, and Africa counts but two species, the fallow-deer and a stag, which inhabit the Mediterranean region. The South American forms, whose domain extends completely across the continent to Tierra del Fuego, have been placed in the genera (or sub-genera) Pudu, Coassus, Furcifer, Blastocerus, and Cariacus, the last of which comprises all the North American deer north of the Mexican boundary, with the exception of the Canada stag or wapiti (Cervus Canadensis), the moose (Alces machlis) and reindeer (Rangifer tarandus), the last two of which are circumpolar, and inhabit the whole northern portion of the Eurasiatic continent, from Norway to China. The European forms are comprised under the three groups Cervus, which includes the stag or red-deer (C. elaphus), whose range embraces, or until recently embraced, the whole of Europe and a large part of Northern Asia; Dama, the fallow-deer, a native of the Mediterranean districts of Europe, Asia, and Africa; and Capreolus, the roe, which was at one time extensively distributed over nearly the whole of Europe, with the exception of the greater part of Russia. Among the better-known Asiatic forms are Axis (peninsula of India, Ceylon, China) and Rusa, the latter containing some of the largest of the cervine tribe, several species of which inhabit the hotter regions of Hither and Farther India, and the islands of the Eastern Archipelago. The muntjacs (Cervulus), which seem to connect the true deer with the musks, inhabit the forest tracts of the Oriental region, from India to China, and from Formosa to the Philippines, Java, and Borneo.

The earliest cervine animals, in the strict sense of the term, with which we are acquainted, are met with in the Middle Miocene deposits of France and Germany, where forms showing evident relationship with the muntjacs, and possessing the simplest kind of horn structure—a simple bifurcated stem—have been variously described as Procervulus, Prox, Dicrocerus, Palæomeryx, and Micromeryx. These are all united by Rütimeyer into the one genus Palæomeryx, to which is also added the supposed differing Dremotherium from the same, and a possibly lower (Lower Miocene), horizon. Of equivalent age is the hornless Amphitragulus. The progressive development from the simple-formed antler to the more complex has been traced through numerous forms of Cervus from the Upper Miocene to the Post-Pliocene, the most complex structure known being that exhibited by C. dicranios, from the Pliocene of the Val d'Arno, Tuscany. Professor Boyd Dawkins thus sums up his observations on this point : " We may gather from the study of the fossil Cervidæ the important fact that in the Middle Miocene age the cervine antler consisted of a simple forked crown only. In the Upper Miocene it becomes more complex, but is still small and erect, like that of the roe. In the Pliocene it becomes larger and longer, and altogether more complex and differentiated, some forms, such as the Cervus dicranios of Nesti, being the most complicated antlers known either in the living or fossil state. These successive changes are analogous to those which are to be observed in the development of the antlers in the living deer, which begin with a simple point and increase their number of tines until their limit is reached. It is obvious, from the progressive diminution in size and complexity of the antlers in tracing them back from the Pliocenes into the Upper and Middle Miocenes of Europe, that in the latter period we are approaching the zero of antler development. In the Lower Miocenes I have failed to meet with evidence that the deer possessed any antlers." [150]

The roe, stag, elk, and reindeer occur fossil in the Quaternary deposits, together with a giant form, the Irish stag (Cervus megaceros), whose extinction appears to have been effected long after the region inhabited by it was also inhabited by man. The stag (wapiti), elk, and reindeer also occur fossil in the American Post-Pliocene deposits, the last, as in Europe, in latitudes very much lower than it now occupies. The genus Cervus dates from the Pliocene. A Quaternary form intermediate in many points of structure between the true deer and the elk, and originally referred to the latter, has recently been re-described by Scott as Cervalces (C. Americanus). Its remains have thus far been met with only in New Jersey and Kentucky. The hornless Miocene genus Leptomeryx, which by Leidy and Marsh is placed near the Cervidæ, is by Rütimeyer considered to more nearly approach the camels.

Of undeterminable position among the Ruminantia, but more or less closely related to each other, and showing certain analogies of structure with the Tragulina, are the Old World Lower and Middle Tertiary genera, Anoplotherium, Xiphodon, Dichobune, and Cainotherium. An equally aberrant type of American ruminants, the "ruminating hogs" (Oreodontidæ) of Dr. Leidy, whose remains are exceedingly abundant in the Western Territories, appears to have been nearly related to the Anoplotheridæ. Thirty-five species, with three exceptions (Merychyus—Pliocene), all belonging to the Miocene period, are referred to this family by Professor Cope. [131]

One of the genera of the family, Agriochœrus, seems to be also represented in the deposits of the Siwalik Hills.

Carnivora.—The members of this order may be conveniently classed under four primary groups, defined in their broadest sense as the cats (Æluroidea), dogs (Cynoidea), bears (Arctoidea), and seals (Pinnipedia). The first of these embrace the cats proper (Felidæ), civets (Viverridæ), the South African aard-wolf (Proteles Lalandii), and the hyenas (Hyænidæ).

The true cats, which by many authorities are considered to be comprised within the single genus Felis, have an almost world-wide distribution, but are most abundantly developed in regions of elevated temperature. No species occurs in either the Australian region or Madagascar. The better-known American forms are the jaguar (F. onça), whose range comprises the entire region included between Patagonia and Texas; the cougar or puma (F. concolor), with probably the most extended north and south range of any mammalian species—Patagonia to the sixtieth parallel of north latitude ; the ocelot (F. pardalis), which, in one or other of its several varieties, ranges from Arkansas through Texas and Mexico to Patagonia; the nearly equally distributed margay (F. tigrina—Mexico to Paraguay), and several allied species of small intertropical "tiger-cats;" the jaguarundi (F. yaguarundi) and eyra (F. Eyra), unspotted cats ranging from Paraguay to the northern boundary of Mexico, the Chilian colocollo (F. colocollo), the pampas-cat (F. pajeros), and the lynx. The last, of which several species or varieties have been described, in whole or in part identical with the common European species (F. lyncus or rufa), inhabits the greater part of the American continent north of Mexico.

Of the Old World cats, besides the lion and tiger, whose range has been specially considered in the zoogeographical portion of this work, the better-known forms are the Felis pardus, leopard or panther, which may represent several distinct species, inhabiting the greater part of Africa and the warmer regions of Asia, from Palestine to Japan; the ounce or irbis (F. uncia), of about the size of, and somewhat resembling, the leopard, a native of the elevated mountain-tracts of Central Asia (Thibet—Siberia), where it ascends to heights of from 15,000 to 18,000 feet; the spotted or clouded tiger (F. macroscelis), an arboreal species, indigenous to the forest regions of Southeast Asia and the adjoining islands of Formosa,

Sumatra, Java, and Borneo; the serval (F. serval), from the greater part of the African continent; and the cheetah or hunting leopard (F. or Cynælurus jubatus), whose domain covers nearly the whole of the African continent and a very considerable part of Southern and Western Asia. The lynx or lynxes range from the polar regions to the Mediterranean, whence they are continued by an allied form, the caracal, over a large part of both Asia and Africa. Felis catus, the wild-cat proper, which is met with in both insular and continental Europe, is not, as is frequently supposed, the ancestor of the domestic animal; this place is now generally conceded to the Egyptian and West Asiatic F. maniculata.

Numerous species of the family, referable in considerable part to the genus Felis of most authors, and differing but little from forms still living, are found fossil in the Post-Pliocene, Pliocene, and Miocene deposits of Europe, Asia, and America. Among the best known of these is the cave-lion (F. spelæa), a species but barely if at all distinguishable from the F. leo, whose remains are abundantly met with in the Post-Pliocene cave deposits of continental Europe and England. During the same period the existing lion appears to have hunted its prey as far north as Yorkshire and on the frontiers of Poland, and the leopard or panther among the Mendip Hills. Felidæ, allied to the panther and lynx, have been discovered in the Pliocene strata of the Siwalik Hills of India. Felis angustus, from the North American Pliocene deposits, was intermediate in size between the jaguar and tiger, while the later F. atrox, which may be considered to represent in the New World the European cave-lion, appears to have surpassed in this respect both the lion and the tiger. In association with the modern type-forms of Felidæ there occur others which depart very widely from these, the most remarkable of which, as representing the most highly specialised forms of the family, and as strictly the most carnassial of all known Carnivora, were the so-called sabre-tooths. The animals of this group are characterised by a prodigious development of the upper canines, which in some instances appear to have measured as much as seven or nine inches in length. The best-known genus is Machairodus (Drepanodon), whose remains have been found in both the Miocene and Pliocene deposits of Europe (Pliocene of India), and whose immediate American representative appears in Smilodon (Pliocene or Post-Pliocene of Buenos Ayres and Texas), a contem-

porary of the giant sloths and glyptodons. A supposed species of
the last (S. gracilis) has also been described from a cave deposit in
the State of Pennsylvania. Other allied forms are the American
genera Dinictis, Nimravus, Pogonodon, and Hoplophoneus, from
the Lower Miocene deposits of the Western United States, whose
members were intermediate in size between the lynx and tiger.
Eusmilus bidentatus, from the phosphorites of France (Upper
Eocene or Oligocene), although the oldest-known form, is, singularly enough, in many respects also the most specialised. Contemporaries with it were Pseudælurus and Ælurogale, the former the
representative of the group denominated by Professor Cope the
" primitive " non-specialised cats.

The Viverridæ, or civet-cats, comprise some one hundred or
more species of moderate-sized Carnivora, which are in the main
restricted to the Ethiopian and Oriental regions. The better-known
types are the true civets (Viverra), from North and Tropical Africa,
India, China, and the Malay Peninsula ; the genets (Genetta), from
Africa (the entire continent), Southwest Asia, and Southern Europe
(France, Spain); and the ichneumons or mongooses (Herpestes),
which are widely distributed over the continent of Africa and Indo-
Malaysia. One species of the last is also found in Spain. Among
other genera of the family are Viverricula, the rasses, and Paradoxurus, palm-civets (both from Indo-Malaysia), Cynogale, the
otter-civets (Borneo), and Cryptoprocta, the last sometimes considered the type of a distinct family, and the largest of the Madagascan carnivores.

Numerous genera, more or less closely allied to recent forms,
carry this family back to the early Tertiary period, where (in the
phosphorites of Quercy, France) we find two or more species representing the genus Viverra itself. Others of the same genus (or
possibly Genetta) are found in the French Miocene, and in the
Pliocene of the Siwalik Hills. A remarkable viverrine form, showing intermediate relationships between the civets and the hyenas,
has been described by Professor Gaudry, from the Middle Tertiaries of Greece and elsewhere, as Ictitherium.

Three well-differentiated species of hyena are recognised by
zoologists—the striped hyena (H. striata), from Africa generally
and Southern Asia; the spotted hyena (H. crocuta)—Africa, south
of the desert, sometimes placed in a distinct genus, Crocuta; and

the brown hyena (H. brunnea), from South Africa. Although now restricted to the continents of Asia and Africa, the numerous remains found in the European Post-Pliocene deposits indicate that this animal, as well as the lion and other semi-tropical species, was an abundant form in the north temperate regions at a comparatively recent period, and that from those parts the Ethiopian realm has drawn much of its existing distinctive fauna. The widely distributed cave-hyena (H. spelæa), whose range embraced a part of the British Isles, was most nearly related to, if not identical with, the H. crocuta, and was without doubt its direct ancestor. The striped hyena may be traced back to the older (Pliocene) H. Arvernensis of Central France, and the brown form not improbably to the Miocene (or Pliocene) H. eximia of Pikermi, Greece. The aberrant form Hyænictis, described by Gaudry from Pikermi, and showing certain viverrine relationships, is considered by Lydekker to represent at most only a sub-genus of hyena. No representatives of this family, either recent or extinct, have thus far been discovered in America, unless, indeed, the Miocene Ælurodon prove to be a distant relative.

The Cynoidea, or canine division of the Carnivora, comprises but a single family, the dogs (Canidæ), whose numerous representatives enjoy a nearly world-wide distribution. Apart from the hunting or hyena dog of South Africa (Lycaon picta *), the long-eared fox (Otocyon megalotis), from the same region, and the bush dog (Icticyon venaticus), from Brazil, all the species—some fifty or more —may be conveniently grouped in the single genus Canis, whose range would then be coextensive with that of the family. If the dingo, or wild-dog of Australia, be proved to be indigenous to that continent, then the genus will be the most nearly cosmopolitan of any of the terrestrial Mammalia. Two clearly defined sections of the genus may be recognised, the lupine and the vulpine, to the former of which belong the wolves, jackals, dogs proper, and a number of not readily classifiable forms which have a general canine aspect; and to the latter the foxes and fennecs. The origin of the various breeds or races of the domestic dog is involved in much uncertainty, and whether their progenitors are to be sought in a

---

* A fragment of a jaw from the Post-Pliocene deposits of Glamorganshire, Wales, has been referred to the genus Lycaon by Lydekker (L. Anglicus).

single one of the feral forms now living, as the wolf or jackal, or in several such forms as are denominated wild-dogs, or in the union of both, still remains to be determined. The researches of Nehring seem to indicate that a race of wild-dogs, akin to the existing domestic one, inhabited a considerable part of Central Europe during prehistoric times. Various forms of wild-dog, as the dhole and buansuah (sub-genus Cuon), range over the greater part of Asia, from Siberia to Java and Sumatra, where they in great measure replace the wolf of the more strictly northern regions. The last (Canis lupus) is found throughout the whole of Europe and Northern Asia, from the Atlantic to the Pacific, and also in Nova Zembla and Japan. There can be little question as to the identity with this form of the corresponding American species (Canis occidentalis), which, in its numerous varieties, covers the entire North American continent, from Mexico to the Arctic Ocean. The coyote, or American prairie-wolf, is by some authors considered to be intermediate between the wolf and fox. South America is wholly deficient in wolves, as in foxes, their place being taken by the fox-like forms which have been referred to the groups Lycalopex, Pseudalopex, and Thous. The most broadly distributed of these is Azara's dog (C. Azaræ), which ranges over the greater part of Brazil, and southward to Patagonia. The most southerly species of the family is the Antarctic dog (C. Magellanicus), which inhabits Chili, Patagonia, and Tierra del Fuego. Of the remaining lupine forms the most widely distributed are the jackals, of which several species are recognised, whose combined ranges embrace the whole of Africa and much of Southern and Western Asia.

The vulpine section of the Canidæ includes the fennecs and foxes, the former all African, the latter—with ten to fifteen species— spread over the whole of North America and Europe, and largely also over Asia and Africa. There can be no question (as in the case of the wolves) that several of the forms that have generally been recognised as distinct species are in reality only varietal types, whose inter-relationship is made manifest when full geographical suites are made use of for comparison. The identity between the common European fox (C. vulpes) and the American red-fox (C. fulvus) may be considered as established. Recognised as one species, the habitat of the common fox of the Northern Hemisphere may be said to embrace the whole of Europe, North Africa, North and Central Asia, and practically the whole of the North American continent, although it appears to be absent from the immediate Pacific coast. A well-marked variety of this form, by many naturalists considered to be a distinct species, is the broadly distributed silver-fox (C. Virginianus), which alone of the different varieties is represented in Central America. The burrowing-fox (C. velox) is an inhabitant of the interior region included between the (lower) Missouri and Saskatchewan rivers and the Cascade Mountains. Over a considerable part of North-Central Asia—Tartary, Mongolia, Siberia—the common fox is replaced by the corsac or steppe-fox (C. corsac), a closely related species. The most northerly species of fox is the Arctic fox (C. lagopus), a circumpolar mainland form, occurring also in Iceland and Spitzbergen.

The Canidæ appear to date from the (Upper) Eocene period, and not impossibly the genus Canis is itself represented in the form that has been described by Cuvier as Canis Parisiensis. Less doubt attaches to the C. Filholi, from the phosphorites of Central France, whose position, however, still remains somewhat uncertain. Barring these two forms, the oldest representatives of the family are seen in the genera Galecynus and Amphicyon, which appear in Europe in the Upper Eocene and Lower Miocene respectively. In America Galecynus is unknown prior to the Oligocene (or Lower Miocene—White River formation), to which period must be referred the most ancient undoubted remains of the family in the New World. To this genus, whose representatives appear to have been very abundant during the Miocene epoch, the existing species of dog are referred in their line of descent by Filhol and Cope. Canis, which in the Miocene is associated with a number of allied generic forms—Temnocyon, Oligobunis, Ælurodon (with feline and hyænoid relationships), in addition to those above named—becomes the dominating, if not the only, type in the Pliocene, where also we meet with the earliest existing species, the wolf and coyote (Western Territories of the United States). The Post-Pliocene deposits contain the remains of the wolf, fox, and dog.*

* Mr. J. A. Allen has recently described a species of extinct dog (Pachyeyon robustus) from Ely Cave, Lee County, Virginia, which in many respects departs widely from the type of any of the ordinary wild or domesticated races. In its general proportions, the shortness of the legs, &c., it more nearly approaches the badgers. Its geological horizon has not been absolutely determined.

The Arctoidea, or ursine division of the Carnivora, includes the bears (Ursidæ), weasels (Mustelidæ), raccoons (Procyonidæ), and the singular panda (Ailurus fulgens), from the Himalaya Mountains, whose connection with the true bears is established by the Thibetan Ailuropus.

The bears, whose range embraces practically the whole of the continents of North America and Eurasia, comprise some ten or more species, most of which fall under the division Ursus proper. Among the better known of these are the American grizzly (Ursus horribilis), from the Western United States and British Columbia; the circumpolar white or polar bear (U. or Thalassarctos maritimus); the brown bear (U. arctos), the common species of Europe and Asia, which also appears to be identical with the common American or black bear (U. Americanus), and to which the (Himalayan) Isabelline and Syrian bears (U. Isabellinus and U. Syriacus) are nearly related; the Japanese bear (U. Japonicus); the Indo-Malaysian sun-bear (U. or Helarctos Malayanus); and the sloth-bear (Melursus labiatus), from India and Ceylon. A solitary species (U. Crowtheri) is found on the African continent (Atlas Mountains), and likewise but a single one in South America, the spectacled bear (Tremarctos ornatus), from the Peruvian and Chilian Andes. The last is, according to Günther,[132] undistinguishable in its dental characters from a species inhabiting the island of Formosa.

Two species of bear, the Ursus Arvernensis and U. Etruscus, are known from the Pliocene deposits of Europe, which, with the forms described from the Siwalik Hills, constitute the oldest members of the genus with which we are acquainted. The complete absence of ursine remains from the American Tertiary deposits would seem to indicate that the existing New World representatives of the family were a recent introduction, a supposition strengthened by the discovery of the remains of the grizzly, as a contemporary of the great cave-bear (U. spelæus), in the European Post-Pliocene deposits. Of forms most nearly related to the true bears are Arctotherium (Pliocene or Post-Pliocene of Buenos Ayres) and Hyænarctos (Upper Miocene and Pliocene of Europe and India), which last, by way of Dinocyon and Amphicyon, would seem to effect a direct transition to the dogs. It thus appears that Amphicyon lies at the converging point of both family lines.

The Procyonidæ are all members of the New World fauna, and

comprise the raccoons (Procyon), from most parts of North and South America; the coatis (Nasua), Mexico to Paraguay; kinkajou (Cercoleptes), Mexico to Peru and Brazil; and the bassarids (Bassaris), from the warmer regions of the United States and Mexico. The family, as represented in the genus Procyon, dates from the Pliocene period.

The Mustelidæ, whose numerous representatives are spread over the greater part of all the continental areas with the exception of the Australian, comprise among better-known forms: Lutra, the otters, whose range embraces nearly the whole of the continents of Eurasia and North America, with parts of South America and Africa; Enhydris, the North Pacific sea-otter (California—Japan); Nutria, the South American west coast sea-otter (California to Chiloe); Meles, the true badger (North Europe to Japan and China); Taxidea, American badger; Mellivora, the African and Indian ratels; Mephitis, the skunk, whose range comprises the entire tract included between Canada and the Strait of Magellan; Ictonyx, the African zorilla; Mustela, the martens, boreal forms of both the Eastern and Western Hemispheres;* Putorius, the weasels, which are distributed over the greater part of the Northern Hemisphere, and enter into tropical Africa and South America; and Gulo, the wolverine or glutton, whose habitat in America extends from about the fortieth to the seventy-fifth parallel (Melville Island), and in Eurasia from Lithuania to Kamtchatka and the Arctic tundras. In the martens are included the Asiatic sable (Mustela zibellina) and the American sable (M. Americana), the range of the latter extending over the greater part of the American continent north of the fortieth or forty-fifth parallel of latitude. Other well-known forms are the Eurasiatic pine-marten (M. martes), and the pekan or fisher (M. Pennanti), which is still extensively distributed over the American continent north of the fortieth parallel of latitude. Of the weasels proper (Putorius) the true or common weasel (P. vulgaris) and ermine (P. erminea) are held in common by the northern regions of Europe, Asia, and America, the true ferret or polecat (P. fœtidus) is Eurasiatic, and the mink, comprising the two species, P. lutreola and P. vison, both Eurasiatic and American.

The total number of fossil forms referable to the Mustelidæ is

* Martes flavigula, the Indian marten, is distributed from the southern slopes of the Himalayas to Ceylon and Java.

very limited. The glutton, badger, otter, marten, and ermine occur in the Post-Pliocene deposit of Europe, while in the equivalent
American series we have species of Galictis and Mephitis. Taxidea,
Lutra, and Mustela are Pliocene in North America, and Mellivora
in India (Siwalik Hills).

The fourth group of the Carnivora, the Pinnipedia or seals,
comprise three very distinct families: 1. Otaridæ, the eared or fur-
seals, sea-lions, with about nine species, whose habitat is the temperate and cold waters of the southern oceans (as far north as the
Galapagos Islands) and the North Pacific (south to California). No
species is known from the North Atlantic. To this group belongs
the highly-prized northern fur-seal (Callorhinus ursinus), which was
at one time abundant along the greater part of the American coast
between Alaska and Lower California, but is now rapidly disappearing, although still very abundant among the Prybilov or Fur-
Seal Islands; the species appears to be found also along the coasts
of Kamtchatka and the island of Saghalien. Of the sea-lions,
commonly so-called, the best-known species are the northern sea-
lion (Eumetopias Stelleri), whose range extends from Behring
Strait to California and Japan, and the Californian or black sea-
lion (Zalophus Californianus), the familiar animal of the harbour
of San Francisco. The common species of the west coast of South
America is the southern sea-lion (Otaria jubata). Other species of
the family, popularly known as sea-bears, whose domain covers
much of the southern seas, from South America to Africa and New
Zealand, are referred to the genus Arctocephalus. 2. The Trichechidæ, walruses, containing a single species, the walrus or morse
(Trichecus rosmarus), whose habitat is the icy waters of the Arctic
regions of North America (from Labrador), Europe, and Asia.
The animal appears not to exist on the American coast between the
ninety-seventh and one hundred and fifty-eighth meridians of longitude, nor on the Eurasiatic coast between the one hundred and
thirtieth and one hundred and sixtieth meridians. East of the
Yenisei it is very rare. The North Pacific form is by some authors
considered to be a distinct species (T. obesus). 3. Phocidæ, the
earless or true seals, which are almost universally distributed over
the temperate and colder portions of the globe. One species, the
monk-seal (Monachus albiventer) inhabits the Mediterranean and the

waters of the Canary Archipelago, and an allied form (M. tropicalis) the shores of the West India islands and Florida. The species of the Caspian Sea (Phoca Caspica) is by many authors identified with the common harbour seal of the North Atlantic and Pacific oceans (P. vitulina), and the seal of Lake Baikal (P. Sibirica) with the northern ringed-seal, P. fœtida. The greater number of the southern forms are generically distinct from the northern, of which last about one-half have a circumpolar distribution. It has been thus far impossible to determine the exact range of the different species of seal, but it appears that, of the northern forms, the harbour-seal is the most widely distributed. On the American coast it has been noted as far south as New Jersey and the Santa Barbara Islands, California, and is reported to have been also observed near Beaufort, North Carolina. Along the European coast it is not rare off the coasts of Spain and France, and is even said to occasionally enter the Mediterranean. The most northerly species appears to be the ringed-seal, which has been met with considerably beyond the eighty-second parallel of latitude. The Greenland or harp-seal (P. Groenlandica) appears to be a permanent inhabitant of the St. Lawrence River. The more aberrant forms of the family are the hooded-seal (Cystophora cristata), from the colder parts of the North Atlantic,* and the sea-elephants (Macrorhinus), of which there are two generally recognised species, one of which (M. angustirostris) appears to be confined to the coasts of California and Western Mexico, and the other, the southern sea-elephant (M. elephantinus or leoninus), to the waters of the southern oceans (Patagonia, Juan Fernandez, Kerguelen Land, Macquarie Island). The former species is now almost completely exterminated.

With the exception of some doubtful fragments described from European museums, the only unequivocal remains of Otaridæ have thus far been described from the Pliocene deposits of Victoria, Australia, and the Post-Pliocene of New Zealand. The remains of the walrus have been found in the Post-Pliocene deposits of various parts of North America—south to New Jersey and South Carolina —while in Europe its representatives appear to be traced back to the Pliocene, or even late Miocene, period. Most of the so-called trichecoid remains, however, have been shown by Van Beneden to

* An individual of this species was obtained in November, 1888, at Spring Lake, New Jersey (Brown, " Am. Naturalist," xvii., p. 1191).

belong to animals having no close relationship with the walruses. The only undoubted phocine fragment found in any American formation of older date than the Quaternary belongs to a form described from the Miocene of Virginia as Phoca Wymani. In Europe, especially in the Antwerp Basin, remains of the family are abundant, and in living and extinct genera (Phoca, Palæophoca, Callophoca, Gryphoca, Monatherium, Prophoca) extend back through the Pliocene to the later Miocene period. Leith Adams has described a species of Phoca from the Miocene calcareous strata of Gozo, near Malta.

Extinct Carnivora of Uncertain Position.—Under the order Creodonta Professor Cope has united a number of generalised European and American carnivore types, which differ in many essentials— greatly reduced cerebral hemispheres, absence of scapho-lunar bone, ungrooved astragalus—from the true Carnivora, and whose direct affinities are at once with these last, the marsupials and insectivores. They are regarded as the primitive carnivores, inasmuch as from these two at least of the great modern groups—the Æluroidea (cats) and Cynoidea (dogs)—are claimed to be directly derived. Of the four extinct families that are referred here, the Miacidæ (Miacis, Didymictis), Oxyænidæ (Oxyæna, Palæonyctis), Mesonychidæ (Mesonyx), and Hyænodontidæ (Hyænodon), the first three are restricted to the Eocene period (beginning with the Lower Eocene), while Hyænodon is both Upper Eocene and Lower Miocene. The Miacidæ and Oxyænidæ are considered to be the ancestral forerunners of the dogs and cats respectively, a union between the latter and the civets being seemingly effected by the genus Proviverra. Here, perhaps, may also be referred the European Arctocyon, one of the oldest forms of Tertiary mammals known.

Primates (Monkeys and Man *).—Naturalists usually recognise three distinct groups of the quadrumanous section of the Primates : the monkeys or apes of the New World (Platyrhina), the apes of the Old World (Catarhina), and the lemurs or halfmonkeys (Lemuroidea), inhabitants of both continental and insular Asia and Africa. In the broader aspects of their distribution the members of this order may be said to be restricted to a zone included between the thirtieth parallels of north and south latitude, although a limited number of forms pass slightly beyond

---

* The consideration of man is not entered into in this work.

18

these boundaries ; they are, therefore, essentially tropical in habit. To what extent, however, climate alone is efficient in determining this distribution still remains to be ascertained, as it is well known that certain forms, most intimately related to species inhabiting the torrid lowlands, appear to habituate themselves to regions of opposite climatic conditions, or where a fairly rigourous winter prevails. Semnopithecus schistaceus has been observed in the Himalayas at an elevation of 11,000 feet, sporting among the garlands of a winter's snow, while a second species of the same genus, S. Roxellanæ, and a macaque (Macacus Thibetanus), steadily inhabit the snow-clad mountains of Moupin, Thibet, at a nearly equal altitude. The most northerly apes known are the two species last mentioned, a species from Japan (Macacus speciosus), and the Barbary ape of the Rock of Gibraltar (Macacus inuus), but it is a little doubtful whether the last is truly indigenous to the region which it now inhabits. The southern limit in the Old World is the region about the Cape of Good Hope, the home of the chacmas. In the New World no form is positively known to pass north of the twentieth parallel of latitude in Southern Mexico (Ateles vellerosus, a species of spider-monkey), but it is by no means improbable that a more northerly extension may be reached by some species; * the American forms being exclusively arboreal in habit, their southern extension will necessarily be determined by the limit of forest growth, which, excepting along the Andean slopes, is in about the thirtieth parallel of latitude, beyond which line no monkeys are known.

It is a circumstance of some little importance, as bearing upon geographical distribution in general, that certain continental islands, as the West Indies, apparently so well adapted in their natural physical conditions to the development of the members of this group of animals, should be entirely deficient in them ; the same holds true with New Guinea, and, indeed, with the entire continent of Australia. Madagascar, while largely supplied with the lemurs, or half-monkeys, is wholly wanting in the apes proper. Long-continued isolation of the tracts under consideration is, doubtless, the primary, if

---

* According to a statement of M. Sallé, made some twenty-five years ago, monkeys (probably a species of Ateles) were found as far north as the upper Tampico, or up to about latitude 23° (Sclater, "Nat. Hist. Review," 1861, p. 509).

not the only, cause of this deficiency. At the same time, it is not quite as easy to account for the present northern limitation. In Mexico, for example, as far as we are able to judge of the general character of the environment, or of the physical conditions governing it, a much more northerly extension might have been assumed than is actually found; but, possibly, the matter of a particular form of food-supply may have something to do with restriction in this quarter. The semi-continent of Europe, again, whose only simian inhabitant is the Barbary ape already mentioned, in view of the fact that at one time it was the home of various forms of ape, offers another instance of a region apparently suited to the wants of these animals, yet practically entirely deficient in them ; but here, doubtless, the extermination was a part of the general extermination which removed so many of the more distinctive types of Miocene and Pliocene mammals into other regions, whatever the exact cause may have been.

The New World monkeys are generally all included in two families : the Cebidæ, with several sub-families, monkeys with thirty-six teeth, and the Hapalidæ, or marmosets, monkeys with thirty-two teeth. The former comprise the sapajous (Cebus), which may be taken as the representative genus of American monkeys, the woolly monkeys (Lagothrix), spider-monkeys (Ateles, and the related Eriodes), howlers (Mycetes), sakis (Pithecia and Brachyurus), night-monkeys or douroucoulis (Nyctipithecus), squirrel-monkeys or saimiris (Chrysothrix), and the related Callithrix. The total number of species known is between seventy and eighty, of which about twenty are sapajous, fifteen spider-monkeys, ten howlers, and about an equal number members of the genus Callithrix.

The extensive equatorial forests of the Amazon and Orinoco, and their tributaries, constitute *par excellence* the home of the American monkeys, but the majority of the genera have a very extended range, appearing in one or more species throughout the greater portion of the tract covered by the entire family. This is more particularly the case with the sapajous, spider-monkeys, howlers, and callithrixes. The range of the species, on the other hand, is not infrequently very sharply defined, as, for example, when a natural barrier, offering insurmountable obstacles to further migration, suddenly interposes itself. Examples of such limitation, as brought about by the dominant water-courses of the equatorial forests, have already been noticed in treating of specific distribution in general (pp. 23, 24). The number of species found in, and north of, the Isthmus of Panama is ten, of which only one, the spider-monkey already referred to, extends into Mexico ; Mycetes villosus, the Guatemalan howler, or *mono*, has thus far been found only in Guatemala and Honduras. It is a little surprising that the range of only two of the species—the black-faced spider-monkey (Ateles ater) and one of the night-apes (Nyctipithecus vociferans) extends beyond Colombia in South America.

None of the South American monkeys appear to pass west of the Andean chain of mountains south of Ecuador, and even north of the Peruvian boundary the number of such transgressional forms is very limited. Indeed, even among the wooded slopes, a habitation along the basal line of the mountain axis seems to be much preferred. The greatest altitude at which monkeys were observed by Tschudi in Peru was 3,000 feet (Lagothrix Humboldtii) ; Ateles ater and Cebus robustus were found at 2,500 feet. On the other hand, Godman and Salvin state that in the district of Vera Paz, in Guatemala, the *mono* or howler is most abundant at an elevation of 6,000 feet ; and on the volcano of Atitlan, in the same country, Mr. Salvin found troops of the Mexican spider-monkey (Ateles vellerosus) in the forest region of 7,000 feet elevation.

The range of the marmosets and oustitis (Hapalidæ) is nearly coextensive with that of the monkeys proper, but no form is thus far known to pass beyond the Isthmus of Panama;* Midas Geoffroyi alone inhabits the Isthmus. The species, of which there are some thirty or more referable to two genera (or sub-genera), Midas and Hapale, are most numerous in the equatorial forests.

Of the Old World Quadrumana, the anthropoid apes (Simiinæ), which include the gorilla, chimpanzee, gibbon, and orang, acquire special importance by reason of their high structural organisation. In the sum of all their characters, the gorilla probably stands the highest, although by many naturalists this place is conceded to the chimpanzee. Only one species of gorilla (Troglodytes gorilla) has thus far been positively determined, but not impossibly other forms may inhabit the interior of the African continent. The recognised habitat of the species is the west coast of Africa a few de-

---

* Midas rufiventer, erroneously described as coming from Mexico, is a Brazilian species (M. labiatus).

grees on either side of the Equator, or the forest region drained by the Gaboon, Muni, Fernand-Vaz, and Ogowai rivers. Much uncertainty still remains as to the number of species of chimpanzee, but most naturalists seem inclined to unite all the variously designated forms, either actually found living or reported to be such, into a single species, Troglodytes niger, whose habitat extends from the west coast (Gambia—Benguela) through the heart of the continent to the central lake region.

The Asiatic anthropoid apes are the gibbons (Hylobates) and orang (Simia satyrus), the latter restricted to the forests of the islands of Borneo and Sumatra. The gibbons, or long-armed apes, which probably comprise a dozen or more species, are confined to South-Eastern Asia, and some of the larger islands of the Eastern Archipelago. On the continent they range from the Brahmaputra River, in Assam, to the Malay Peninsula, and eastward to the region about Canton, China (Hylobates pileatus); the Chinese species is also found in the island of Hainan. The better-known forms are the siamang (H. siamanga), the largest member of the genus, from Sumatra, hoolock (H. hoolock), the most northern form (Assam, Bengal), and lar (H. lar), from Siam, the Malay Peninsula, and Sumatra.

Of the non-anthropoid Quadrumana of the Old World the most numerous in point of species are the dog-apes (Cynopithecinæ)— green-monkeys, macaques, drills, baboons, &c. The long-tailed forms of the genus Cercopithecus are exclusively African, and comprise all the more graceful monkeys of the continent that have been variously designated guenons, green-monkeys, and white-nosed monkeys. Collectively, they range over the greater part of the tracts included between the Gambia and Congo Rivers on the west and Abyssinia and the Zambezi on the east. The mangabeys, sometimes separated as a distinct genus (Cercocebus), are West African, as is also the talapoin (Miopithecus talapoin). Almost equally distinctive of the African region are the dog-faced baboons of the genus Cynocephalus, which have a very general distribution throughout the continent, extending also into the adjoining tracts of Asia (Arabia). Among the better-known members of this group are the mandrill and drill (C. maimon or mormon, and C. leucophæus), both from the west coast (Guinea) ; the baboons proper (C. babuin), whose habitat extends from Abyssinia and Kordofan into the wilds of the interior ; the nearly related and rock-inhabiting chacma (C. porcarius) and sphinx (C. Sphinx), from the south and west of the continent respectively ; and the hamadryas (C. hamadryas), whose home is constituted principally by the coast mountains of Abyssinia and Southern Nubia, and the littoral of Western Arabia. A somewhat aberrant form, the gelada (C. gelada), differing from the other baboons in the non-terminal position of the nostrils, and hence sometimes constituted into a distinct genus (Theropithecus), inhabits the highlands of Abyssinia at an elevation, according to Schimper, of from 10,000 to 13,000 feet.

The macaques, if we exclude the Barbary ape or magot, whose habitat in the north of Africa and on the Rock of Gibraltar has already been noted, are exclusively Asiatic, ranging on the continent from the Himalayas to Japan, and southward to the extremity of the Malay Peninsula. One or more species of the group are found on nearly all the more important islands of the Malay Archipelago, from Sumatra to Timor. The best-known form is the common macaque (Macacus cynomolgus), whose habitat comprises nearly the whole of Southeast Asia, and the islands of Sumatra, Banca, Java, Borneo, Celebes, Bali, Lombok, Flores, Sumbawa, and Timor. In Java, where it ascends to a height of 5,000 feet, it is one of the commonest of animals, and has been brought into a general condition of domestication. Other well-known forms of the group are the Rhesus-monkey (M. Rhesus), whose home is British India, especially the wooded tracts of the lower Himalayas, and the wanderoo (M. Silenus), from the forest region of Malabar. The former has been observed to ascend the Himalayas to an elevation of upwards of 10,000 feet, and even during the winter it is said to dwell habitually in the snow-clad forests about Simla. A remarkable and somewhat aberrant form of this group, the black macaque (M. niger), whose relationship with the African baboons is more intimate than that of any of the other species, inhabits Celebes (and Batchian ?); it is frequently recognised as the type of a distinct genus, Cynopithecus.

The remaining types of Old World monkeys are usually included in the genera Semnopithecus and Colobus, constituting the subfamily Semnopithecinæ, the former of which, with probably not less than twenty-five species, are exclusively Asiatic, and the latter, considerably less numerous, African. Of the genus Semnopithecus, whose distributional area extends from Ceylon and the snow-bound heights of Thibet (S. Roxellanæ) to the islands of the Malay Archipelago, which properly constitute its headquarters, the better-known species are the sacred entellus, or hoonuman (S. entellus), from the Gangetic provinces, and the proboscis monkey (S. nasicus) of Borneo. A form related to the last, characterised by an excessively up-turned nose, is the proboscis monkey of Thibet above mentioned (S. Roxellanæ).

The African Colobi are slender, long-tailed monkeys like the Semnopitheci, from which they are barely separable, but differ in the complete absence of the thumb. Of probably not more than a dozen species, whose combined habitat embraces the greater part of the African continent, from the west coast to Abyssinia and Zanzibar, the best known, and, at the same time, probably the most graceful and beautiful of all monkeys, is the guereza (C. guereza), whose home appears to be the highlands of Abyssinia, at elevations of from 7,000 to 10,000 feet. A closely related form is Colobus Angolensis.

The total number of apes inhabiting the islands of the Malay Archipelago is, according to Rosenberg,[133] twenty-five, distributed among the different islands as follows : Sumatra, twelve ; Banca, four; Borneo, eleven; Java, five; Celebes, two; and Bali, Lombok, Flores, Sumbawa, and Timor, each one. The rapid diminution in the direction of the Australian continent, which is entirely deficient in the animals of this class, is very marked. Only one form, the common macaque, is common to all the islands; Sumatra holds only one species in common with Java, whereas, surprisingly enough, four of its species are represented in Borneo. The greater number of the species are restricted to individual islands.

No unequivocal remains of true monkeys are known to antedate the Miocene period, and in America they do not appear before the late Pliocene or Post-Pliocene. Several forms, referred to the South American genera Cebus, Callithrix, and Hapale, and one representing an extinct type, Protopithecus, probably allied to the howlers, have been described by Lund from the cavern deposits of Brazil. Protopithecus Bonæriensis is founded upon a number of incisor teeth obtained by Ameghino in the neighbourhood of the city of Buenos Ayres. No quadrumanous remains other than those referable to the lemurs, or to a type, Laopithecus (Miocene), standing intermediate between these and the Cebidæ, have as yet been discovered on the North American continent. To the extent of our present knowledge, therefore, the type of the Old World monkeys appears to have had no representatives in the Western Hemisphere.

Numerous remains of Quadrumana are found in the Tertiary deposits of France, Germany, Switzerland, Italy, and Greece, indicating for these animals a much broader distribution in past periods of time than they now enjoy. Several of these forms are referred to existing genera, such as the Miocene Colobus grandævus, from Steinheim, Würtemberg, and the Pliocene Macacus priscus and Semnopithecus Monspessulanus, from Montpellier, France. A macaque (Macacus Pliocænus) has also been cited from Essex, England, and several forms have been described from the Val d'Arno, Italy. But the greater number of the more ancient species still remain with undetermined relationships. One of the most remarkable of these is the Dryopithecus Fontani, from the Middle Miocene deposits of St. Gaudens, France, and the Swabian Alps, which in stature appears to have rivalled the largest of the existing anthropoid apes, although probably more nearly related to the gibbon than to any other living member of this group. Two other apparently anthropoid forms of somewhat smaller dimensions have been described from the nearly equivalent deposits of Sansan, France, and Elgg, in the Canton of Zürich, Switzerland, as Pliopithecus antiquus and P. platyodon respectively ; these are by some authors considered to be more nearly related to the group of the Semnopithecinæ, or even to the macaques, while by others, as Rütimeyer and Lydekker, they are referred to the modern genus Hylobates. Of still more doubtful relationship are the singular Mesopithecus Pentelici, from the Mio-Pliocene of Pikermi, Greece, which in its cranial and dental features most nearly approaches Semnopithecus, while in the structure of the limbs it approximates the macaques, and the probably still less simian Oreopithecus Bambolii, from Monte Bamboli, Tuscany.

Several species of fossil ape have been described from the Siwalik Hills (Pliocene), and are by Lydekker referred to the genera Palæopithecus, Semnopithecus, Macacus, and Cynocephalus. If the determination in the case of the last-named genus be correctly made, it is interesting as proving the former much further extension of the baboons than we find at the present time. Cynocephalus Atlanticus occurs in the late Pliocene or Post-Pliocene deposits of Algeria.

The lemurs, or half-monkeys, constitute a well-differentiated group of the Primates, differing, indeed, in so many essential points of structure from the type of this order as to have induced many naturalists to elevate them to an order apart by themselves, the Lemuroidea or Prosimiæ. Their non-simian affinities are at once with the Insectivora and Ungulata, to which they appear to be united by many connecting ties, both recent and fossil. Upwards of fifty more or less clearly defined (recent) species, representing a dozen or more genera, have been referred to this group, more than one-half of which, embracing all the typical lemurs of the genera Lemur (some fifteen species), Hapalemur, and Lepilemur, are absolutely confined to the island of Madagascar, where they inhabit the forest region. The indrises (Indris, Propithecus), which are likewise confined to the Madagascan region, comprise some of the largest of the lemurs, Indris brevicaudatus measuring upwards of two feet in length.

The sub-family of the galagos numbers probably not less than twenty species, distributed under the two genera Chirogaleus and Galago, the former of which is restricted to Madagascar, while the latter inhabits the scattered wooded tracts of the interior of the African continent, from Senegambia to Abyssinia, and southward to Natal ; no species is known from Madagascar. All the Asiatic lemurs, if we except the very remarkable tarsier (Tarsius spectrum), the type of a distinct family, which inhabits some of the larger islands of the Malay Archipelago (Sumatra, Borneo, Celebes) and the Philippines, and which differs from the lemurs proper, apart from other general characters, in the large eyes and unusual elongation of the tarsal elements of the foot, belong to the sub-family of the lories. They constitute a limited group of small nocturnal animals, destitute of a tail, and distinguished, as far as their habits are concerned, by their exceedingly slow movements. Hence they are frequently termed the "slow lemurs." Of the two Oriental genera, Nycticebus, the typical slow-lemur, is distributed over Cochin China, Siam, the Malay Peninsula, and the larger islands of the adjoining archipelagos—Sumatra, Java, Borneo. Loris, with a single species, the graceful loris (L. gracilis), is a native of Ceylon. Two other species of the same sub-family, resembling the last in their habits, but provided with a rudimentary tail, and with a greatly reduced index-finger, inhabit the west coast of Africa : the potto (Perodicticus potto) is a native of the Gaboon region and the territory of Sierra Leone, and the awantibo (P. [Arctocebus] Calabarensis) of Old Calabar.

The most aberrant form of lemur is the Madagascan aye-aye (Chiromys Madagascariensis), an animal of about the size of a cat, with a rodent-like dentition, and singularly elongated fingers furnished with pointed claws. For a long time the position of this remarkable animal was misunderstood, it having been placed alternately with the lemurs, insectivores, and rodents. It constitutes the type of a distinct family, Chiromyidæ.

The somewhat anomalous distribution of this group of animals, taken as a whole—their headquarters in Madagascar, with a thinning out towards the west on the African continent, and their reappearance in Ceylon and the mainland of Asia—has suggested to some naturalists the notion that at a former, and fairly ancient, period of the earth's history direct land connection existed between these various points, bridging over the chasms that now separate them in the way of water, and permitting of ready migration from one region to another. For this hypothetically assumed, now sunken, continent, Mr. Sclater has proposed the name "Lemuria." In how far such a connecting land-mass may have existed in fact, or in how far, if it actually existed, it was directly concerned with the present distribution of the lemurs, still remains to be determined.

The earliest lemuroid remains of the Old World are probably those of Cænopithecus lemuroides, described by Rütimeyer from the Eocene deposits of Egerkingen, Jura Mountains, and supposed by their discoverer to represent an animal of intermediate relationships between the true lemurs and the American monkeys. This form is not unlikely identical with a species described from the gypsum deposits of the Paris Basin and the phosphorites of Quercy (Oligocene) as Adapis Parisiensis—for a long time supposed to represent an ungulate—with which, also, the Palæolemur Betillei, from Béduer, France, has been identified. The animals here referred to appear to have their nearest analogues among the African lories or galagos. From certain peculiarities in the structure of the cranium, which are supposed to represent similar structures seen in the Ungulata, Filhol recognises in these animals an extinct zoological type, designated the Pachylemur, which stands intermediate between the true lemurs and the pachyderms. Necrolemur antiquus and N. Edwardsi, on the other hand, from the phosphorites of Quercy, are considered to be true lemurs. Lemuroid forms do not appear to be represented in any of the Tertiary formations newer than the Lower Miocene or Oligocene.

Numerous forms, referable to the same group of animals, have been described from the Lower Tertiaries of the Western United States (Lemuravus, Limnotherium, Microsyops, Hyopsodus, Mixodectes, Anaptomorphus, &c.). These in part indicate a transition to the hoofed animals, while others, again, are so closely linked with the Insectivora that they are barely, if at all, separable from them. A reference to some of these forms will be found in the section following the discussion of the Insectivora. As in Europe, no lemuroid forms are known from either the American Miocene or Pliocene formations.

REFERENCE NOTES.

1. On the authority of Wallace. It would appear, however, from the observations of Taczanowski ("Ornithologie du Pérou," vol. i, p. 321, 1884), that the bird is less rare in the region indicated than has been generally supposed.

2. Jeitteles, "Verhandl. d. zool. bot. Gesell. Wien," 1862, p. 262.

3. To this list might also be added the chipmunk, Arctic hare, lynx, wolf, walrus, several seals, &c.

4. Allen, "North American Pinnipeds," pp. 609 *et seq.*

5. Allen, "North American Rodentia," "U. S. Geol. Survey," vol. xi.

6. "Monograph of the Strepomatidæ," p. xli, Smithson. Misc. Pub., 253.

7. "Island Life," pp. 20-22.

8. Seebohm, "Catalogue of Birds," British Museum, v, p. 328.

8a. Since the preparation of the text a large number of additional species, and several genera, of paradise-birds have been described from New Guinea by Finsch, Meyer, Forbes, and others.

9. The family comprises some thirty-five or more species (Beddome, "Ann. Mag. Nat. History," January, 1886).

10. Gray, "Catalogue of Edentate Mammalia," British Museum, 1869, p. 389.

11. Brehm, "Thierleben," i, p. 391.

12. "Ceylon," ii, p. 287.

12a. "Proc. Zool. Soc.," London, pp. 221, 222.

13. Newton, "Encycl. Brit.," article "Humming-Bird," ninth ed., xii, p. 359.

14. Mosenthal and Harting, "Ostriches and Ostrich Farming," p. 28.

15. "Geograph. Distrib. of Animals," ii, p. 330.

16. Lyell, "Principles of Geology," eleventh ed., ii, p. 369.

17. "Encycl. Brit.," ninth ed., iii, p. 461.

18. Lyell, "Principles of Geology," eleventh ed., ii, p. 366.

19. Baird, "Am. Journ. Science," 1866 (xli), p. 340.

20. Baird, "Am. Journ. Science," 1866 (xli), p. 346.

21. Sharpe, "Catalogue of Birds," British Museum, ii, p. 238.

22. Wallace, "Island Life," p. 296.

23. Wallace, "Island Life," p. 253.

24. Wallace, "Geographical Distribution of Animals," i, p. 32. The occurrence was noted by Mr. Lowe, who communicated the facts to Sir Charles Lyell.

25. Wallace, "Geograph. Distrib.," i, p. 180.

26. Wallace, "Geograph. Distrib.," ii, p. 13.

27. Coues, "Key to North American Birds," p. 317.

28. Gould, "Birds of Australia," ii, p. 1.

29. "Ibis," 1884, p. 471. Two hundred sheep are said to have been killed in a single night by a flock of these birds, at a station on Wanaka Lake.

30. "Proc. Zool. Soc.," London, 1864, p. 456.

31. Elwes, "Proc. Zool. Soc.," 1873, p. 648.

32. Forsyth Major, "Zoogeographische Übergangsregionen," "Kosmos," 1884, p. 106.

33. Forsyth Major, "Kosmos," 1884, p. 109. Forsyth Major is in error in quoting from Böttger that twenty-seven out of the forty reptilian and amphibian species inhabiting Morocco are also found in Spain; the number stated is twenty-two ("Abhandl. d. Senckenb. Naturf.-Gesellsch.," xiii, 1863, p. 146).

34. Cope, "Bull. U. S. National Museum," 1875; Heilprin, "Proc. Acad. Nat. Sciences," Philadelphia, 1882.

35. Moseley, "'Challenger' Reports," "Zoology," ii, pp. 188, 189.

36. Wyville Thomson, "Depths of the Sea," p. 454.

37. Lyman, "'Challenger' Reports," "Zoology," v, p. 327.

38. A. Agassiz, "'Challenger' Reports," "Zoology," iii, p. 30.

39. A. Agassiz, op. cit.

40. Milne-Edwards, "Contes Rendus," December 17, 1883.

41. Wyville Thomson, "Voyage of the 'Challenger,'" ii, p. 350.

42. "Nature," March 20, 1884.

43. Günther, "Study of Fishes," p. 305.

44. Wyville Thomson, "Voyage of the 'Challenger,'" ii, pp. 352, 353.

45. "Annals and Magazine of Natural History," January, 1883; "Ver handl. d. k. k. geol. Reichsanstalt," 1882, No. 4.

46. "Nature," xxvi, p. 560.

47. "Bull. Museum Comp. Zoology," Cambridge, vi, p. 153.

48. "Nature," September 3, 1885.

49. "Bull. Soc. Vaud.," xiv, p. 211, 1876.

50. "Rendic. R. Istit. Lomb.," ser. 2, xii, p. 694.

50a. Asper and Heuscher have quite recently shown, through the use of a "pelagic" net, that the pelagic faunas of lakes are far more prolific in microscopic animal forms than has hitherto been supposed. A drop of water from Lake Zürich was estimated to contain ten individuals of Anuræa foliacea, eight of Anuræa longispina, sixty of Ceratium hirundinella, and millions of Dinobryon forms and Asterionellæ, besides various heliozoans, rotifers, and crustaceans. Identical results were obtained under all the most varied conditions of light (darkness) and water, in the open lake and along the shallower shore-line ("Zoologischer Anzeiger," June 19, 1886).

51. "Nature," June 11, 1885.

52. "Am. Journ. Science," 1871, p. 161.

53. "Bull. Soc. Vaud.," xiii, xiv, 1874, 1876. Dr. Henri Blanc enumerates the following twelve species of Rhizopoda as entering into the composition of the deep-water fauna of Lake Geneva (seventy to one hundred and twenty metres): Amœba proteus, A. verrucosa, A. radiosa, Difflugia pyriformis, D. urceolata, D. globulosa, Hyalosphenia cuneata, Arcella vulgaris, Centropyxis aculeata, Pamphagus hyalinus, Actinophrys sol, and an undetermined large Difflugia. All or most of the above forms have been observed by Leidy in the surface-waters of the United States, and it is remarked that the species indicated to be rare by Leidy are also rare in the deep waters of the lake ("Bull. Soc. Vaud.," ser. 2, xx, p. 287).

54. "Am. Journ. Science," 1871.

55. "Anniversary Address, London Geol. Soc.," 1881.

55a. Probably the most striking and convincing evidence indicating convergent modification is presented by the Australian fauna, where, among the numerous implacental forms, we have such remarkable reproductions of the distinctive types seen among the Placentalia, although based upon an entirely different type of structure, and arising independently of the other.

56. "Paleontographical Soc. Reports," 1884.

57. "Ann. Mag. Nat. Hist.," 1874, xiii, p. 222.

58. Heilprin, "Proc. Acad. Nat. Sci.," Philadelphia, March 4, 1884.

59. "Anniversary Address, London Geol. Soc.," 1881.

60. Medlicott and Blanford, "Geology of India," part i, p. 282.

61. Seeley, "Q. Journ. Geol. Soc.," London, 1883.

62. Dawson, "Am. Journ. Science," third ser., xx, pp. 403 et seq.

63. Anodonta Jukesii, from the Old Red Sandstone of Ireland. Professor Hall recognises in the Cypricardites Catskillensis of Vanuxem, from the Oneonta Sandstones of the State of New York (Middle Devonian), a

fresh-water mussel having the general characters of Anodonta ("Science," New York, December 11, 1880)

64. "Review of the Non-Marine Fossil Mollusca of North America," "U. S. Geol. Surv.," third annual report, 1881–'82.

65. Smith, "Proc. Zool. Soc.," London, 1881 ; Crosse, "Journal de Conchyliologie," 1881 ; Bourguignat, "Mollusques Terr. et Flur. du Lac Tanganyika," August, 1885.

66. Marshall, "Ann. Mag. Nat. History," December, 1883.

67. Richthofen, "China," iv.

68. "Q. Journ. Geol. Soc.," London, 1878, pp. 568 *et seq.*

69. "Manuel de Conchyliologie," p. 144 (1881).

70. Tryon, "Structural and Systematic Conchology," i, p. 159.

71. Carpenter, on the authority of Fischer, *op. cit.*, p. 168.

72. "Palæontologia Indica," ser. ix, 1875.

73. "Palæontologia Indica," 1865.

74. "Kreidebildungen von Texas," 1852.

75. "Zeitschr. d. deutsch. geol. Ges.," 1870, pp. 191–251.

76. "Jahrb. d. k. k. geol. Reichsanstalt," 1871, p. 524.

77. "Palæontologia Indica," "Cephalopoda of Kutch," ser. ix, p. 237.

78. "Kreideb. v. Texas," pp. 22–25.

79. "Smithsonian Miscell. Pub.," vii.

80. "Jahrb. d. k. k. geol. Reichsanstalt," 1878, p. 44.

81. "British Assoc. Reports," 1884, p. 555.

82. "Anniversary Address, Geol. Soc.," London, 1862, p. xiv.

83. Bronn's "Klassen und Ordnungen d. Thier-Reichs, Protozoa."

84. From the data furnished by Brady, "'Challenger' Reports," "Zoology," ix.

85. From the data furnished by Brady, "'Challenger' Reports," "Zoology," ix.

86. "'Challenger' Reports," "Zoology," vi, p. 132.

87. "'Challenger' Reports," "Zoology," ii, "Corals," pp. 132, 133.

88. Duncan, "Q. Journ. Geol. Soc.," London, xxvii, p. 437.

89. Duncan, "Q. Journ. Geol. Soc.," London, xxix, p. 561.

90. Duncan, "Q. Journ. Geol. Soc.," London, xxvi, p. 313.

91. "Denkschr. d. k. k. Akad.," Vienna, 1872, p. 199.

92. "Q. Journ. Geol. Soc.," London, xxvi, p. 311.

93. "Q. Journ. Geol. Soc.," London, xxxii, p. 343.

94. Zittel, "Handbuch der Palæontologie," i, pp. 717–720.

95. "Manuel de Conchyliologie," pp. 173 *et seq.*

96. Ellsworth Call, "Bull. U. S. Geol. Survey," xi (1884), p. 43.

97. Thesaurus Siluricus ; Thesaurus Devonico-Carboniferus.

98. " Handbuch der Palæontologie," ii, p. 320.

99. " Encycl. Brit.," ninth ed., vi, p. 663 (" Crustacea ").

100. Bronn's " Klassen und Ordnungen des Thier-Reichs," v, p. 1073.

101. Hall, " Pennsylvania Second Geol. Survey," 1884, PPP, p. 29.

102. McLachlan, in Nares's " Voyage to the Polar Sea," ii, p. 234.

103. " Trans. Asiatic Soc. of Japan," November, 1885.

104. " Nature," December 24, 1885.

105. " Mem. Boston Soc. Nat. History," April, 1885.

106. Scudder, " Mem. Boston Soc. Nat. History," April, 1885, p. 355.

107. " Nature," January 29, 1885, p. 297.

108. " Voyage to the Polar Sea," ii, p. 1220.

109. " Study of Fishes," pp. 307-311.

110. Günther, " Study of Fishes," pp. 270, 271.

111. " Science," May 23, 1884.

112. Filhol, in " Science," May 23, 1884.

113. " Science," May 23, 1884.

114. " American Naturalist," March, 1885, p. 289.

115. " Journ. Asiatic Soc. Bengal," 1864, p. 544.

116. Leydig, " Abhandl. d. Senckenb. Naturf.-Gesell ," xiii, 1883.

117. Böttger, " Abhandl. d. Senckenb. Naturf.-Gesell.," xiii, 1883.

118. Bosca, " Bull. Soc. Zool. de France," 1880.

119. Bedriaga, " Bull. Soc. Natural.," Moscow, 56, 1881.

120. " Ann. Mag. Nat. Hist.," January, 1886.

121. Woodward, " Geol. Magazine," November, 1885.

122. " Palæontologia Indica," ser. x, iii, p. 146.

123. " Q. Journ. Geol. Soc.," London, August, 1885.

124. " Encycl. Brit.," ninth ed., xv, p. 399.

125. Dobson, " Monograph of the Insectivora," 1882.

126. " American Naturalist," May, 1885, p. 467.

127. Dobson, " Catalogue of the Cheiroptera, British Museum," p. xxx.

128. " Am. Journ. Science," April, 1886.

129. " Zoologischer Anzeiger," June, 1883.

130. " Q. Journ. Geol. Soc.," London, 1878, p. 419.

131. " Trans. Am. Philos. Society," January 18, 1884.

132. " Proc. Zool. Soc.," London, p. 443.

133. " Zoologischer Garten," xxiii (1882), pp. 111-115.

Astartidæ, 148, 166, 271, 272.
Asteroidea (deep-sea), 111.
Asterolepis, 302.
Astræa, 247–249.
Astræidæ, 144.
Astrangia, 248.
Atalapha, 349, 351.
Atax, 126.
Ateles, 394–396.
Atherines, 104.
Atherura, 362.
Athyris, 145.
Atlanta, 120, 122.
Atlantosaurus, 161.
Atrypa, 145, 214.
Auchenia, 375, 376.
Aurelia, 122.
Aurochs, 379.
Australian fauna, anomalies of, 9.
Australian realm, 97.
Aviculidæ, 148, 271.
Axis, 381.
Axolotl, 306.
Aye-aye, 402.
Azores, birds of the, 7, 48.

Babbling-thrush, 72, 94, 107.
Baboon, 397.
Babyrousa, 374.
Baculites, 168, 267.
Badger, 390.
Baikal, Lake, fauna of, 212.
Bairdia, 207, 273, 274.
Balæna, 341.
Balænoptera, 341, 344.
Balænotus, 345.
Baltimore bird, 66, 79.
Bandicoot, 99.
Barbary ape, 394, 395.
Barbel, 68, 291.
Barbus, 68.
Barracuda (Sphyræna), 294.
Barramunda, 103.
Barriers affecting migration, 41.
Bascanium, 323, 324.

Basilisk, 317.
Bass (Labrax), 294.
Bassalian province, 299.
Bassaris, 390.
Bathmoceras, 192.
Bathyactis, 110, 243.
Bathycrinus, 111.
Bathyergus, 357.
Bathygnathus, 160.
Bathyophis, 298.
Bats, distribution of, 349.
Battocrinus, 151.
Bear, 4, 27, 389.
Beaver, 360, 364, 365.
Bee-eater, 87, 94, 106.
Bees, 279, 284.
Beetles, 150, 279, 280, 283, 284.
Belemnitella, 168.
Belemnites, 137, 156, 168, 268.
Belemnosepia, 138, 166.
Belideus, 99, 100.
Bell-bird, 78.
Bellinurus, 278, 279.
Bellows-fish (Centriscus), 295.
Beloteuthis, 166.
Beluga, 344.
Bermudas, birds of the, 48, 51.
Bernissartia, 329.
Beroe, 240.
Beyrichia, 272.
Beryx, 294, 303.
Bibos, 379.
Big-horn, 379.
Birds, American, common to Europe, 46; European, common to America, 48; dispersal of, 45; migrations of, 47; oceanic journeys of, 47; Arctic, 70; geological distribution of, 330; of the Holarctic realm: of the Eurasiatic division, 65; of the North American, 66; of the Neotropical realm, 77–79; of the Ethiopian realm, 87, 88; of the Oriental realm, 94, 95; of the Australian realm, 100–102;

THE END.

NATURAL SCIENCES IN AMERICA

An Arno Press Collection

Allen, J[oel] A[saph]. **The American Bisons,** Living and Extinct. 1876

Allen, Joel Asaph. **History of the North American Pinnipeds:** A Monograph of the Walruses, Sea-Lions, Sea-Bears and Seals of North America. 1880

American Natural History Studies: The Bairdian Period. 1974

American Ornithological Bibliography. 1974

Anker, Jean. **Bird Books and Bird Art.** 1938

Audubon, John James and John Bachman. **The Quadrupeds of North America.** Three vols. 1854

Baird, Spencer F[ullerton]. **Mammals of North America.** 1859

Baird, S[pencer] F[ullerton], T[homas] M. Brewer and R[obert] Ridgway. **A History of North American Birds:** Land Birds. Three vols., 1874

Baird, Spencer F[ullerton], John Cassin and George N. Lawrence. **The Birds of North America.** 1860. Two vols. in one.

Baird, S[pencer] F[ullerton], T[homas] M. Brewer, and R[obert] Ridgway. **The Water Birds of North America.** 1884. Two vols. in one.

Barton, Benjamin Smith. **Notes on the Animals of North America.** Edited, with an Introduction by Keir B. Sterling. 1792

Bendire, Charles [Emil]. **Life Histories of North American Birds** With Special Reference to Their Breeding Habits and Eggs. 1892/1895. Two vols. in one.

Bonaparte, Charles Lucian [Jules Laurent]. **American Ornithology:** Or The Natural History of Birds Inhabiting the United States, Not Given by Wilson. 1825/1828/1833. Four vols. in one.

Cameron, Jenks. **The Bureau of Biological Survey:** Its History, Activities, and Organization. 1929

Caton, John Dean. **The Antelope and Deer of America:** A Comprehensive Scientific Treatise Upon the Natural History, Including the Characteristics, Habits, Affinities, and Capacity for Domestication of the Antilocapra and Cervidae of North America. 1877

Contributions to American Systematics. 1974

Contributions to the Bibliographical Literature of American Mammals. 1974

Contributions to the History of American Natural History. 1974

Contributions to the History of American Ornithology. 1974

Cooper, J[ames] G[raham]. **Ornithology.** Volume I, Land Birds. 1870

Cope, E[dward] D[rinker]. **The Origin of the Fittest:** Essays on Evolution and **The Primary Factors of Organic Evolution.** 1887/1896. Two vols. in one.

Coues, Elliott. **Birds of the Colorado Valley.** 1878

Coues, Elliott. **Birds of the Northwest.** 1874

Coues, Elliott. **Key To North American Birds.** Two vols. 1903

Early Nineteenth-Century Studies and Surveys. 1974

Emmons, Ebenezer. **American Geology:** Containing a Statement of the Principles of the Science. 1855. Two vols. in one.

Fauna Americana. 1825-1826

Fisher, A[lbert] K[enrick]. **The Hawks and Owls of the United States in Their Relation to Agriculture.** 1893

Godman, John D. **American Natural History:** Part I — Mastology and **Rambles of a Naturalist.** 1826-28/1833. Three vols. in one.

Gregory, William King. **Evolution Emerging:** A Survey of Changing Patterns from Primeval Life to Man. Two vols. 1951

Hay, Oliver Perry. **Bibliography and Catalogue of the Fossil Vertebrata of North America.** 1902

Heilprin, Angelo. **The Geographical and Geological Distribution of Animals.** 1887

Hitchcock, Edward. **A Report on the Sandstone of the Connecticut Valley,** Especially Its Fossil Footmarks. 1858

Hubbs, Carl L., editor. **Zoogeography.** 1958

[Kessel, Edward L., editor]. **A Century of Progress in the Natural Sciences:** 1853-1953. 1955

Leidy, Joseph. **The Extinct Mammalian Fauna of Dakota and Nebraska,** Including an Account of Some Allied Forms from Other Localities, Together with a Synopsis of the Mammalian Remains of North America. 1869

Lyon, Marcus Ward, Jr. **Mammals of Indiana.** 1936

Matthew, W[illiam] D[iller]. **Climate and Evolution.** 1915

Mayr, Ernst, editor. **The Species Problem.** 1957

Mearns, Edgar Alexander. **Mammals of the Mexican Boundary of the United States.** Part I: Families Didelphiidae to Muridae. 1907

Merriam, Clinton Hart. **The Mammals of the Adirondack Region,** Northeastern New York. 1884

Nuttall, Thomas. **A Manual of the Ornithology of the United States and of Canada.** Two vols. 1832-1834

Nuttall Ornithological Club. **Bulletin of the Nuttall Ornithological Club:** A Quarterly Journal of Ornithology. 1876-1883. Eight vols. in three.

[Pennant, Thomas]. **Arctic Zoology.** 1784-1787. Two vols. in one.

Richardson, John. **Fauna Boreali-Americana;** Or the Zoology of the Northern Parts of British America, Containing Descriptions of the Objects of Natural History Collected on the Late Northern Land Expeditions Under Command of Captain Sir John Franklin, R. N. Part I: Quadrupeds. 1829

Richardson, John and William Swainson. **Fauna Boreali-Americana:** Or the Zoology of the Northern Parts of British America, Containing Descriptions of the Objects of Natural History Collected by the Late Northern Land Expeditions Under Command of Captain Sir John Franklin, R. N. Part II: The Birds. 1831

Ridgway, Robert. **Ornithology.** 1877

Selected Works By Eighteenth-Century Naturalists and Travellers. 1974

Selected Works in Nineteenth-Century North American Paleontology. 1974

Selected Works of Clinton Hart Merriam. 1974

Selected Works of Joel Asaph Allen. 1974

Selections From the Literature of American Biogeography. 1974

Seton, Ernest Thompson. **Life-Histories of Northern Animals: An Account of the Mammals of Manitoba.** Two vols. 1909

Sterling, Keir Brooks. **Last of the Naturalists:** The Career of C. Hart Merriam. 1974

Vieillot, L. P. **Histoire Naturelle Des Oiseaux de L'Amerique Septentrionale,** Contenant Un Grand Nombre D'Especes Decrites ou Figurees Pour La Premiere Fois. 1807. Two vols. in one.

Wilson, Scott B., assisted by A. H. Evans. **Aves Hawaiienses:** The Birds of the Sandwich Islands. 1890-99

Wood, Casey A., editor. **An Introduction to the Literature of Vertebrate Zoology.** 1931

Zimmer, John Todd. **Catalogue of the Edward E. Ayer Ornithological Library.** 1926